Experimental Fluid Mechanics

More information about this series at https://link.springer.com/bookseries/3837

Klaus Hufnagel

Wind Tunnel Balances

Klaus Hufnagel
Wind Tunnel
Technische Universität Darmstadt
Darmstadt, Germany

ISSN 1613-222X ISSN 2197-9510 (electronic)
Experimental Fluid Mechanics
ISBN 978-3-030-97765-8 ISBN 978-3-030-97766-5 (eBook)
https://doi.org/10.1007/978-3-030-97766-5

This Springer imprint is published by the registered company Springer Nature Switzerland AG
The registered company address is: Gewerbestrasse 11, 6330 Cham, Switzerland

I would like to dedicate this book to Prof. Bernd Ewald, who was my teacher and doctoral advisor. From him, I acquired most of the knowledge in the field of wind tunnel balances. In the beginning of the 1980s, I was hired by him for a research project with the aim to develop internal balances for the emerging cryogenic wind tunnels. He was aware that the very promising aim of having the same Reynolds number in a wind tunnel test as on subsonic cruise flight could not be achieved without a force measuring technique that has the same accuracy as the systems built for conventional tunnels. The project was successfully brought to completion, but there was no interest in the industry to take over this knowledge of cryogenic balances for industrial production, and so, he decided to offer such balances directly from the university. He was an engineer that showed us how engineering creativity could solve totally new problems when it is coupled with solid engineering knowledge and art. He analyzed lot of problems by simple models and so delivered the basis and the direction for the knowledge transfer to practical applications. His impact on the development on present status of wind tunnel balances is documented in numerous papers, the study of which is highly recommended if somebody wants to enter this field. Prof. Ewald passed away in June 2019.

Preface

The motivation to write this book came to me while sorting out material which the head of our institute, Prof. Bernd Ewald, had produced and collected on the subject of balances during his career. Following his retirement in 1998, he continued to work on balance design, and although he had intended on summarizing his life experience with balances in the form of a book, he became side-tracked—he dedicated his time to rebuilding a famous flying wing aircraft, the Horton IV. On every visit to the office, he therefore brought with him a trunk full of wind tunnel balance documents, rather than throwing them away. Together with the documentation I had collected at this time over my own 15 years of working with balances, I was therefore faced with deciding the fate of all this accumulated knowledge. The documentation comprised over six hundred articles related to wind tunnel balances and calibration machines and addressed such issues as operation over a wide temperature range or under large temperature gradients. Especially while reading through Prof. Ewald's notes, it became clear to me that we had overcome innumerable design problems in the past that represented important solutions and experience that others could benefit from.

It eventually became apparent that the only solution to preserving this knowledge for future reference was to write a book, condensing this combined experience obtained over a period of 36 years. Although I started this endeavor well in advance of my own retirement, this 14 year lead time was still not sufficient to finish the project. It was also never evident to me whether the book would be outdated before publication. The eventual role of computational fluid dynamics (CFD) was and is still not clear—to what extent wind tunnel testing would remain an important design tool for the aviation industry? In comparison with complicated and costly experiments employing contemporary techniques such as particle image velocimetry (PIV) or pressure/temperature sensitive paints (PSP/TSP), CFD appeared in many respects to be advantageous. However, in the end, both CFD and wind tunnel can be viewed as aerodynamic simulation tools that have their own specific uncertainties in the prediction of the airflow around an airplane. Since the development of cryogenic wind tunnels, which are able to close the Reynolds number gap between model testing and reality, wind tunnel testing now delivers extremely precise data for the performance of an airplane. So, time has revealed that wind tunnel balances remain an important tool in the overall design of aircraft, and therefore, it is my hope that this book will be of use to present and future generations of engineers and technicians dealing with this measurement technology.

Darmstadt, Germany Klaus Hufnagel
April 2021

Acknowledgements Writing a book is a one part of the story, the other is to get it published. First, I have to thank my longtime colleague Matthias Quade for the first review of the text. For bringing the book to a scientific standard, I want to thank very much Prof. Dr.-Ing Cameron Tropea for the time he spent in proofreading and his support in the layout of the book. Finally, I would like to extend a word of thanks to numerous of my colleagues throughout the world, who have graciously sent me pictures of their balances and instruments and allowed me to use them in this book.

Contents

List of Figures

List of Tables

Chapter 1
Historical Review

1.1 Introduction

Before embarking on a detailed description of wind tunnel balance design, it is instructive to first review the evolution of force measurements using such balances. The justification for such a review in an engineering handbook is quite simple and comes from years of experience.

Despite having access to innumerable articles written on a subject, the chronological sequence of these articles can be particularly revealing about why and when certain inventions were made. Circumstance often lies behind the saying that 'necessity is the mother of invention'. This is very well exemplified by the parallel development of two sectors eventually merging to yield what we now know as strain gauge based wind tunnel balances. On the one hand very basic physical laws and effects of elasticity and electricity were being developed, eventually culminating in the fundamentals of measuring strain with a strain gauge. On the other hand, the need for aerodynamic force measurements was rising rapidly with the advent of wind tunnel testing. It is this sequence of events and steps of progress which is summarized in the following sections.

1.2 From Fundamental Physics to the First Force Transducers

The basic research to build a force transducer with a metal spring and a wire strain gauge was conducted by *Robert Hooke* (1635–1703), *Georg Simon Ohm* (1789–1854), *Charles Wheatstone* (1802–1875) and *William Thomson, 1st Baron Kelvin of Largs* (1824–1907).

Robert Hooke was a famous physicist and an architect. Among other things he formulated in 1678 the basic theory of elasticity [9] and this is also the reason why the elastic relationship between stress and strain is known as *Hooke's Law*.

K. Hufnagel, *Wind Tunnel Balances*, Experimental Fluid Mechanics,
https://doi.org/10.1007/978-3-030-97766-5_1

Georg Simon Ohm was professor for physics at the universities of Cologne, Nuremberg and Munich. In 1827 he published his work on the correlation between voltage, current and electrical resistance. His relation between voltage current and resistance became the basic principle for an electrical circuit and the measurement of electrical resistance [16]. Using his name as the unit for the electrical resistance honored this work.

Charles Wheatstone contributed work on the electrical telegraph and published work on the "*Wheatstone Bridge*" circuit in 1843 [22]. Although he never claimed to have invented the special electric circuit to determine small electrical resistances that was named after him. The first description of the bridge circuit was given by *Samuel Hunter Christie* (1784–1865), from the Royal Military Academy, who published it in 1833 [4].

William Thomson was knighted in 1866 and became *1st Baron Kelvin of Largs* in 1892. His most famous works were his contributions that made the first transatlantic telegraph cable a success, but he also published numerous articles on physics, among these in 1856 an article on the relationship between mechanical stress and electrical resistance of metals [21], which had been already mentioned by *Wheatstone*.

This small excursion to early developments shows, that the basics for metal strain gauges resulted from the work of several scientists working sequential to one another with only minimal temporal overlap. For instance, after *Thomson's* contribution, another 61 years went by until the first use of the effect was reported for a sensor application. The first person who actually used the effect of the change of electrical resistance under mechanical stress for measurements was *Walther Nernst* (1864–1941), who built a pressure gauge in 1917 to measure the pressure fluctuations inside a piston engine. This application, and the resulting pressure diagram were published in 1928 [12, 13].

Edward. E. Simmons (1911–2004) was an assistant at the *California Institute of Technology* when he invented a material testing apparatus for measuring the percussive force in 1936. At that time *Simmons* and others he worked with probably did not realize the importance of their invention and that is one reason that it was patented only several years later [19, 20]. The patent for this apparatus [17] was granted in 1942 (Fig. 1.1). This is the first application where a wire strain gauge was used to measure a force.

At nearly the same time on the east coast at *Massachusetts Institute of Technology*, *Arthur Claude Ruge* (1905–2000) used a meandering wire on a piece of paper to measure the strain on the surface of a containment vessel model to predict the stresses in the real vessel (Fig. 1.2). This was the first use of a resistance strain gauge in experimental stress analysis.

The aerospace industry was developing an urgent need for simple and inexpensive strain measurement sensors at that time and so the first industrially produced foil strain gauge, ***SR4***, immediately became a success. The sensor was named ***SR4*** because *Simmons* and *Ruge* together with four people (*de Forest, Tatnall, Clark and Hathaway*) negotiated the terms of the corporate patent of *Simmons* and *Ruge* for the wire strain gauge.

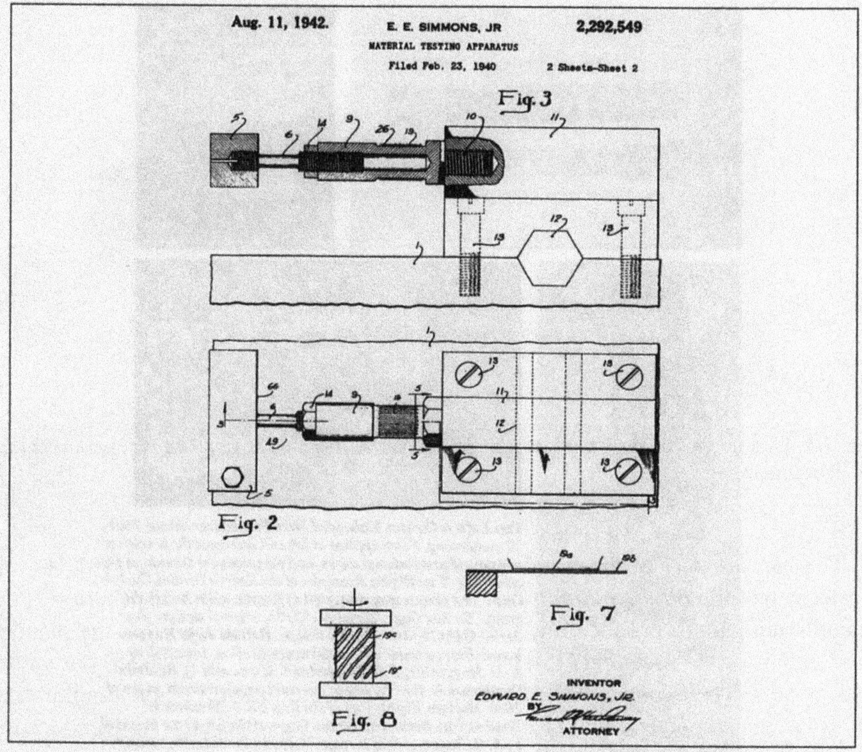

Fig. 1.1 Extract from Simmons' material testing apparatus patent document (HBM Hofmann)

Fig. 1.2 Prof. Ruge with a model of a containment vessel (HBM Hofmann)

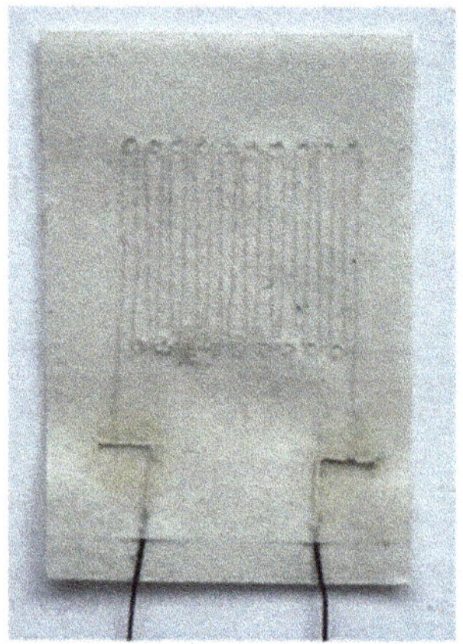

Fig. 1.3 Photographs of early wire gauge and the patented strain gauge SR4 (with permission by Dr. Stockmann)

The success story of the metal strain gauge was described by *Tatnall* [8]. He was a salesman of the *Baldwin-Southwark Company* and promoted the distribution of the paper strain gauges produced by the *Ruge/de Forest Company*, so that the planned production for the first year (1941) of 50,000 gauges (Fig. 1.3) was sold out within two months. The Second World War and the associated rapidly expanding aircraft industry created great demand for these strain gauges for material testing and testing of aircraft structures. The paper strain gauge dominated the area of experimental stress analysis very rapidly. The strain gauge production company of *Ruge* and *de Forest* was sold to *Baldwin Lima Hamilton (BLH)* in 1955, a company which today still produces gauges designated ***SR4***.

In the field of force measurement some developments can be traced back to the beginning of the 20th century using electro dynamometers as force transducers. In the USA *Burton McCollum* and *O.S. Peters* published an article about a "new electrical telemeter" in 1924 [15]. They used a stack of carbon plates as a strain sensitive element and so, were the first to use a semiconductor gauge for strain measurement. Another example was a piezoelectric dynamometer that was used in a test by *Max Kramer* at the *Aachen University of Technology* in 1932 to determine the dynamic lift force of a two-dimensional wing generated by a quick change of angle of attack [14]. In 1932 *Fred Scoville Eastman* from the *University of Washington* reported on a weigh beam for an external balance using an electromagnetic dynamometer [6].

All of these applications were used or proposed for use in a wind tunnel, but eventually metal strain gauge sensors dominated this application for wind tunnel strain gauge balances. One main reason for this was the next significant step in the development of the strain gauges made by *P. Jackson* in Great Britain with the invention of the foil strain gauge, where the grid was no longer a wire [10]. The production of the gauge was performed using a photo-chemical etc.hing process, similar to that of the printed circuit technology by the *Technograph Inc. Company* under the license of the *Saunders-Roe Company (UK)* in 1952, where *P. Jackson* worked as a test engineer.

This process resulted in the mass production of the gauges, which made high quality gauges much less expensive and more reliable. In 1954 the *Baldwin Company* bought the license for the production of foil strain gauges from *Technograph Inc.* (UK) and the *SR-4 foil strain gauge* was produced.

In the 1950s several companies emerged to produce strain gauges in large numbers and in increasingly varied shapes. *Hottinger Messtechnik* started production of foil strain gauges in Darmstadt in 1955 [11] and in 1963 this company merged with *Baldwin Lima Hamilton Corp.* and formed *Hottinger Baldwin Messtechnik (HBM). HBM* still exists today, but is now owned by other companies. Since 2001 *BLH* is part of the *Vishay Company,* which was founded at the beginning 1960s and is now the largest strain gauge manufacturer worldwide. Numerous other strain gauge manufacturers were established over the years, but none of these play a role in the market comparable to *BLH*, *HBM* and *Vishay Micro Measurements*.

Up to now in this historical review the story of the semiconductor strain gauges was not mentioned. One reason for this is that semiconductor strain gauges do not play a major role in the production of wind tunnel balances. This is because of their nonlinear characteristic and their strong temperature sensitivity. However, for some applications they are applicable, where a quick response to a sudden or dynamic change in load must be measured. Then their high sensitivity and the likely higher stiffness of the balance body are decisive. Semiconductor strain gauges use the piezoresistive effect, the change of the electrical resistance caused by the change of density in the crystal structure of a semiconductor under stress. *P.W. Bridgman* conducted comprehensive experiments on the electrical resistance of metals and crystals and in 1932 he published an article on the effect of homogeneous mechanical stress on the electrical resistance of a crystal [3]. He tested the change of resistance on crystals by the influence of static pressure. Under his supervision *Mildred Allen* performed first experiments on the effect of tension on crystals in 1932 [1]. She tested the change of electrical resistance on Bismuth crystals under tension, related to different crystal orientations. However, her measurements did not directly lead to the development of a strain sensor.

About 20 years later *Charles S. Smith* at the *Bell Laboratories* discovered the strain sensitivity of Germanium and Silicon. This research was used to develop the semiconductor strain gauge [18]. In 1958 *Honeywell* offered the first commercially produced semiconductor strain gauge [18]. In the coming years numerous developments arose using semiconductor strain gauges for pressure transducers (*Kulite, Honeywell*). Since 1962 *Baldwin Lima Hamilton (BLH)* has offered bondable semiconductor strain gauges, similar in use to the metal strain gauges they produce. However productions of these strain gauges was terminated some years ago. The use of bonded semiconductor strain gauges was described in detail by *James Dorsey* from *BLH* in his *Semiconductor Strain Gage Handbook* [5] in 1964 and *BLH* offered semiconductor strain gauges until 2004. Nowadays, semiconductor strain gauges are available from *Kulite, Kyowa and Micron Instruments*.

1.3 Force Measurement in Wind Tunnels

Apart from the development of strain gauges, there was also the development of force transducers and wind tunnel balances. With the acceleration of aerodynamic research and the use of wind tunnels in the early 20th century, the measurement of aerodynamic forces on test specimens was of paramount importance.

Benjamin Robins (1707–1751) and later on in 1804 *Sir George Cayley* were the first to perform experiments with a whirling arm to determine lift on plate segments [2]. Also *Otto Lilienthal*, around 1888, used a whirling arm apparatus to obtain lift and drag for different profiles (Fig. 1.4). Lift was measured by the weights, which balanced the thrust of the "propeller" and the drag was proportional to the time which the weights needed to reach the ground. The disadvantages of such a system are obvious. There are only short or no moments with steady state conditions during the experiment. This is likely one of the reasons why *Frank H. Wenham* (1824–1908) built a wind tunnel in 1871, which used a steam engine to drive a fan upstream of the model to generate a constant airflow through a wooden box of 3.7 m length and a cross section of 45 cm × 45 cm. This wind tunnel is the first documented wind tunnel and it was built for the Royal Aeronautical Society. It is reported that *Wenham* used a device to measure the forces on profiles by compensating the forces with weights outside the tunnel section. This device looked like a balance which was usually used to measure weight, and so the designation "Wind Tunnel Balance" may be traced back to this force measuring instrument. Later on the *Wright Brothers* employed a small wind tunnel with an external balance for their experiments with airfoils.

One of the first larger wind tunnels in which experiments with models of airplanes were conducted was built by *Gustave Eiffel* in 1910 [7]. His principle of a flow-through wind tunnel, sucking air through a nozzle, test section, collector and a diffuser with a fan drive at the end of the diffuser, is still in use today and tunnels built according to this design are called *Eiffel* type wind tunnels (Fig. 1.5). He also used an external wind tunnel balance according to the compensation principle (Fig. 1.6). So the use of external balances, working according to the compensation principle, prevailed as the wind tunnel force measurement system.

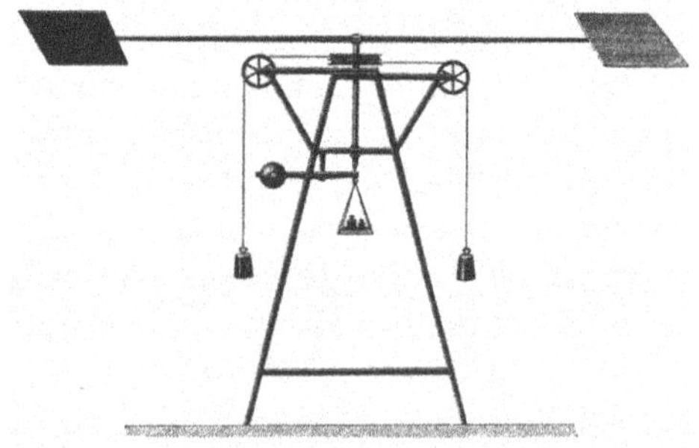

Fig. 1.4 Test rig of Lilienthal for airfoil testing

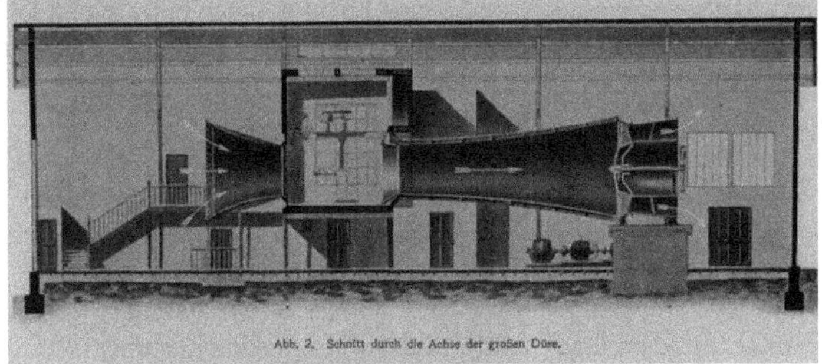

Fig. 1.5 Schematic picture of Eiffel's wind tunnel in Auteuil

Fig. 1.6 Eiffel's wind tunnel balance with model

As remarked above, some key developments took place in the early 1930s and in the mid 1940s, when electrical sensor based force transducers were used to measure dynamic forces in a wind tunnel, or in the first pressure transducer of *Nernst*. However the real breakthrough for transducer based force measurements in a wind tunnel was the invention of the strain gauge. It is understandable that scientists and the engineers immediately adopted this inexpensive and precise sensor for the development of force transducers. The high sensitivity of the strain gauge allowed the development of force transducers with a very high stiffness and precision.

In all countries with an aircraft industry, numerous wind tunnels were being built. These tunnels required precise, multi-component force measurements. Following the Second World War a large number of high *Mach* number tunnels were also built. To achieve high *Reynolds* numbers at supersonic speed these tunnels were pressurized and the model loads increased, caused by the higher density of the gas. This circumstance, and the relatively low interaction afforded using a back sting support, made the development of the compact sting balance necessary and the development of the internal sting balance was only possible by using strain gauges.

The first report of such an internal sting balance is the report of Wingham [23] from 1945. *Wingham* used strain gauges to measure lift and pitch on a model in a high-speed wind tunnel. This balance was a sting balance with two bending sections (Fig. 1.7).

In this report a reference to an earlier report from 1944 by members of *Vickers-Armstrong Ltd.* was mentioned, but this report was not published. Thus, although it is not absolutely clear who and when the first sting strain gauge balance was built, it is clear that shortly after the strain gauge was commercially available, wind tunnel engineers started to design and build wind tunnel balances using strain gauges as sensors. In the early 1950s numerous developments in the area of sting balances and external balances with force transducers are reported. Along with the construction of new wind tunnels, the development of new balances was necessary to achieve precise and reliable results for the aerodynamic force measurement.

After a long period of development, the emergence of cryogenic wind tunnels (around 1980) set new requirements for the temperature stability of internal wind tunnel balances. The balances for these tunnels were required to measure with the precision of balances at ambient temperature, but over a much larger temperature range. Without this precision, the advantage of the high *Reynolds* number achieved using the cryogenic temperatures was useless. The challenges generated by the cryogenic wind tunnels were responsible for the latest developments of strain gauge based wind tunnel balances.

Optical strain gauges with higher sensitivity than the metal foil gauges did not have a major impact on the development of the wind tunnel balances. Their capabilities can only be fully exploited when new balance materials with a much higher *Young's modulus* than that of steel are available.

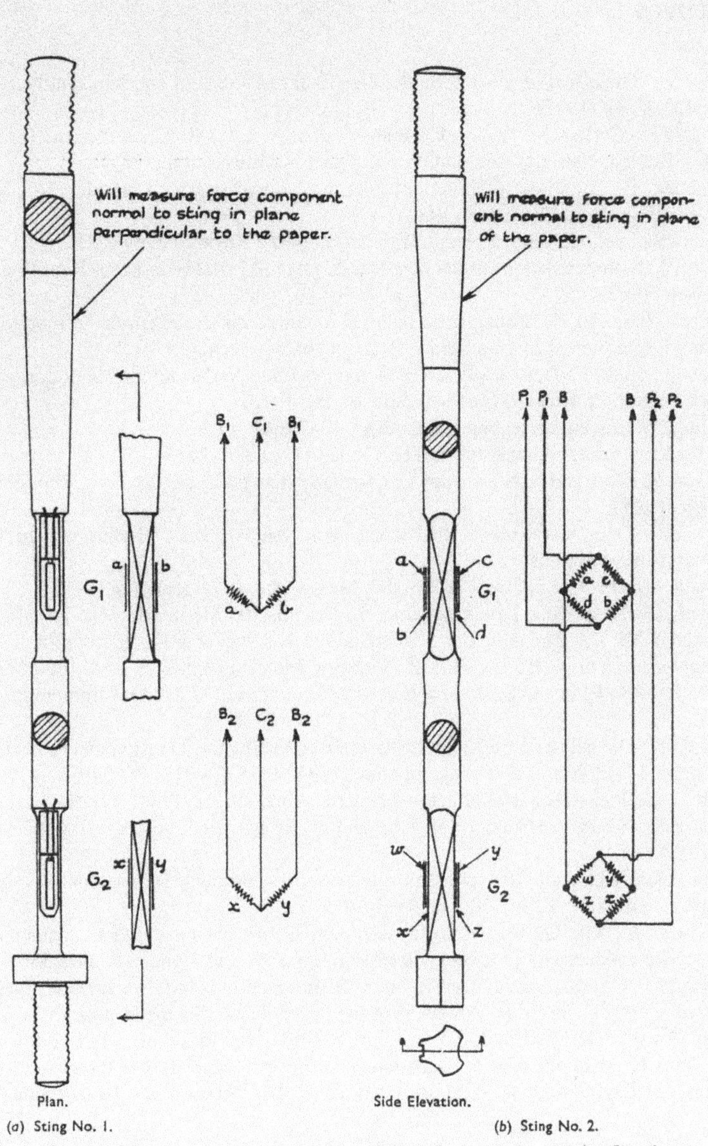

Fig. 1.7 Schematic drawing of P. Wingham balance in his report from 1945

References

1. Allen, M.: The effect of tension on the electrical resistance of single bismuth crystals. Phys. Rev. **42**(6), 848 (1932)
2. Baals, D.D., Corliss, W.R.: Wind Tunnels of NASA, vol. 440. Scientific and Technical Information Branch, National Aeronautics and Space Adminstration (1981)
3. Bridgman, P.: The effect of homogeneous mechanical stress on the electrical resistance of crystals. Phys. Rev. **42**(6), 858 (1932)
4. Christie Samuel, H.: Philosophical Transaction of the Royal Society (1833)
5. Dorsey, J.: Semiconductor Strain Gage Handbook, p. 17. Baldwin-Lima-Hamilton Electronics Division (1963)
6. Eastman, F.S.: An electromagnetic balance for force measurement or current control. Engineering Experimental Station Series, Bulletin No. 60. Univ. (1932)
7. Eiffel, G., Huth, F.: Der Luftwiderstand und der Flug: Versuche: im Laboratorium des Marsfeldes ausgeführt. Richard Carl Schmidt (Berlin, 1912)
8. Gosling, W.: Tatnall on testing. Strain **2**(4), 43 (1966)
9. Hooke, R.: De potenitia restituvia. TH Riemann (London, 1678)
10. Jackson, P.: Electrical resistance devices. British patent no.: 728,606. Tech. Rep. 728,606, Great Britain (1955)
11. Keil, S.: Eine historische Rückschau aus Anlass des 50-jährigen Jubiläums des Dehnungsmessstreifens
12. Keinath, G.: Die Technik der elektrischen Messgeräte. R. Oldenbourg (1921)
13. Kelnath, G.: Elektrische Druckmessung. tm-Technisches Messen **13**(JG), 180–183 (1932)
14. Kramer, V.M.: Die Zunahme des Maximalauftriebes von Tragflügeln bei plötzlicher Anstellwinkelvergrösserung (Boeneffekt). Z. Flugtech. Motorluftschiff **23**, 185–189 (1932)
15. McCollum, B., Peters, O.S.: A New Electrical Telemeter, 247. US Government Printing Office (1924)
16. Ohm, G.S.: Die galvanische Kette, mathematisch bearbeitet. TH Riemann (1827)
17. Simmons, J.E.E.: Material testing apparatus (1942). US Patent 2,292,549
18. Smith, C.S.: Piezoresistance effect in germanium and silicon. Phys. Rev. **94**(1), 42 (1954)
19. Stein, P.K.: Strain gage history and the end of the twentieth century. Experi. Techn. **25**(2), 15–16 (2001)
20. Stein, P.K.: A banner year for strain gages and experimental stress analysis, an historical perspective. Experi. Techn. **30**(1), 23–41 (2006)
21. Thomson, W.: Xix. On the electro-dynamic qualities of metals: Effects of magnetization on the electric conductivity of nickel and of iron. Proc. R. Soc. London **8**, 546–550 (1857)
22. Wheatstone, C.: An account of several new instruments and processes for determining the constants of a voltaic circuit. In: Abstracts of the Papers Printed in the Philosophical Transactions of the Royal Society of London, vol. 4, pp. 469–471. The Royal Society London (1843)
23. Wingham, P.: An application of strain gauges to the measurement of normal force and moments in high speed wind tunnels. Tech. Rep. No. 2316, ARC Reports and Memoranda (1945)

Chapter 2
Basics

In this book the basic terms of metrology are based mainly on common standards and definitions given by an international agreement on terminology, prepared as a collaborative work of experts appointed by BIPM, IEC, IFCC, ISO, IUPAC, IUPAP, OIML and published by the International Organization for Standardization (ISO). The vocabulary is published in "*The International vocabulary of basic and general terms in metrology*" [8]. This vocabulary covers subjects related to measurement and includes information on the determination of physical constants and other fundamental properties of materials and substances. Most of the metrological expressions are defined in the *Guide to the expression of uncertainty in measurement* (GUM) [7] and some other standards are listed in the references for this chapter.

However a wind tunnel balance is a rather complicated force measurement system and sometimes the common definitions do not cover special definitions related to the balance characteristics and uncertainties. In this chapter the common definitions are modified or supplemented for balance specific applications. They do not claim to be international definitions, but based on extensive experience, they should be appropriate.

In Europe and North America different definitions are used in defining a coordinate system, making international understanding unnecessarily difficult. For instance, for the United States there are the AIAA standards written down in "*Assessment of Experimental Uncertainty with Application to Wind Tunnel Testing*" [1] and the "*Calibration and Use of Internal Strain Gauge Balances with Application to Wind Tunnel Testing*" [5].

Nevertheless, in this book SI units are used. SI units are more practical than other unit systems because no complicated conversion factors are required.

K. Hufnagel, *Wind Tunnel Balances*, Experimental Fluid Mechanics,
https://doi.org/10.1007/978-3-030-97766-5_2

2.1 Basics of Wind Tunnel Balance System

Authors of a technical book are confronted with the question, to whom should the book be addressed—the absolute beginner wishing to learn how to build a balance or the expert already familiar with balances and wishing to have a reference to refer to when in need of specific information? In the former case, explanations must be much more encompassing and thorough, in the latter case, many details can be presumed known and basics can be dispensed with. The present book attempts to accommodate both audiences. In these first chapters the basics are reviewed, possibly superfluous information for the experienced user, but essential for newcomers to the field. Here some basic vocabulary and concepts are introduced.

The basic question is: What is a wind tunnel balance used for?

One might think to ask this question is unnecessary because it is straightforward and the easy and correct answer is: To measure the aerodynamic loads on a model. But often the real consequences from this easy answer are not well understood.

The first and most significant consequence is that the axis system used to determine the loads is the ***model axis system*** and not an axis system that is fixed on the balance. When referring to loads acting on the balance, forces and moments that are given relate to the model axis system. Every balance, even the stiffest, measures some deformation of a spring to determine the load and so the balance reference point moves away from the model reference point.

The only part of the balance that remains inline or in the same position related to the model is the model joint, where model and balance are fixed together. All other parts of the balance change their position related to the model under load.

From this definition another problem occurs. When the model determines the position of the axis system, how can its position be known during design and manufacturing of the balance? In this phase even the model to be measured with the balance may not be known. To solve this problem a virtual reference point related to the model or metric end of the balance is defined as the model reference point. This model reference point is identical to the balance reference point in the unloaded condition (Fig. 2.1). All calculations and calibrations are performed relative to this point.

If a new model is designed for a wind tunnel test, the location of the model on the balance has to be defined. Usually the center of gravity of the original is defined as

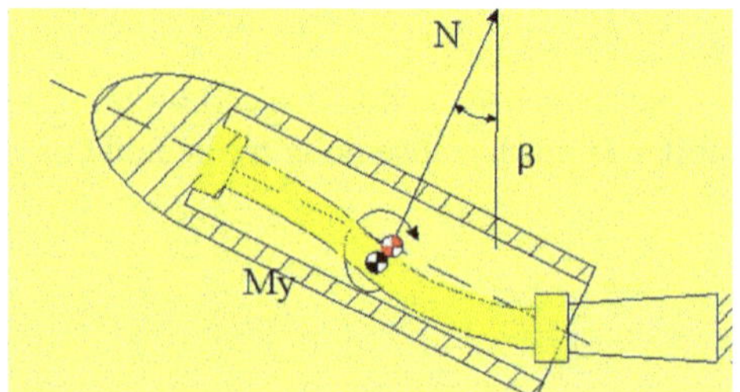

Fig. 2.1 Model reference point (red) and balance reference point (black)

the model reference point. Since the balance is usually mounted inside the model, the model reference point (center of gravity) is at the same position as the balance reference point in an unloaded condition. If this is not possible, the distances between the two points have to be precisely determined, otherwise the moment measurement will be incorrect. This applies for all types of balances. However, it is possible to use the displacement between the two points to adapt the moment range of the balance to moments that are expected during the measurement, as long as the range for the force is sufficient for the test.

Another basic question is: How do we get the loads from the balance signals?

The relation between the electrical output and the measurand of a balance is determined by a calibration, as with normal transducers. For a single transducer this is a simple procedure, where well-known calibration quantities are applied to the transducer and the readings of the transducer are listed. The slope of the relation between readings and calibration quantity is the sensitivity of the transducer. Once calibrated, the transducer can be used as an instrument to measure unknown quantities.

Unfortunately, the electrical outputs of a balance are seldom directly proportional to a particular load, as with a force transducer measuring a single load. With balances it is typical that load components other than only the desired component influence the signal. These interactions are systematic errors and must be taken into account by factors determined during the calibration. The number of these interactions depends on the number of components the balance measures and the balance design and characteristics. It can vary from only a few interactions up to 198. These factors form a matrix called the *Evaluation Matrix*. This matrix defines the relation between the signals of the balance and the loads to be measured. This matrix is also defined in the model axis system. So the outcome of a calibration is not only one calibration factor, it is a set of factors. That makes the calibration of a wind tunnel balance much more complicated and laborious than that for a single transducer.

2.2 Basic Terms of Balance Metrology

Full Scale Load, Nominal Load [N, Nm]

Full Scale Load can have different definitions; usually it defines the maximum combined load, i.e.. the maximum of one load when all other loads are also a maximum. It is also possible to use the maximum single loads. The maximum single loads are usually much higher than the maximum combined loads, so uncertainties related to these values are much smaller than those related to the maximum combined loads. This has to be taken into account if balances from different manufacturers are compared. However, in a wind tunnel test, single loads do not occur and the use of maximum combined loads or signals as full scale reference provides much more realistic information about the capability and accuracy of the balance.

Full Scale Output, Nominal Output [mV/V]
Full Scale Output is the signal measured by the equipment under full scale load defined above.

Systematic Errors or Bias of an Instrument [% of F.S.]
Systematic Errors are repeatable errors that occur for every measurement. If the reason for a systematic error can be detected, it can be eliminated by calibration or compensation. If the sources of a systematic error cannot be detected under repeatable conditions, it is the difference between the measured value and the true value, despite the fact that the true value may be unknown.

Random Errors [% of F.S.]
Random Errors are the difference between the mean value of an infinite number of measurements under repeatable conditions and the single measurement. The value of the random error is equal to the difference between the total errors minus the systematic error.

Resolution [% of F.S.]
Resolution is the smallest value that can be detected by a balance, or the smallest difference between two loads that can be detected. The maximum resolution should be in the range of 0.005% of full scale load. For a normal wire strain gauge, resolution in this range is not a problem and the remaining measurement equipment determines the limit of the resolution.

High resolution equipment today offers a resolution of 0.0003% or less of full scale. Thus resolution nowadays is only a problem of the investment available for the equipment.

Repeatability [% of F.S.]
If a certain load situation on a balance is exactly reproduced, the repeatability is the difference of the signals produced by the balance. This is a very important characteristic for a balance, because many tests in a wind tunnel only compare different model configurations. In most cases in a wind tunnel test the flow situation related to the original object cannot be reproduced exactly. So the test itself does not represent reality and engineers look at the difference between two designs under test condition and the effects measured at the model are extrapolated to the original. Repeatability can be as good as the resolution, but is usually worse. The repeatability of a good balance is in the range of 0.005% of full scale.

Repeatability depends also on time, so one distinguishes between short term repeatability, for example between two wind tunnel runs, and long term repeatability, which means the difference between two wind tunnel campaigns.

Both requirements are in the same regime because the differences between the designs of a new aircraft and the data of a well-known aircraft must be detected with the same accuracy as the differences between new model configurations. The reason for this is the gap between the flow parameters, especially Reynolds number, in the wind tunnel test and the original environment. Most of the uncertainties for the prediction of the forces on the real airplane are caused by this gap and therefore, relative measurements to the model of a well-known real airplane are used to close this gap.

Interactions, Interference [N/(mV/V); Nm/(mV/V)]
One of the major systematic errors of a wind tunnel balance is caused by the Interactions or Interference of the load to be measured with all the other remaining components. This means, for example that the sensor which measures the axial force will also yield a signal which is caused by the other components and this signal can be in the order of several percent of the measured axial force signal. This error is a systematic error that can be corrected by calibration. The corrections are given by non-diagonal elements in the evaluation matrix. These elements determine the sensitivity of the component to be measured to other moments and forces, therefore the dimensions of these elements are [N/(mV/V); Nm/(mV/V)].

Balance Zeros [mV/V; % of F.S.]
Balance Zeros or zero signal seems to be a relative simple definition. At first sight, the zero signals are the signals of the balance components measured with no load on the balance itself. This requirement is correctly fulfilled only at Zero-G conditions, which is rather difficult to achieve. That is why the balance zeros are taken when the balance is mounted to the earth side or sting side with no adapters or model on the model side. This setup is achieved during the preparation for a new wind tunnel test and in this situation the balance zeros can be taken for a check of the balance. To check the balance during a wind tunnel campaign, the wind off zero can be used. The balance zeros should be in the range of ± 10% of the full scale output, because some measurement equipment has a limited range and the full scale output of the balance itself must be lower than the maximum permissible input for the measurement equipment.

Wind Off Zeros [mV/V]
The Wind Off Zeros are the signals of the balance in a precisely reproduced position of the model with no wind in the tunnel. It is not necessary to have the model and balance in horizontal position or at zero angle of attack. The important thing is to find a position in the wind tunnel that can be reproduced very accurately. This position depends on the repeatability of the drive and the control system of the model support, which is sometimes more accurate in the end position.

Zero Drift [(mV/V)/K; %/K of F.S.]
Zero Drift is the change of the balance zeros with temperature. In this case the balance zeros can be one of the balance zeros defined above. The reasons for this change are a combination of several influences.

The first is the so-called "*Apparent Strain*" of the strain gauge itself. Apparent strain is caused by the sum of a thermal and a mechanical effect inside the strain gauge when the temperature changes. The thermal effect is the change of resistance of the gauge grid by temperature. The mechanical effect is caused by the different heat expansion coefficients of gauge and base material. Caused by this differences the gauge grid can be stretched or compressed because the thermal expansion of the gauge does not follow exactly the thermal expansion of the base material.

In an ideal *Wheatstone* bridge, the sum of all the different individual apparent strains of the gauges plus the influence of the resistance changes of the wiring, results in the *zero drift* of the sensing element.

Sensitivity [(mV/V)/N; (mV/V)/Nm; N/(mV/V); Nm/(mV/V)]
The Sensitivity of a component is the ratio of the signal for this component per unit load or vice versa. Normally the sensitivity of a balance component is very constant and it can be used to determine the signal generated by a load or vice versa. But for accurate determination of a load by a signal the use of the sensitivity is not enough. Small non-linearity can have a influence on the result and if they are not taken into account, significant uncertainties occur. So strictly speaking the sensitivity of the balance is not a constant but a function of the load itself.

Sensitivity shift [(mV/V)/NK); (mV/V)/(NmK); % of Nominal Value]
Sensitivity Shift is the change of the above defined sensitivity with temperature. This shift is also caused by different sources. One is the change of the gauge factor (sensitivity of the gauge itself), the change of the geometry of the measuring element by thermal expansion and the change of *Young's Modulus* of the base material with temperature. Adapting the shift of the gauge factor to the shift of the geometry and the shift of the *Young's modulus* can compensate sensitivity shift.

Temperature Gradient Sensitivity [mV/(VK)]
A Temperature Gradient inside a balance means that the distribution of the temperature inside the balance structure is not uniform. This non-uniform temperature distribution leads to internal deformation of the balance. These internal deformations may cause strain in the measuring elements that are sensed by the strain gauges and causing a signal too. So this signal is a real response to a deformation and is completely different from other thermal influences like zero drift or sensitivity shift. If temperature gradients cannot be avoided by the test setup, the compensation of their influence has to be achieved through the design of the balance. Only if always the same gradient occurs, compensation by thermal controlled measures is possible. This is mostly not the case, because different models always generate different heat flux inside the model leading to different gradients inside the balance.

Accuracy [% of F.S.]
The term Accuracy covers numerous different effects and is generally used to describe the correlation between certain data. For wind tunnel test data it is normally the quality of the correlation between the aerodynamic data taken in the tunnel and the true value that will be subsequently measured in a flight test. Numerous factors influence this difference, like the measurement of static and dynamic pressure, wall and sting interactions, model geometry precision, flow quality, angle of attack and balance uncertainty. This makes it very difficult to specify the accuracy for a wind tunnel test. Balance uncertainty has to be taken into account by the use of the uncertainties determined by the calibration of the balance.

Absolute Error; Uncertainty [% of F.S.]
The Absolute Error is the difference between the real load acting (true value) and the load detected by the balance. Normally the true value is not known so in practice a "conventional true value" is used. [8]. Besides all the error sources in the instrumentation and the balance, the error of the load measurement is strongly influenced by the calibration of the balance, because in this process the relation between the signals of the balance and the true loads is determined. This error includes all the errors of the measurement and the calibration. According to the "*Guide to the Expression of Uncertainties in Measurement*" [7] shortly *GUM*, this error is defined as uncertainty of the measurement and can be given as the experimental standard deviation of the entire process.

For a wind tunnel test the uncertainty of the measured data is not only dependent on the measurement of the loads. The uncertainty is also influenced by the measurement of the angle of attack, the dynamic pressure, model geometry, flow quality and all the instruments which are used to determine the aerodynamic performance and derivatives. Regarding the desired precision of the test result, minimum requirements for the load measurement can be estimated from flight test data of an existing aircraft.

Accuracy Requirements
The Requirement of the Accuracy is also connected to the price of the balance. If a balance is specially designed for one test it needs to be only as accurate as this test requires. This is unfortunately not the usual case. The high cost of a balance allows a tunnel only to have a small number of balances to cover the test capabilities of the tunnel. An estimation of the maximum required accuracy for the development of a transport aircraft [3, 4] can be made by analyzing the difference recognized between the design of a well-known aircraft and the new aircraft design in a wind tunnel test. The outcome for the accuracy requirement is that the balance has to measure a difference of 1–2 drag counts, which is equivalent to an accuracy requirement for the balance of better than 0.07–0.1% of full scale of the balance output. This value is a very approximate one and should be specified in more detail for every new test setup.

Creep mV/(Vsec); [% of F.S.]
Creep is a time or temperature dependent change of the transducer signal without a change of load. For force transducer applications usually the time to determine the creep is between 5 and 30 min after applying the load, while the temperature is constant. Values for this period are also called short-term creep. For long-term creep the inspection time can vary from several hours to one month at constant load without having a constant temperature situation.

Hysteresis [mV/V; % of F.S.]
If a force transducer yields different signals depending on the time history of the loading, the difference in signal at one load is called Hysteresis. This must be explained in more detail and the most expedient way to do this is to use a simple diagram (Fig. 2.2).

According to Fig. 2.2, a calibration load cycle is performed, if hysteresis occurs only full scale values are the same for the up and down path. Therefore the measured

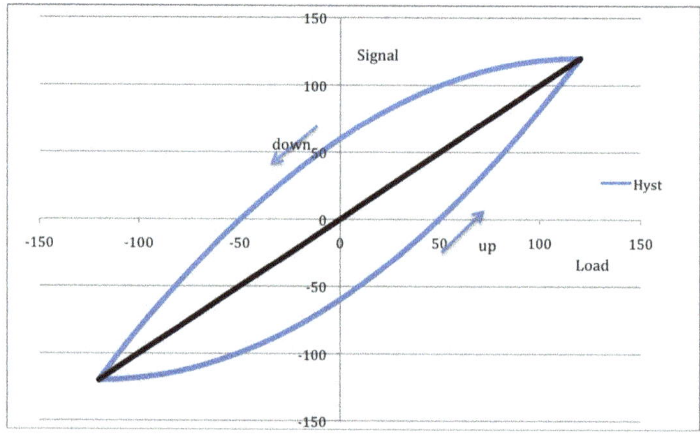

Fig. 2.2 Calibration load cycle with hysteresis

signal depends on the history of loading. In a real force transducer the values for the hysteresis are very small and cannot be observed by plotting the calibration curve (signal versus load). To make hysteresis visible, the differences between an approximation and the measured signals must be plotted.

2.3 Definition of Axis Systems

An axis system fixed to the wind tunnel is known as the wind axis system, because the main flow direction in the tunnel does not change and so these two axis systems, the flow axis system and the wind tunnel axis system, are identical.

The lift force is defined as the force on the model acting vertical to the main flow direction and according to this, drag is defined as the force acting in the main flow direction. This definition is common all over the world. Not uniform is the definition of the positive direction of the forces and moments. Due to historical reasons the definitions vary from country to country. The problems and advantages of such definitions are exemplary discussed for the definitions in Europe and the USA.

Whereas lift (normal force) and drag (axial force) are defined positive in the USA (Fig. 2.3), in Europe (Fig. 2.4) weight and thrust are defined positive in the wind axis system.

To form a right-hand axis system the side force in USA has to be positive in the starboard direction. The definitions of the positive moments do not follow the sign rules of the right-hand system. The pitching moment is defined positive turning right around the Y axis, but yawing and rolling moment turn left around their corresponding axis. This makes this system non consistent in a mathematical sense, which can cause problems when data from different tunnel are processed by common software.

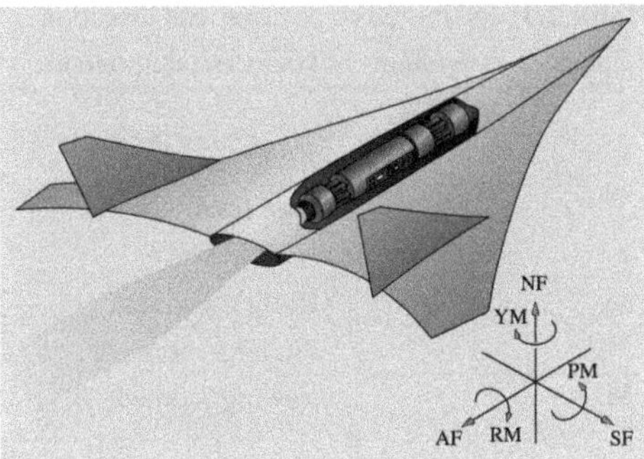

Fig. 2.3 Definition of wind axis system in USA

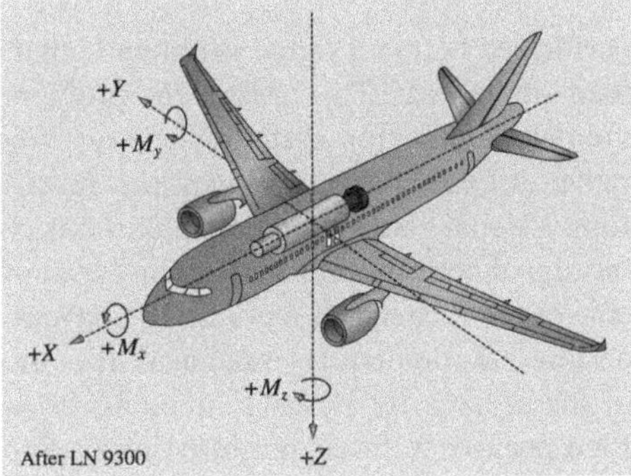

Fig. 2.4 Definition of model axis system in Europe

The European axis system is a consistent right-hand system and the definition is based on a standard written down in *DIN 9300* or *ISO 1151*.

A balance, staying fixed to the tunnel with respect to the wind axis system, always provides the pure aerodynamic loads on the model.

For the model fixed axis system, the balance does not measure the aerodynamic loads directly. The balance yields the loads acting on the model and the pure aerodynamic loads must be calculated from these components by using the correct yaw, roll, and pitch angles. The difference between the American and European definition for the positive direction remains the same in this case (Table 2.1).

2.4 Signal Conventions

For the so-called *Direct Read Balances*, where the output is directly proportional to the component to be measured, the sign convention is very easy. A positive load or moment should result in a positive output.

However numerous balances do not provide directly an output for every component. On the contrary, for most balances, to obtain the relation between output and the component to be measured some calculations of signal combination must be

Table 2.1 Definition of positive axis direction

Balance axis system	Name of component	European	USA
		Positive direction	Positive direction
X	Axial force	In flight	In wind
Y	Side force	To starboard	To starboard
Z	Normal force	Down	Up
Mx	Rolling moment	Roll to starboard	Roll to starboard
My	Pitching moment	Turn up	Turn up
Mz	Yawing moment	Turn to starboard	Turn to starboard

performed before a signal is obtained that is proportional to the value of the desired load component. This creates the problem that after the connection of a balance to the data acquisition system, checking whether the balance works correctly can be rather difficult. Several bridges may react, given an applied force or moment. Therefore, a signal convention that makes this check easier is proposed. On the other hand the signal convention for different types of balances can be different, but a practical convention instead of a common convention should be preferred. For example for an internal moment type balance, it is more practical to define the positive output of the sensors for positive moments around the X, Y, Z axes and positive signals for a positive X force, because these components can be generated by hand without installing a loading device. With four simple hand loadings, all six signals of such a balance can be checked for correct sign and function.

2.5 Relevant Wind Tunnel Standards

There is no common international standard for wind tunnel balances. There is also no agreement between the USA and European standards. The main differences are the definition of the axis systems and the unit dimensions used. In this book only SI units are used, because most of the countries in the world agreed to the SI unit convention. Even the USA signed the contract for the use of SI units [m] Meter, [kg] Kilogram and [s] Second in 1875(!), but did not establish its use up to now.

In Table 2.2 a collection of publications relevant for the use of wind tunnel balances is given. One can hope that there may be only one someday! Further references to additional norms and standards can be found in the bibliography of this chapter.

Table 2.2 Wind tunnel standards

Subject	Number	Date of publication	Region
Axis system	DIN 9300	1990	D
	ISO 1151	1982–1996	EU, (World)
Terms	ISO 5725	1994–2002	EU, (World)
Calibration	AIAA R-091-2003	2003	USA
Laboratories	DIN EN ISO/IEC 17025	2000	World
Statistics	ISO 3534	2006–2009	World

References

1. AIAA: Assessment of experimental uncertainty with application to wind tunnel testing. AIAA S-017A Washington DC **6**, 1–15 (1999)
2. EN, D.: DIN EN 45020 2007–03 Standardization and related activities; General vocabulary; (ISO/IEC Guide 2 2004); Trilingual version EN 45020 2006. Beute Verlag, Berlin (2004)
3. Ewald, B.: Balance Accuracy and Repeatability as a Limiting Parameter in Aircraft Development Force Measurements in Conventional and Cryogenic Wind Tunnels. Technische Hochschule Darmstadt, Darmstadt, Germany (1987)
4. Ewald, B., Krenz, G.: The accuracy problem of airplane development force testing in cryogenic wind tunnels, p. 776. In: 14th Aerodynamic Testing Conference, West Palm Beach (1986)
5. Internal Balance Technology Working Group: Recommended Practice: Calibration and Use of Internal Strain-Gage Balances with Application to Wind Tunnel Testing (AIAA R-091-2003) (2003)
6. ISO: ISO 5725-1:1994 Accuracy (Trueness and Precision)of Measurement Methods and Results—Part1: General Principals and Definitions. International Organization for Standardization (1994)
7. ISO: Guide to the Expression of Uncertainty in Measurement (GUM). International Organization for Standardization, Genève, Switzerland (1995)
8. ISO: International Vocabulary of Basic and General Terms in Metrology (VIM), pp. 09–14. International Organization for Standardization, Genève, Switzerland (2004)
9. ISO: Vocabulary and Symbols—Part 1: General Statistical Terms and Terms Used in Probability (ISO 3534-1: 2006). International Organization for Standardization (2006)
10. International Organization for Standardization: ISO 8402: 1994: Quality Management and Quality Assurance-Vocabulary. International Organization for Standardization
11. International Organization for Standardization: ISO 10012 Measurement Management Systems-Requirements for Measuring Processes and Measurement Equipment. International Organization for Standardization (2003)

Chapter 3
Balance Types

Balance types are differentiated by the number of components they measure simultaneously: one to six is possible. Balances placed inside the model are called "*Internal Balances*" and if they are located outside the model or the wind tunnel, they are called "*External Balances*".

Another criterion to differentiate balances is the way they are built. If they are manufactured from one piece they are called "*One Piece Balance*" or "*Mono Piece Balance*". If they are built from various parts that are assembled together they are called "*Multi Piece Balance*". Some mixed constructions exist, where parts are brazed or welded together, or where only a few parts are screwed into a single piece block.

Last but not least balances can be differentiated by the relationship between the loads and the signals. For the "*Direct Read Balances*" one signal is direct proportional to one load and in contrast with "*Indirect Read Balances*" more than one signal is affected by the load to be measured.

In every category of balance, numerous different sub-types exist and there is no common classification for all variations, so there are no doubt balances which may not fit into the above classification scheme.

3.1 External Balances

"*External Balances*" are always placed outside the wind tunnel test section and the reference axis system of the balance can be either the model axis system or the wind axis system or even a combination of both. Which system is most appropriate depends on the position where the model positioning system is integrated into the entire setup.

If the model positioning system is placed between the balance and the model, the balance always remains in the wind axis system. If the model positioning system is integrated between the balance and the earth, the balance measures in the model

K. Hufnagel, *Wind Tunnel Balances*, Experimental Fluid Mechanics,
https://doi.org/10.1007/978-3-030-97766-5_3

axis system. There are combinations where the angle of attack is adjusted by a mechanism between balance and model and the yaw positioning is performed by rotation of the whole support. In this case the loads measured by the balance must always be transferred into one of two axis systems.

Two types of external balances exist. The first is the "*One Piece External Balance*", which is made from one piece of material and is equipped with strain gauges (Fig. 3.2). Another expression for this kind of balance is also "*Sidewall Balance*" or "*Half Model Balance*". The main characteristic of a sidewall or half model balance is that they usually measure only 5 components. The force vertical to the wall is not measured. In a half model setup, as shown in Fig. 3.1, it can be seen that a measurement of force vertical to the wall makes no sense. Moreover, to add the 6th component the effort to build the balance is much higher.

The other type of balance is the "*Multi Piece Balance*". It is formed of single force transducers that are connected by a framework (Fig. 3.3). Such a design can be built very stiff, but it requires significantly more space compared to the one-piece design. However, outside the wind tunnel more space is usually available and thus, the construction can be optimized with respect to measurement requirements, like optimized sensitivity, stiffness and decoupling of load interactions. Usually such balances are designed to measure six components. The capability to use it for full models or half models makes the correct definition of the load ranges difficult, because a half model setup requires a much larger rolling moment range than a full model setup.

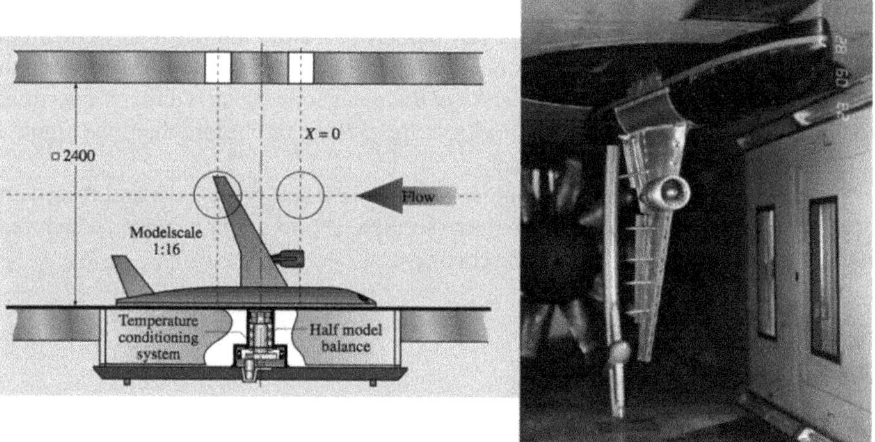

Fig. 3.1 Half model balance test setup

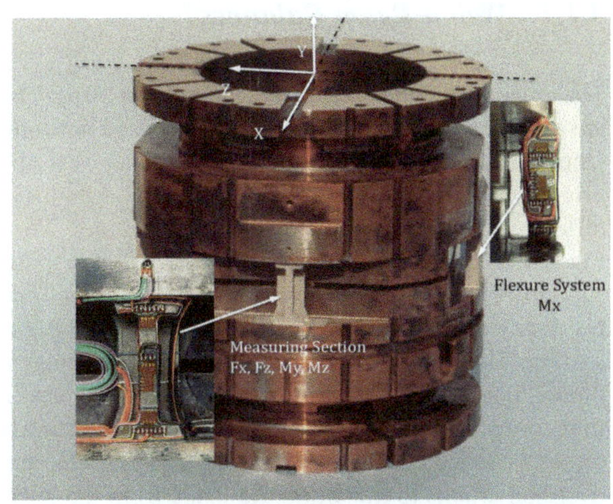

Fig. 3.2 Half model balance

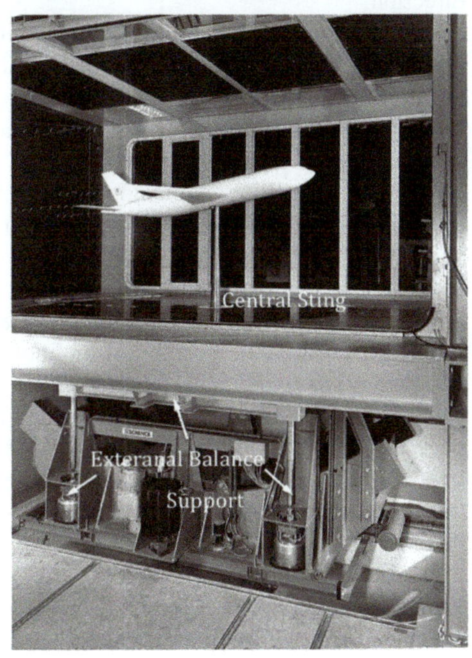

Fig. 3.3 External balance and support with a full model

3.1.1 Weigh Beam Balances

The archetype of the external balances is the "*Weigh Beam Balance*". These balances are designed like the real beam balances used in former times on every marketplace. One example of such a weigh beam balance was already shown in Chap. 1 in the original Eiffel wind tunnel (Fig. 1.6). Six weigh beams are used to measure the different components on a wind tunnel model. The principle design is shown in Fig. 3.4. Characteristic for this type of balance is the overhead mounting and the support of the model by wires. This arrangement makes the use of large pre-loads necessary, which are mostly damped in a water basin beneath the wind tunnel test section. The wire support system has the advantage of very low interaction on the model, but the entire system has a very low natural frequency that requires a long time for balancing and averaging the data. The weigh beams for lift, pitch and roll are directly connected to the model, while side force, drag and yaw must be connected over bell cranks to transform the loads from the horizontal direction to the vertical direction.

In the example shown in Fig. 3.4 the weigh beams are numbered with W_i for the drag and A_i for the lift direction. Side force is the only component direct proportional to the measurement of the weigh beam S. All other components are not measured directly by one weigh beam, but are obatined by summing up or subtracting different force measurements. Drag is proportional to $W_1 + W_2$, lift is equal to the sum of $A_1 + A_2 + A_3$, rolling moment is equal to $A_1 - A_2$, pitch is equal to $A_1 + A_2 - A_3$ and yaw is proportional to the difference of $W_1 - W_2$ (Figs. 3.5 and 3.6).

Originally the sliding weights on the weigh beam were balanced manually and the data was read off the beam. So practically six people were needed and had to read the data simultaneously. Later the balancing was performed automatically and a sensor measured the position of the motor-slid weight. That reduced the time for one data point significantly and increased accuracy. However the very low natural frequency and the long time that is needed for the stabilization of the sliding weight on the weigh beam still required a very long time for one measurement. Therefore, in some balances a load cell on each weigh beam replaced the sliding weight and thus, the compensation measurement principle of the balance was changed to the deflection measurement principle, albeit with relatively low deflections.

Throughout this step-by-step evolution some very old external balance mechanisms survived in some wind tunnels and are still in use today.

These balances are in most cases relatively accurate, but due to the high balance mass and the flexures for the bell cranks etc., the natural frequency does not change very much and so the measurement is usually slow. The above-mentioned "relative small deflections" of the balance can be in the range of some millimeters for the lateral deflections and about $0.1°$ for the angular deflections. For precision measurements the lateral deflections are not of interest, but angular deflections greater than $0.02°$ must be corrected to obtain reliable results. For this reason usually the angle of attack is measured directly in the model itself. The magnitude of the deflections depends on

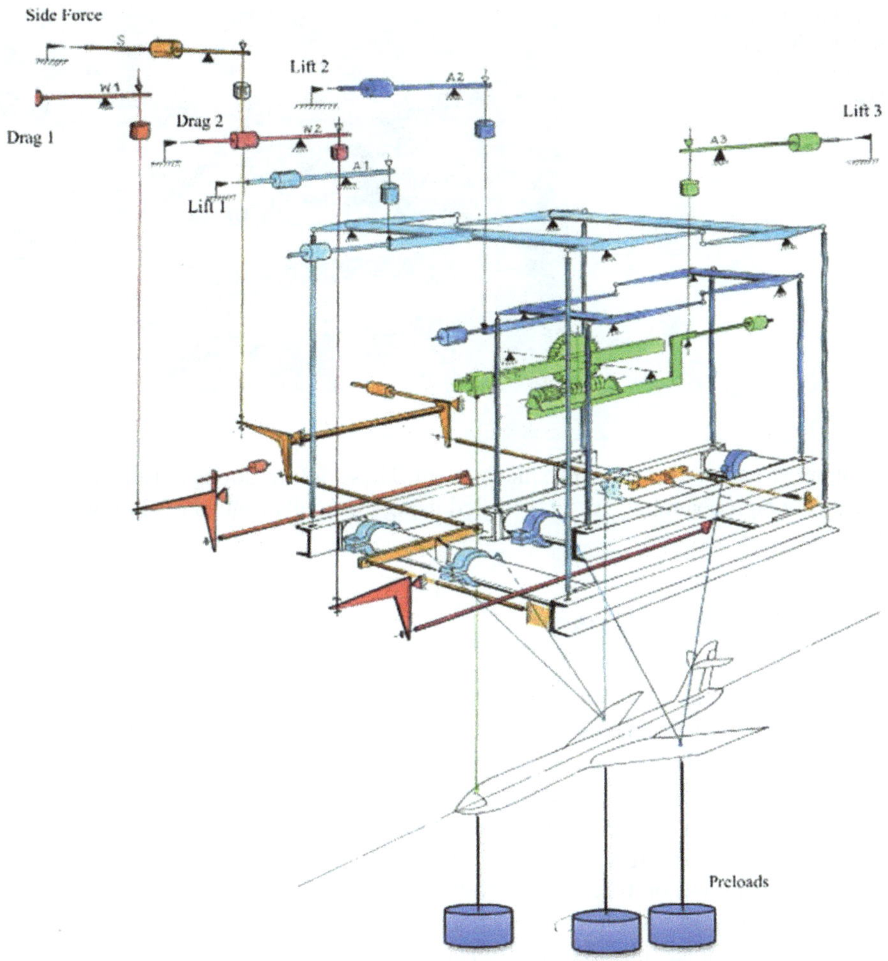

Fig. 3.4 Schematic drawing of a weigh beam balance

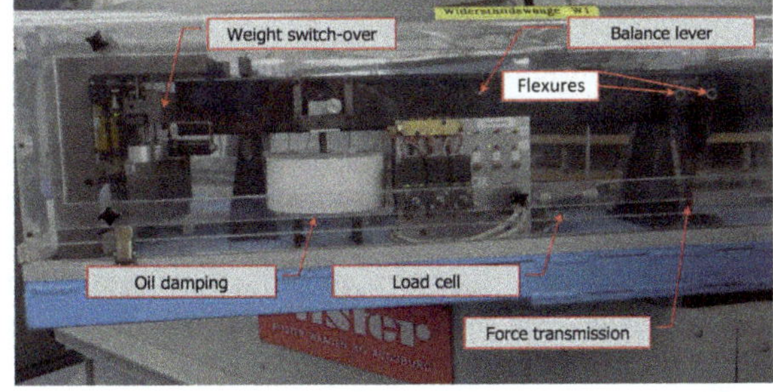

Fig. 3.5 Load cell suspended weigh beam system

Fig. 3.6 Overhead balance with mechanical controlled weigh beams (TU Darmstadt)

the inner structure of the balance mechanism. These deformations must be estimated before a weigh beam balance is modified using load cells instead of slide weights.

3.1.2 Pyramidal Balance

A principle design of a "*Pyramidal Balance*" is shown in Fig. 3.7. The reason why it is called a pyramidal balance is that the extensions (a, b, c, d) of the four inclined linking rods from the first plate and the second plate intersect exactly in the reference point (O) of the balance. The linking rods have pivots at each end so that the two platforms can be moved relative to each other.

If a force is acting in X, Y or Z direction exactly in this point there will only be tension or compression forces in the rods $(a,\ b,\ c,\ d)$ and the supports A, B, C, D. The flexible system stays in place without any movement. The forces $X,\ Y,\ Z$ can be measured by load cells that are aligned to the axis of the rods $(a,\ b,\ c,\ d)$. X is proportional to the signals of $(A + E) - (C + D)$, Y is proportional to $(A + B) - (C + D)$ and Z is proportional to $(A + B + C + D)$.

If a moment (M_x, M_y, M_z) is applied in the reference point, the two platforms try to move relative to each other, except this movement is prevented by the supports in (E, F, G). So the product by the forces measured in the support points E, F, G multiplied by the distance to the relevant rotation axis (e, g) is only proportional to the moments M_x, M_y, M_z and not to any force applied to the reference point ($M_x \sim F * e$; $M_y \sim E * e$; $M_z \sim G * g$). The signals for the moments can be adjusted to the load range of the load cells, for example by varying the length of the levers.

This is the very basic principle of a pyramidal balance and the great disadvantage can be easily seen by the equations to obtain signals from the forces. The signals of all load cells must be used to compute any one force. So even if the aligning of rods and the sensitive axes of the load cells is perfect, there are inherent interactions among the forces and the full scale range of every load cell is determined by the sum

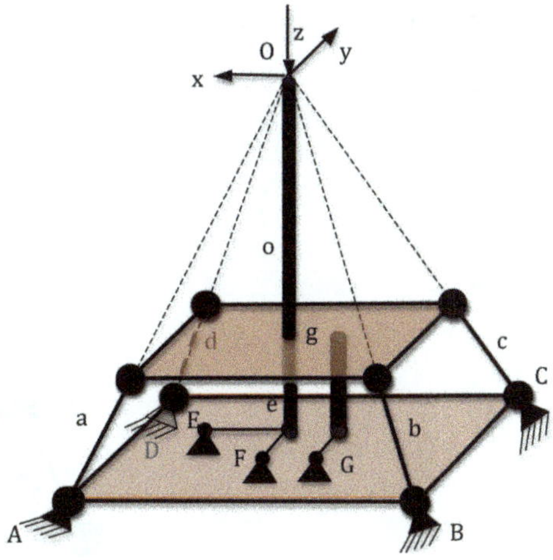

Fig. 3.7 Principle design of a pyramidal balance

of the maximum single loads. For each component the full scale output will be only one third of the full scale signal of one load cell for the ideal case that all full scale loads are the same. Normally the full scale values for X, Y, and Z differ in the order of 3–5 and so the resolution for the most sensitive load is 3–5 times lower than for the maximum single load. This leads to the problem that for the load with the lowest load range the sensitivity is the poorest.

To separate the influence of the forces on each other and to adjust the ranges of a load cell to the range of the load the principle design from Fig. 3.7 can be extended by additional rocking piers, as shown in Fig. 3.8. Now the X force is measured by the load cell in M and the Y force is measured by the load cells L_1 and L_2. Two load cells are needed to stabilize the whole arrangement with respect to yaw. The vertical force Z is directly proportional to the sum of signals in $H,\ I,\ J,\ K$.

With such a system the decoupling of forces and moments is nearly perfect and this is the big advantage of such a pyramidal balance. There is no summing up or subtracting of different signals to obtain readings that are proportional to the components. Thus, the pyramidal balance can be also called "*Direct Read External Balance*". The advantage of obtaining direct readings for every component is achieved by combining a pyramidal balance with a platform balance, which will be described later. Seen from the technical standpoint, the "*pyramidal balance*" is a relatively complicated external balance, because numerous mechanisms have to be built and a several load cells are needed (see Fig. 3.8). The precision in aligning the rods has to be perfect, otherwise the effect of a perfect decoupling of forces and moments cannot be achieved and the effort to determine the interaction through calibration increases (Fig. 3.9).

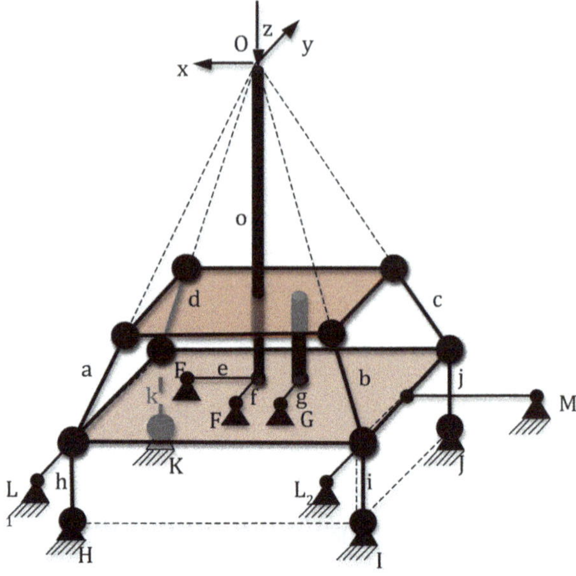

Fig. 3.8 Extended principle design of a pyramidal balance

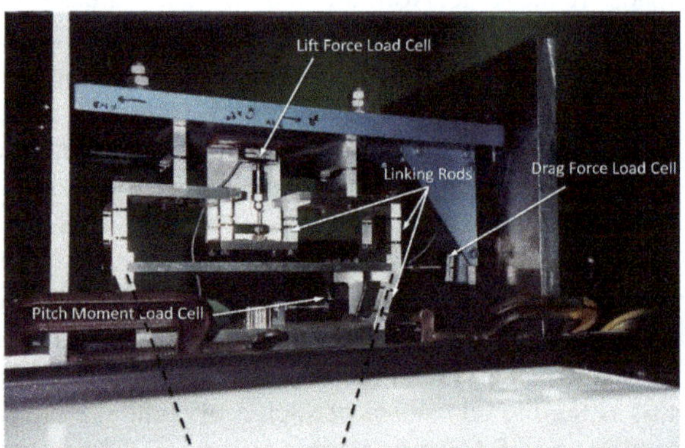

Fig. 3.9 Three component pyramidal balance overhead on a small wind tunnel at the University of Wichita

3.1.3 *Platform Balance*

The "*Platform Balance*" is now the most common type of external balance. In the principle illustration in Fig. 3.10 it can be seen that the design is much simpler than the design of a pyramidal balance. The platform (red triangle) is supported by load cells at the points B, C, D, E, F, and G and the load cells are connected to the platform by rods b, c, d, e, f, and g with elastic hinges at both ends. These rods are used to decouple the load cells from loads other then the force acting directly in the axis direction. The adaption of the load cell range to the moment range of the balance is achieved by the choice of the distance from the load cell to the balance axis and

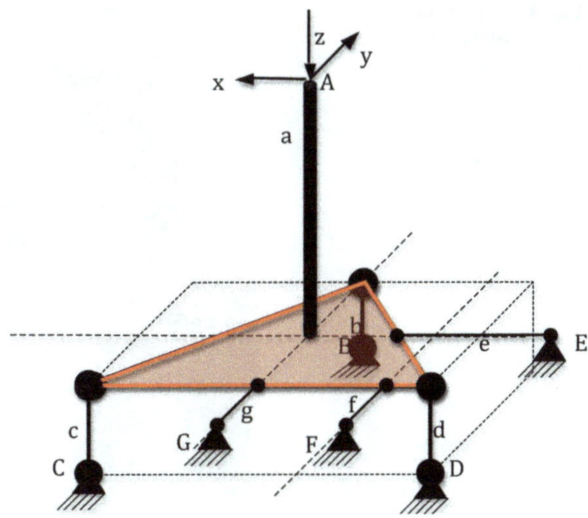

Fig. 3.10 Principle design of a platform balance

the distance relative to the load cells (GF; CD; CB; CD). Load cells to measure only the forces are placed directly in the axes of the balance.

In Fig. 3.10 the load cell at point E measures the axial force F_x, the load cell at point G measures the side force F_y. The normal force F_z is given by the sum of the signals of the three load cells in the points $(B + C + D)$. The yawing moment M_z is measured at point F. Pitch M_y is the combination of $(C - D)$ and roll M_x is proportional to B.

There are other possibilities to obtain the loads out of the six signals of the load cells. For example, side force can also be proportional to the sum of $(G + F)$, while the yawing moment M_z is proportional to the difference $(G - F)$. When the normal force F_z is again the sum of $(B + C + D)$, the combination of $(C + D - B)$ is proportional to roll M_x and $(C - D)$ is again proportional to pitch M_y. The load cell at point E gives a signal only related to axial force. Using these combinations the load cell G does not have to be aligned with the Y axis.

This description shows the main difference between the platform balance and a pyramidal balance. While in the pyramidal balance the signal of one load cell is almost proportional to one component, in a platform balance the signal that is proportional to one component must be calculated from different readings of several load cells. This means that there is a built in interaction between components, because some load cells react to more than one component. Some of these interactions can be up to 50%. Consequently, these interactions must be precisely measured during calibration and taken into account in the evaluation matrix of the balance. To decouple the measurement of loads as good as possible, rocking piers are used. The longer they are the lower are the influence of moments or orthogonal forces.

Figure 3.11 shows the external platform balance of the TU Darmstadt low speed wind tunnel. The entire balance is mounted on a support that can be rotated by $\pm 180°$ in yaw direction, so that side wind and cross wind can be simulated. The

Fig. 3.11 External platform balance of TU Darmstadt low speed wind tunnel

angle of attack is adjusted by a mechanism on the balance itself. For a full model and pure changes of angle of attack, the balance measures in the wind axis system. If additionally the yaw angle is changed, some of the loads are measured in the wind axis system and some are measured in the model axis system.

If a half model is mounted on the balance, the forces are always measured in the model axis system, because now the yaw system is used to change angle of attack.

3.1.4 Coaxial Balances

Another interesting type of balance is the "*Coaxial* or *Column Balance*". This balance is especially designed for low lateral space on any side of the test section. It requires a small long cylinder vertical to the roof or bottom of the test section and it looks like an extension of the central sting with a larger diameter. An example of such a balance can be seen in the Kirsten Wind Tunnel of the University of Washington (Fig. 3.12).The principle design of a column balance can be seen in Fig. 3.13.

This balance is also a balance where the signal for a component has to be calculated from signals of different load cells, with the exception of the normal force Z, which is directly proportional to the signal of the load cell at point G. The axial force X is proportional to the sum of $(C + F)$ and the side force is given by the sum of $(B + D + E)$. The moment around the X axis is proportional the difference $(B - E)$ and the moment around the Y axis is proportional the difference $(C - F)$. M_z is proportional to the signal measured in E. The fact that the most sensitive component of the balance has to be calculated by signals out of two load cells that measure the M_y moment also, is the most notable disadvantage of this balance type. The measurement of the axial force is influenced by another load that generates much

Fig. 3.12 Column balance of Kirsten wind tunnel; University of Washington

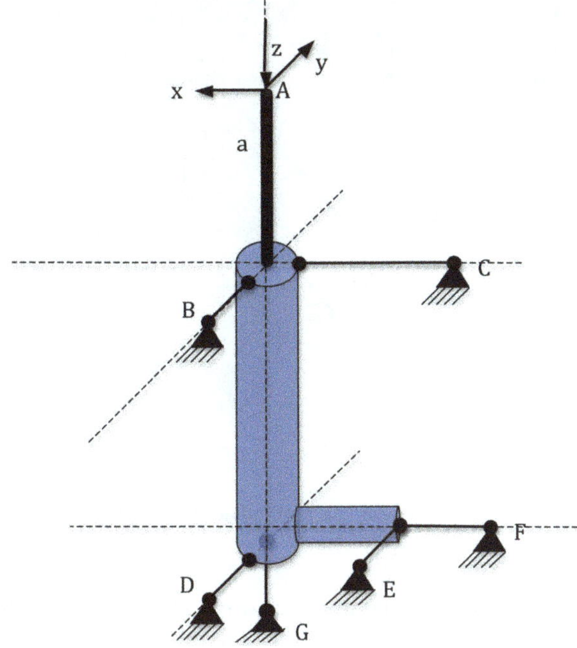

Fig. 3.13 Principle design of a column balance

higher forces than the axial force full scale and so load cells with a higher full scale range are needed. The result is a much lower sensitivity for axial force and a higher interaction of M_y on the axial force X.

Some problems of such a column balance are overcome by the design of the external balance for the Kirsten wind tunnel of the University of Washington in Seattle. The balance is designed in such a way that the forces measured by electromagnetic load cells are almost only proportional to a single force or moment. This must to be explained in more detail.

The sketch of the balance is shown in Fig. 3.14. The basic design consists of two concentric tubes. The inner tube (orange) is connected to the model. In the real

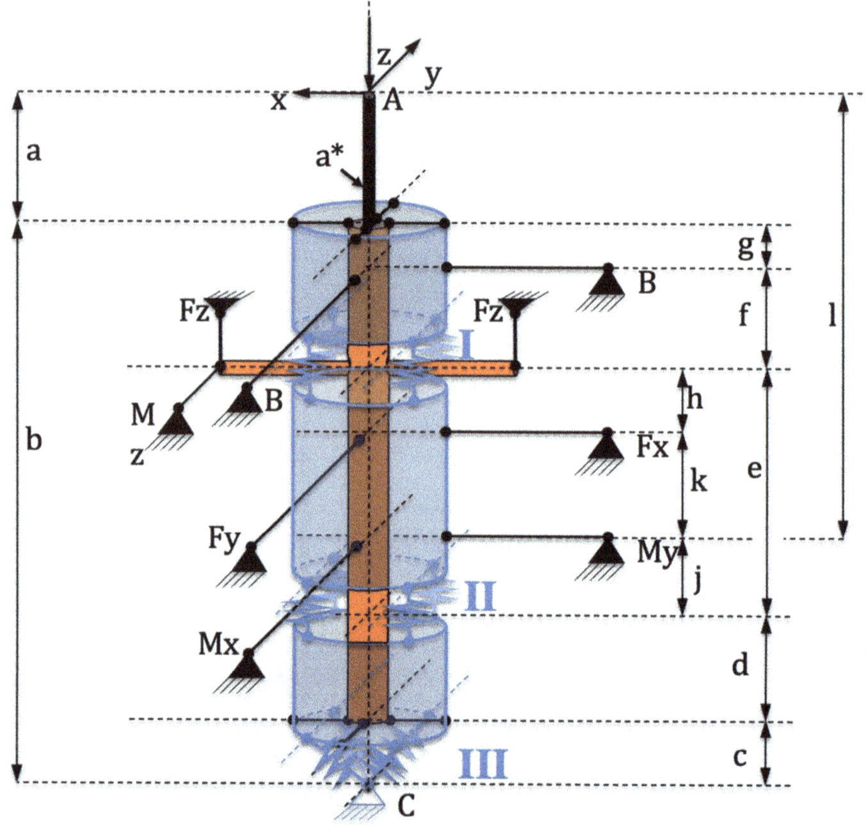

Fig. 3.14 Sketch of Kirsten wind tunnel external balance

balance the yaw and the pitch drive are integrated in the upper end of the main tube between the main tube and a^*, but this mechanism is not shown in the sketch.

The main tube is connected via rocking piers to the upper and lower outer tube (blue). Upper and lower tube are connected in their axis by the two elastic gimbals I and II. All three are fixed by another gimbal III to the earth, which acts in the sketch coexistent as the lower side support in point C.

The inner tube transfers the lift force F_z via two cantilevers directly to the lift load cells, and in the same manner the yawing moment (M_z) is transferred to the yaw load cell. All other loads, drag (F_x), side force (F_y), pitching moment (M_y) and rolling moment (M_x) are transmitted by the rocking piers from the main beam over the upper and lower outer coaxial tubes to the middle one. This relatively complicated arrangement is chosen to separate axial force F_x and pitching moment M_y in the X-Z plane and in the same way to separate side force F_y and rolling moment M_x in the Y-Z plane. By a calculation shown in [6] the dimensions h, k, and j can be calculated for given dimensions a, b, c, d, e, f, and g so that a very good separation of load measurement is possible. The equations for the calculation can be obtained by balancing the moments and the forces in the gimbals I, II und III. To separate loads in this complicated way is due to the fact that in 1952 no computer was available to perform a matrix operation. So a numerical correction of the interactions was not possible and it had to be integrated into the design of the balance. In this design yaw and pitch adjustments are achieved by mechanisms placed between the inner balance tube and the model, so this balance always measures in the wind axis system (Fig. 3.15).

3.1.5 *Yoke Balance*

In some cases the space outside the test section of a wind tunnel is not large enough for a platform balance. So the "*Yoke Balance*" can be a solution to solve such space constraints. On two sides of the test section there are two separate three-component balances that are connected by a yoke underneath the bottom. Both three-component balances are used to measure the five components X; Y; Z; M_x; M_z by adding or subtracting signals of the load cells B; C; D; E: $X \sim$ Sum (B, E); $Y \sim$ Sum (D); $Z \sim$ Sum (C); $M_x \sim$ Difference (C); $M_z \sim$ Difference (B) and M_y is measured by the adding the signals of the load cells in B subtracting it from the signal in E.

The extra support in E is necessary because the connections between the load cells and the yoke must be rocking piers to minimize the interactions. If the point A is part of the center axis of the three-component balances (Fig. 3.16) the signal of E is direct proportional to M_y and the sum of the signals in B are direct proportional to X.

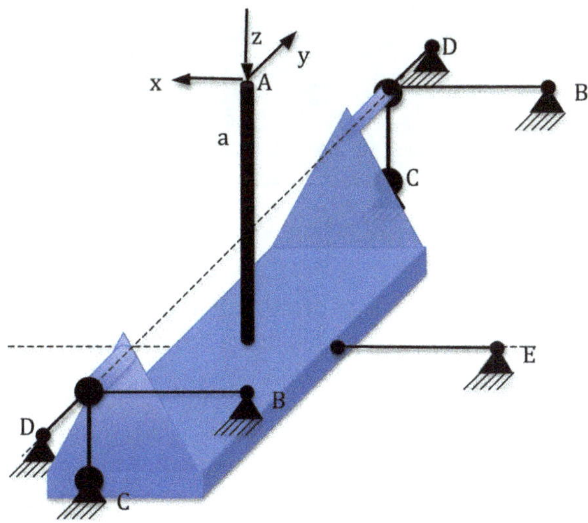

Fig. 3.15 Principle design of a yoke balance

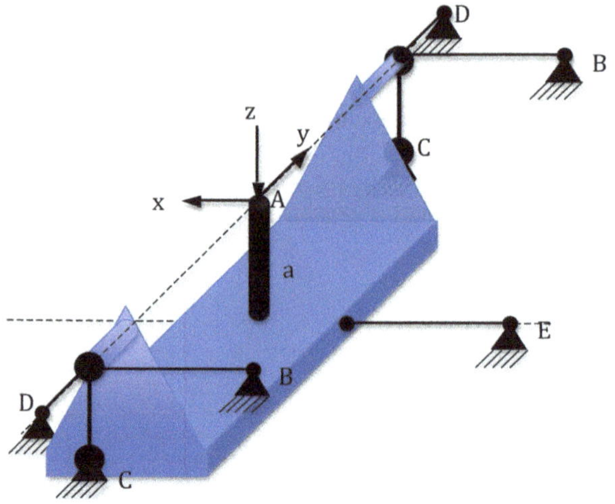

Fig. 3.16 Yoke balance with centered balance reference point

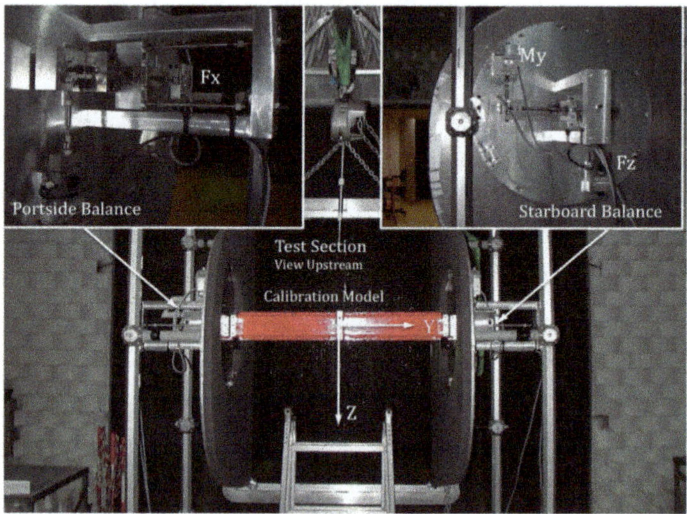

Fig. 3.17 Yoke balance of DLR acoustic wind tunnel

The external balance of the DLR acoustic wind tunnel in Braunschweig (Germany) is a yoke balance type, although it does not look like one (Fig. 3.17). On each side of the large plates that form a kind of nozzle, a full three component balance is mounted. These two balances are connected by the model, which assumes the function of the yoke. This balance was built only to measure lift, pitch and drag on profiles and to allow acoustic measurements at the same time.

In the picture shown in Fig. 3.17, the balance is shown during calibration and a red calibration bar is installed instead of an airfoil model.

3.1.6 Spiral Spring Type Balance

A very interesting type of external balance was designed by a student of the University of Applied Sciences of Regensburg. The balance is designed like a *spiral spring*, where the forces are measured by bending beams and the moments are measured by torsion bars (see Fig. 3.18). If one follows the flux of forces from the metric end to the non-metric end this flux forms a spiral, which is why this type can be designated as a "*Spiral Spring Type Balance*". Attached to the metric end there is a torsion bar to measure the yawing moment Mz, followed by a bending beam parallelogram to measure the side force Fy. After that again there is a torsion bar to measure pitch and a bending beam parallelogram to measure drag. Finally there is another torsion bar to measure roll and another bending parallelogram to obtain lift.

This design leads to a very compact balance. The disadvantage is that there is no decoupling of the components. Every sensor is loaded with all components at the same time and the sensitivity to the desired component is managed by the orientation of the sensor and the kind of strain that is measured. Also the adaption of the sensitivity to the required load range is very difficult, especially if the variation in full scale loads is large.

3.1.7 Half Model and Side Wall Balances

"*Half Model or Sidewall Balances*" are external balances that are mounted into one sidewall or in the top or bottom of the tunnel test section. The reason for sidewall mounting is mostly given by the test setup that requires free access to the top or the bottom of the model. The problem related to a sidewall balance is to achieve a stiff mounting platform. Most sidewall balances are half model or box type balances that are designed as external balances. A typical half model setup and balance has already been shown in Figs. 3.1 and 3.2.

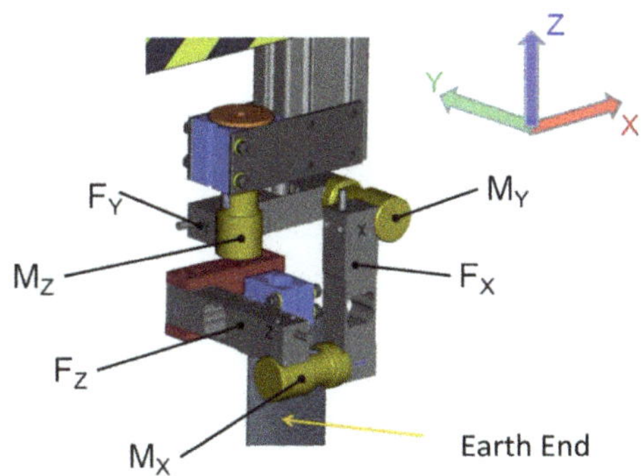

Fig. 3.18 Spiral spring balance from Regensburg University of Applied Science wind tunnel

3.1.8 Three-Flange Balance

The "*Three-Flange Balance*" (Fig. 3.19) is a combination of a half model balance with a compact external balance. Normally half model balances are not able to measure the normal force and so for full model measurement an additional six-component balance is needed to perform testing with a central sting mount. The integration of a normal force measurement section into a half model is very difficult, because this section has to carry the high rolling moment generated by the asymmetrical half model configuration. On the other hand, the high rolling moment needed for the half model makes the balance very insensitive to roll for the full model configuration.

The idea of the three-flange balance is to have two flanges on the metric side. One is used for the half model setup and the other one is used for the full model configuration. With this configuration of sort a two range balance is designed and the load ranges can be switched by mounting the model to either one of the two flanges.

The additional second flange is connected via spokes to the original flange of the half model balance. The spokes are designed for normal force and pitching and rolling moment of the full model. Side force and axial force and yawing moment are measured by original half model configuration.

The advantage of this configuration is to use the sidewall balance as a full functional external balance without the disadvantage of the unadapted load ranges for normal force roll and pitch in one balance setup.

3.2 Internal Balances

"*Internal Balances*" designates all balances that are placed inside the model. They always move with the model and their reference axis system is always the model axis

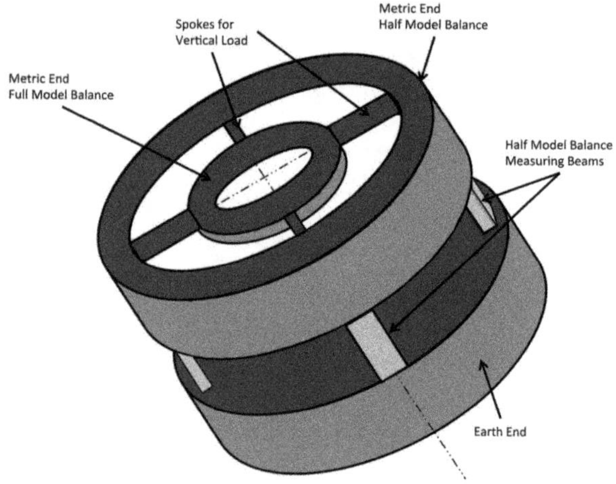

Fig. 3.19 Three flange external balance

system. Inside the model the available space is restricted, so internal balances have to be relatively small compared to external balances. Generally two different types exist: the monolithic type, where the balance body is a single piece of material, which is designed in a way that some areas are stressed mostly by the load they have to measure. The other type uses small transducers that are orientated with their sensing axis in the direction of the load they should measure. They are combined within a solid structure.

An internal balance measures the total model loads, it is mostly placed in or near the center of gravity of the model.

Two general groups of internal balances exist. The first group consists of the so-called "*Box Balances*". They can be built either from one piece of material or they can be assembled out of many different parts. Their main characteristic is that their outer shape looks like a cuboid and the loads are transferred from the top to the bottom of the balance, as will be detailed in Sect. 3.2.2.

The other group is known as *sting balances*. These balances have a cylindrical shape and the loads are transferred from one end of the cylinder to the other end in longitudinal direction (Fig. 3.20). Sting balances can also be built from one piece or out of several pieces.

Another possibility to characterize internal balances is given by the manner in which they measure the components. The *force balances* use the deformation measurement caused by tension and compression to determine the loads and the *moment balances* use the effect of bending or torque to determine the loads. Also types that use all effects are conceivable. This differentiation is possible for both sting balances and box balances.

The next characterization of balances is the relationship between signals and loads. First there are the *Direct Read Balances*. The output of each sensor of such a balance is directly proportional to one load component. In contrast, the *Indirect Read Balance* uses more than one signal for the determination of a load and more than one load affects a particular sensor.

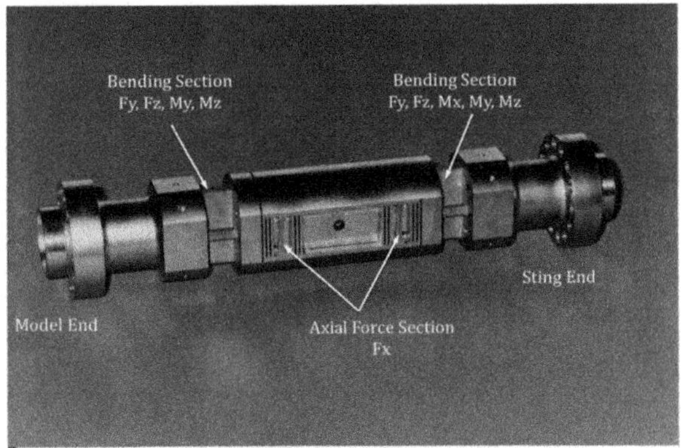

Fig. 3.20 Internal sting balance for a cryogenic wind tunnel with double axial force section

3.2.1 *Sting Balances*

"*Internal Sting Balances*" can be divided into two different groups. One group is the "*Force Balance*" type and the other group is the "*Moment Balance*" type. If the bridge output of every bridge is directly proportional to one load component these balances are called "*Direct Read Balances*".

Typical for all groups is that axial force and rolling moment are directly measured with one bridge. The measurement of lift force and pitching moment or of side force and yawing moment is achieved in different ways that characterizes every group.

Moment type balances and force type balances do not naturally give a direct output proportional to lift/pitch and side/yaw. The signals, which are proportional to each of these loads have to be calculated by summing or subtracting a signal from each other before they are passed to the data reduction scheme. The advantage of this is a very concentrated wiring on each section that is less sensitive to temperature effects. Moment type balances can be built as direct read balances, but in this case the inner bridge wiring has two short and two long wires. The long wires must be installed across the active length of the balance and so their influence on the bridge output, especially due to temperature sensitivity, is very high. (see Sect. 8.6).

3.2.1.1 Force Balances (Tension/Compression)

This type of balance uses two measurement sections that are placed in the forward and the aft section of the balance. In these measurement sections a forward and aft force is measured using tension and compression transducers. These forward and aft force components are used to calculate the resulting force in the plane and a moment around the axis that is orthogonal to the force axis. Typical force balances are the balances built by *Able* Corporation (see Fig. 3.21). The axial force and the moment around the X axis are measured in a separate section at the forward end. The rolling moment is the only component that is measured by a moment sensor directly.

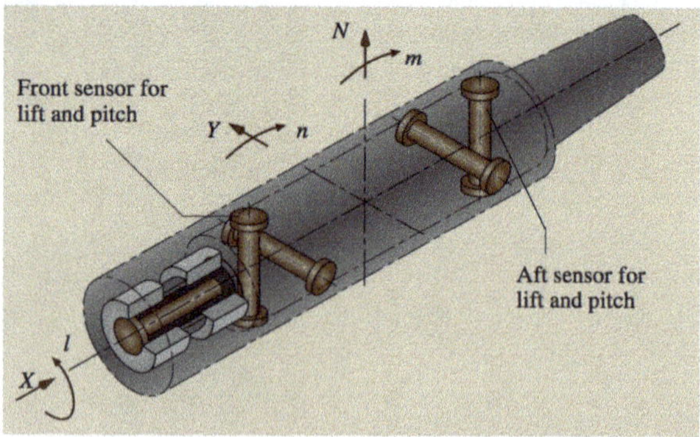

Fig. 3.21 Force balance with tension transducers in forward and aft section

The model adapter is the entire outer cylinder, while the inner part is arranged concentric with the sting (see also Sect. 3.2.3). This design makes this kind of balance very attractive for testing of high performance aircraft models like fighters, that normally have their engine inside the fuselage. The space that is occupied by the engine can be used for the balance and the model adapter leaves sufficient space for the air duct. A second advantage is the fact that the two measuring springs for force and moment are connected in parallel, while in a moment balance these two elements are connected in series. For the same deformation/output, the total deformation of the balance at the end is smaller and less clearance at engine outlet is required. Advantages are usually accompanied by some disadvantages and so it is for this type of balance. The inner cylindrical bar is relatively small and so the stresses inside this part of the balance are rather high relative to those of the outer cylinder, because the loads are the same but the section modulus is much lower. The tension and compression elements must be quite sophisticated because a signal due to tension or compression is about 2/3 of a signal due to bending. To achieve the same output, the stress in the sensor must be about 35% higher. As a consequence, some stability problems may occur in such a balance.

3.2.1.2 Moment Type Balances

"*Moment Type Balances*" have a bending moment measuring section in the front and aft part of the balance (S_1; S_2 in Fig. 3.22).

The measurements S_1 and S_2 and of the two bending moments are used to compute a signal that is proportional to the force in the measurement plane and another one that is proportional to the moment around the axis that is vertical to the measurement plane. The stress distribution shows that the moment $M_y(M_z)$ is proportional to the sum of S_1 and S_2. However the force $F_y(F_z)$ is proportional to the difference of the signals and S_1 and S_2.

To obtain the rolling moment M_x, a separate part of one bending section must be applied with shear stress gauges to detect the shear stress $\boldsymbol{\tau}$.

The most complicated part of the balance is the axial force section that consists of flexures that enables axial movement and carries the other loads, as well as a bending beam to detect axial force (Fig. 3.23).

3.2.1.3 Direct Read Balances

A "*Direct Read Balance*" can be either of the force balance type or of the moment balance type. Instead of measuring the bending moment or a force at each section separately, there are half bridges on every section directly wired to a moment bridge and another set of half bridges directly wired to a force bridge. So the difference between this type and the others is only in the wiring of the bridges.

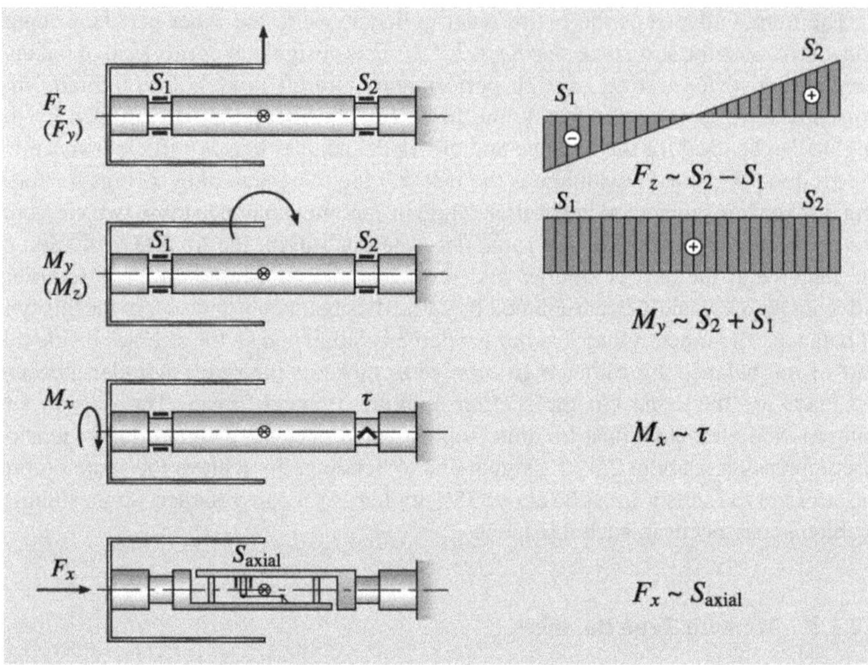

Fig. 3.22 Principles of a moment type balance

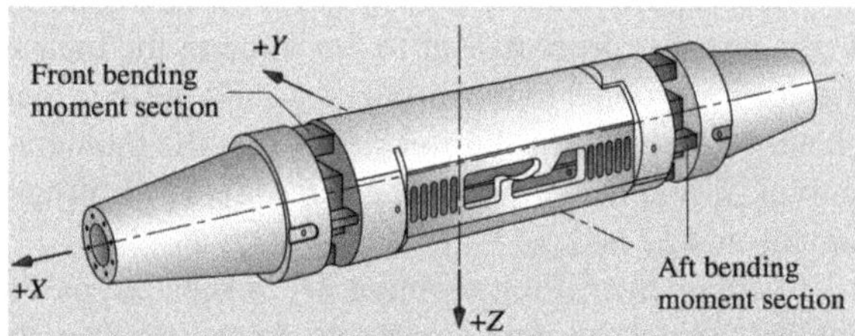

Fig. 3.23 Moment type balance

The disadvantage of such a wiring scheme are the long wires from the front to aft end that belong to the sensitive part of the bridge. All temperature changes inside these wires will cause errors and changes in the zero signal. For details of wiring scheme see Sect. 8.6.

3.2.1.4 Sting Integrated Balances

A *sting integrated balance* is in most cases a moment type balance, where the bending sections are not inside the model and the balance and sting are made out of one piece of material. If models are small and highly loaded, the balance joints become increasingly more the area of stress concentration. So it will be increasingly complicated to build in a mechanical interface to separate the balance from sting or model. In this case it can be necessary to skip the interface and build sting and balance out of one piece. Sometimes even model, sting and balance must be made out of one piece.

3.2.2 Box Balances

The main differences of the "*Box Balances*" to the "*Sting Balances*" are the areas of model and sting attachment (see Fig. 3.24). The transfer of load is from the top to the bottom along the vertical Z axis. So these balances are used with a central sting arrangement for an airplane configuration (Fig. 3.3) or inside a vehicle model.

The "*Mono Piece Box Balance*" is made out of one single piece of material. The advantages of this relative expensive manufacturing process are the low hysteresis and a good creep behavior, which are also basic requirements for a good repeatability of the balance.

The multi-piece box balances are built with different parts that can be manufactured separately. The load transducers can be either integrated into the structure or separate load cells can be used. This enables a parallel manufacturing process with a final assembly of the parts at the end. It makes the whole process much faster than that for the mono-piece balance, where one step after the other has to be performed.

The box type balances are internal balances, but they have more in common with the half model balances. Especially their temperature behavior is similar to that of semi span balances (Figs. 3.25 and 3.26).

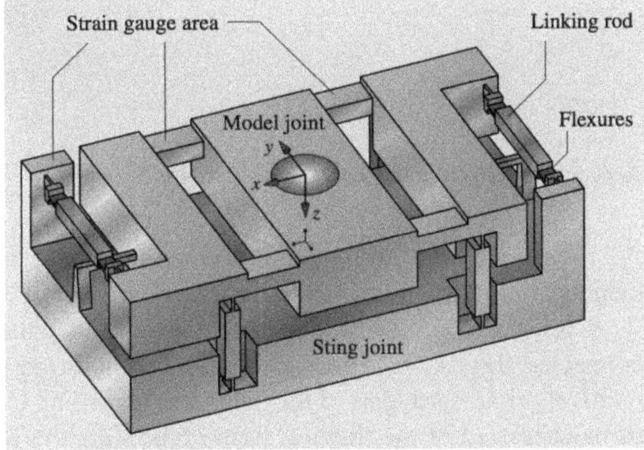

Fig. 3.24 Mono piece box balance

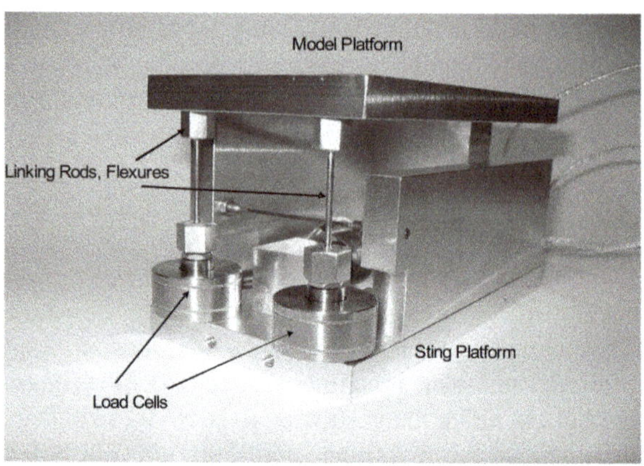

Fig. 3.25 Small box balance with load cells

Fig. 3.26 Vehicle fixed wheel load balance

3.2.3 Floating Frame Balances

A "*Floating Frame Balance*" describes not a special type of internal or external balance, but rather it characterizes a design type.

A relatively rigid frame is the metric end of the balance where the model is connected (Fig. 3.27). This rigid frame is connected via the sensing elements to the relatively rigid earth part of the balance. According to this definition, a large number of balances are of the floating frame type, like the force balance in Fig. 3.21 as an example for an internal balance, as well as most external balances.

Fig. 3.27 Compact floating frame balance (Carl Schenck Company Darmstadt)

3.2.4 Rotating Balances

In some cases it is not possible to fix the non-metric end to ground or a static base. For example to measure the forces on a propeller or a wheel it can be necessary to let the balance rotate with the propeller or wheel. In this case there is first the major problem of getting the supply voltage into the sensor and to extract the signals out of the sensor. This problem can be overcome by the use of slip rings or non-contact inductive devices.

For the balance itself the major problem is that beside the forces to be measured, centrifugal and inertial force are present. These forces are also a function of the rotational speed and must be separated from the other loads (Figs. 3.28, 3.29, 3.30 and 3.31).

To obtain the thrust of a propeller it is not sufficient to measure only one component, because the thrust vector is not always aligned with the rotational axis. To always obtain the exact thrust in the direction of rotation axis, a minimum five components must be measured.

Beside the problem of the inertial forces, also strong temperature variations and temperature gradients occur in the balance because the drive motors for the propellers are powerful and the generated heat is transferred through the shaft and through the balance to the propeller, which then acts as a active cooling system. The fundamental design principle to overcome the thermal problems and the effect of the centrifugal forces is an absolute symmetrical structure with measurement elements having a large area of constant stress in the area of the strain gauge application. Symmetry is

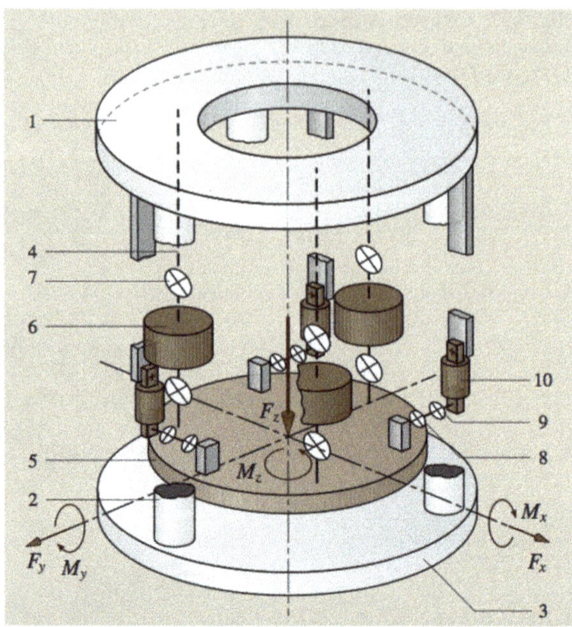

Fig. 3.28 Principle of a compact external floating frame balance

Fig. 3.29 Six-component rotating balance (DNW, EADS)

required in terms of geometry and mass distribution. Large constant stress areas in the area of strain measurement will improve the symmetrical distribution of sensitivity for the different sensing elements. For equal sensitivity the parasitic effects cancel out by summing up or calculating the difference of the correspondent sensors.

Special efforts must also be made to minimize the influence of thermal effects and inertial forces on the wiring. Asymmetrical wiring or moving cables under inertia loads affect the measuring result significant.

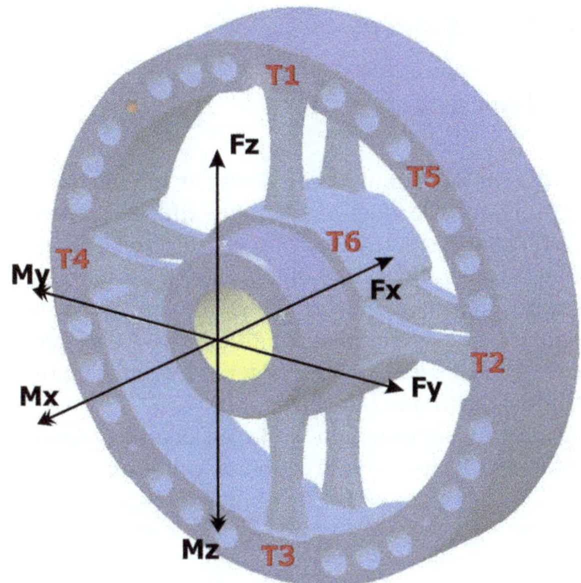

Fig. 3.30 Axis system and temperature measurement locations on EADS/DNW rotating balance

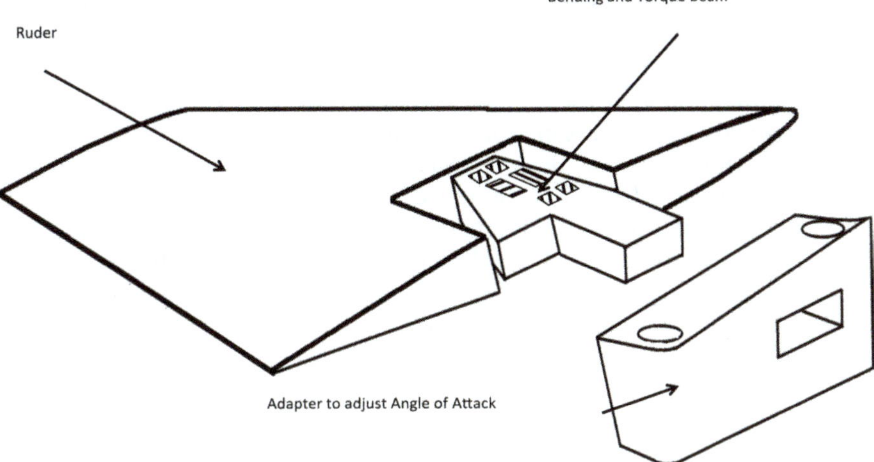

Fig. 3.31 Rudder balance (change of angle of attack by different adapters)

3.2.5 *Hinge Moment Balances*

3.2.6 *Rudder Balances*

To obtain the forces on a moving tail the device is too thin to build in a separate balance. To obtain the root bending moment, the lift and the torque of the rudder, a three-component balance is integrated into the root connection of the rudder. On a trapezoidal shaped beam at the front and at the end, full bridges are installed to measure the bending moment at these sections. The sum of the signals of these two bridges is calibrated to measure lift and the difference of the signals is calibrated to measure the root bending moment or the rudder. Between these two bridges another full bridge to measure the torque is installed.

Such designs are very compact and so the uncertainties due to interactions are relatively high. However, it is possible to calibrate such an arrangement with sufficient accuracy.

3.2.7 *Missile Balances*

Missile balances or in general balances for outboard extensions are in most cases very small balances and often only measure five components (without drag), because the drag measurement makes the balance very expensive (Figs. 3.32 and 3.33).

For missile separation tests the drag of the missile itself is not relevant, because missiles should normally have more thrust than drag. When they are fixed to the model or shortly after their release, the total change of drag for the entire configuration is needed and the internal balance of the model measures exactly that. Missile balances are required in order to obtain information about the effect on the aircraft during separation and the forces on the missile while flying near the aircraft. With the latter information the trajectory of the missile after separation can be determined. This is a rather complex procedure because usually missiles must be releasable in almost every flight condition of the aircraft.

3.3 Magnetic Suspension Balances

Measuring aerodynamic loads with balances in a wind tunnel always poses the problem that some kind of support is needed to fix the model inside the tunnel. Strut and stings always cause interactions that do not only influence the force measurement, but generates aerodynamic loads themselves. They also influence the flow around the model. This influence can be so significant that the flow completely differs from the flow without support. The results of such measurements are often useless. To completely eliminate the influence of the support one idea to fix the model in the

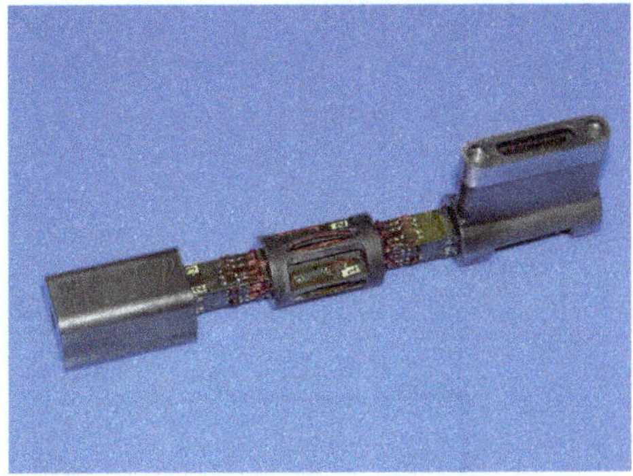

Fig. 3.32 Missile balance built by Cassidian (Germany) for Eurofighter

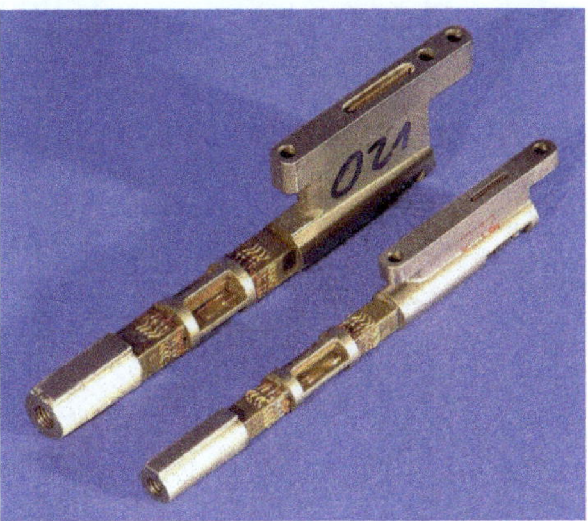

Fig. 3.33 Missile balances of Cassidian (Germany)

middle of the wind tunnel is to use magnetic fields. To compensate the reaction of the aerodynamic loads by electromagnets, an array of six different magnets has to be arranged round the test section to keep the model in a stable position. The energy used in the electromagnets is in some way proportional to the loads and can be used to determine the inertia and the aerodynamic loads. See [2–4, 7]. The measurement of aerodynamic loads by magnetic suspension has not been very successful in commercial testing and has primarily been only used for academic research.

3.4 Electromagnetic Balances

An electromagnetic balance in principle does not differ from an external balance that uses strain gauge based force transducers. Instead of strain gauge based load cells electromechanical force transducers are used. Figure 3.34 shows the external balance of the Kirsten Wind tunnel at University of Washington. The load cells of this balance

Aug. 3, 1954 H. G. SLOTTOW ET AL **2,685,200**

ELECTROMAGNETIC WEIGHING DEVICE

Filed March 27, 1952 2 Sheets-Sheet 1

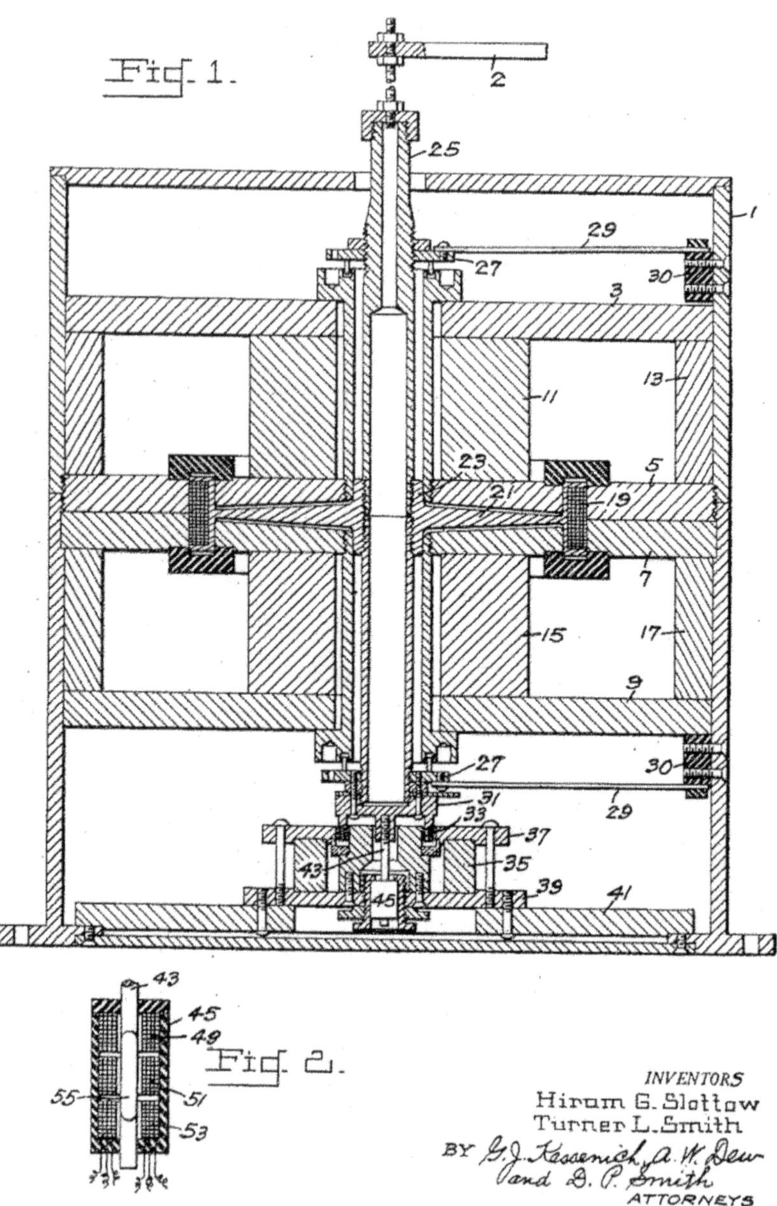

Fig. 3.34 Electromagnetic force transducer

are electromagnetic load cells. The working principle is an electromagnet compensating the deformation of a flexure. The current that is needed for compensation is proportional to the applied load.

References

1. Bazin, M., Bret, J., Fetet, T., Jacquemmoz, M.: Rotating balances for direct propeller performance evaluation on motorised models, 2nd edn. In: International Symposium on Strain Gauged Balances, Bedford, United Kingdom, May 4-7, 1999, ONERA, TP, 1999-56 (1999)
2. Boom, R.W., Abdelsalam, M.K., Eyssa, Y., McIntosh, G.: Magnetic Suspension and Balance System Advanced Study-Phase II. NASA Langley Research Center (1990)
3. Britcher, C.: Progress toward magnetic suspension and balance systems for large wind tunnels. J. Aircraft **22**(4), 264–269 (1985)
4. Covert, E.E.: Magnetic suspension and balance systems for use with wind tunnels. In: ICIASF'87-12th International Congress on Instrumentation in Aerospace Simulation Facilities, pp. 283–294 (1987)
5. Giesecke, P., Polansky, L.: Six-Component Underfloor Balance. Development Stages of External Wind Tunnel Balances. Schenck Company Report, Darmstadt (1984)
6. Gratzer, L.: Design of a New Balance System for the Kirsten Wind Tunnel, vol. 2 (1952)
7. Li-Ming, Y., Long-Hua, S., Quan-Ling, Y.: The technology research of 15 cm × 15 cm magnetic suspension and balance system (MSBS). In: NASA Conference Publication, pp. 711–726. NASA (1999)
8. Philipsen, I., Hoeijmakers, H.: Dynamic checks and temperature correction for six-component rotating shaft balances. In: 4th International Symposium on Strain-Gauge Balances (2004)
9. Rae, W., Pope, A.: Low Speed Wind Tunnel Testing. John Wiley, New York (1984)

Chapter 4
Model Mounting

Whenever a model is placed in a wind tunnel the forces and moments must be carried by some kind of support. There are only two exceptions to this rule. The first exception is the vertical wind tunnel, when spin tests are performed. In most vertical tunnels the model is spinning in free flight and the measurements are taken by an imaging system and no balances can be used. The other case is the magnetic suspension system that has been described in Sect. 3.3. All other support systems are mechanical. These mechanical support systems have to fulfill two main requirements:

1. The stiffness should be as high as possible to prevent the model from moving relative to the wind axis system.
2. The influence of the support system on the flow around the model should be as low as possible to assure the highest possible similarity with the flow around the original.

In many cases these two requirements are conflicting, demanding a compromise. The problem and associated solution can be quite different for external and internal balances.

External balances require in most cases relatively low deformations to generate a signal, or, if they work according to the compensation principle, they do not even have deformation at all and most of the relative movement is caused by the support system.

The deformation of internal balances is of course inherently required to generate the signals and this deformation directly influences the model position. Therefore, for a precise measurement, a model attitude measurement system is required to monitor the correct angle of attack and the correct yaw angle.

It must be emphasized that without knowledge of the angle of attack and the yaw angle any information from a balance is virtually useless. So either a deformation free setup is required or the knowledge of the sting system deformation must be available. To illustrate the gravity of the situation an example will be given, working from the sketch given in Fig. 4.1. In this sketch, one can imagine a model of a

K. Hufnagel, *Wind Tunnel Balances*, Experimental Fluid Mechanics,
https://doi.org/10.1007/978-3-030-97766-5_4

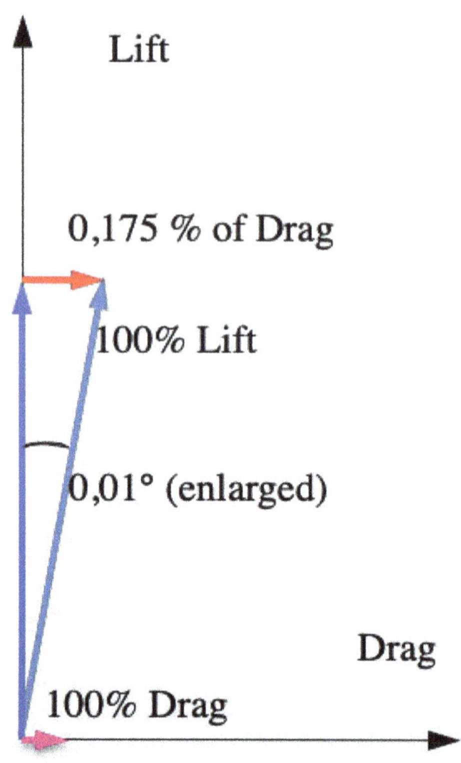

Fig. 4.1 Effect of lift on drag by an error in angle of attack

transport aircraft is being tested. The glide ratio of this model is 10:1, which means that lift is ten times higher than drag. The error on drag created by the lift force at a change in angle of attack of 0.01° is 1.75‰. The specified uncertainty for a balance in drag is often below 1.0‰. Thus, achieving this uncertainty specification with the balance makes no sense without an accurate measurement of the angle of attack.

The aerodynamic loads on a model are always affected by the presence of the model mounts. Sometimes the sting system is so rigid, that it is difficult to eliminate its influence on the measurement. However, if only comparative tests are required, it may be sufficient to work with a support system that always generates the same absolute value of interaction on the model flow. In this case, the obtained aerodynamic data is not correct in absolute terms, but the differences between two configurations are measured very precisely, if the two configurations do not differ substantially.

If absolute aerodynamic data are required, the aerodynamic interaction problem of the support system is more serious. The mounting loads themselves are subtracted from the model loads by performing tests without the model in place.

The second effect to be considered is the influence of the mounts on the flow field around the model and the influence of the model on the flow field around the mounts. A complete separation of these effects is not possible. Therefore, it is not possible to entirely eliminate the influence of the model-mount interference. Nevertheless, there are two main methods to resolve this problem:

1. An experimental approach, where several measurements with and without dummy stings in normal and upside down positions are performed, and so at least the influence of the support system cancels out.
2. The entire wind tunnel test is simulated in a numerical calculation. In this case,"entire" means the discretization of the model in the test section with nozzle and diffuser to insure correct aerodynamic data. The advantage of the numerical simulation is that it can be performed with and without considering the model sting. The difference of these two calculations can yield the influence of the sting on the flow to a very good approximation.

Both methods demand significant effort and the decision about which approach is more appropriate depends on the capabilities of the tunnel and available simulation tools.

A central motivation for including a chapter on model-mounting in a handbook on balances is that one question continually arises: what should be stiffer, the balance or the support? At the latest, this question arises when a model begins to oscillate during a wind tunnel test. In most cases the sting flexibility is responsible for the vibrations, so a sting damping device can solve the problem. This can be a simple oil damper or an active piezoelectric damping device, as it is used in the ETW *(European Transonic Wind Tunnel).*

4.1 Mounting of Models to External Balances

When using external balances, there are many different possibilities for mounting the models. For example, an aircraft full model can be supported by a three-strut arrangement (Fig. 4.2) or a central sting (Fig. 4.3). For cars and buildings, more than three supports are sometimes necessary. The external balance support system should provide for as many arrangements as possible. For aircraft models, the struts on the wings support the model at the quarter-chord positions and carry most of the load, except for the pitching moment that is balanced by the strut at the tail of the model. All links to the model are joined around the Y axis, so that a vertical movement of the tail strut allows for easy variation of the angle of attack. Yaw angles are set by turning the whole support and balance. The change of length of the tail strut while changing the angle of attack has to be taken into account for the strut correction. Therefore, a tare measurement without the model must be performed for every strut position.

Using the central-strut mounting, the model is fixed inside the fuselage in all directions. The adjustment of the angle of attack is achieved through a Y axis mechanism inside the fuselage. All loads must be carried by this support and therefore it must be very rigid to minimize dynamic movements of the model during testing.

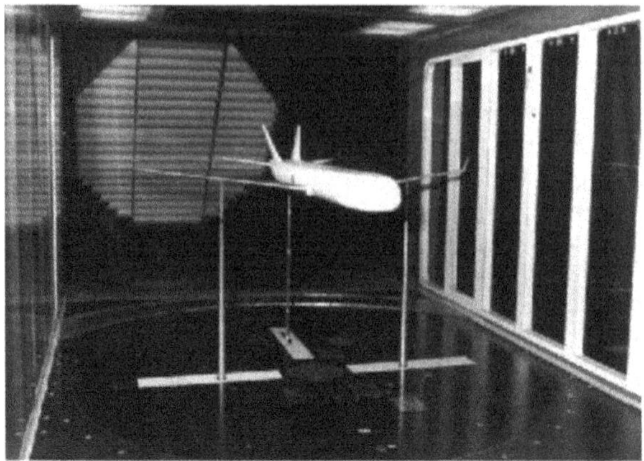

Fig. 4.2 Three-sting mounting on external balance

Fig. 4.3 Centre-strut mounting on external balance

Semi-span models (Fig. 4.4) are used to increase the effective Reynolds number of the tests by increasing the geometry of the model. Aside from achieving a higher Reynolds number, the larger model size allows models with variable flaps and slats to be constructed much easier; thus, semi-span models are often used for the testing of takeoff and landing configurations.

The support of the model by wires is seldom used anymore (Fig. 4.5). In such cases the balance has to be an overhead external balance and the model hangs from the balance by wires. To keep the model stable in the tunnel, the system has to be pre-loaded by weights (S_p in Fig. 4.5), which are usually dampened in a water basin under the test section. The advantage of such a model support system is the very low interference on the flow around the model. Especially if small variations on low drag configurations must be measured, a wire suspension arrangement is the

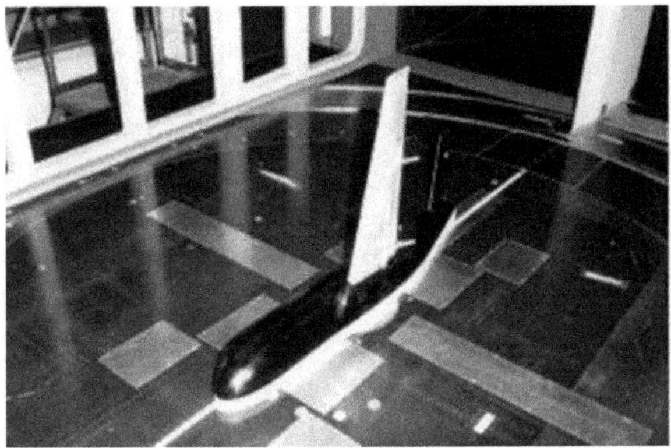

Fig. 4.4 Semi-span model on external balance

simplest possibility to reduce interactions so that they do not contaminate the drag force measurement.

In Fig. 4.5 the model hangs upside down in the tunnel, minimizing the pre-loading. This is helpful since the balance cannot measure loads smaller than the pre-loading. Sometimes modern wind tunnels test the model upside down in order to pre-load the balance in the lift direction. By doing this, the lift of the model contributes to the pre-loading such that the balance signal does not pass through zero as in a normal setup. So the additional non-linearities associated with the zero-load regime can be avoided.

4.2 Mounting Models to Internal Balances

The most common setup using an internal balance is the back-sting arrangement (Fig. 4.7), in which the sting is attached through the afterbody of the fuselage. In some cases the vertical fin (Fig. 4.6) is also used to support the model.

For tests that require the free flow around the after body, two sting setups (Fig. 4.8) can be used. In such cases, one balance is needed inside each sting. To determine the influence of the tail sting, two measurements are performed, one with a dummy tail sting and one without the dummy tail sting in place.

Another possibility to achieve relatively undisturbed flow around the afterbody is the vertical fin sting mounting (Fig. 4.9) which has been successfully used in the $S1$ tunnel (ONERA) by Airbus.

To connect the balance with the sting there has to be a sting adapter inside the fuselage and of course a slot around the fin. The advantage of a relative undisturbed flow around the afterbody causes a low lateral stiffness in side force direction of the mounting. Thus, the entire setup is sensitive to model vibrations in the lateral direction and this limits the use of this setup to low variations in angle of attack and to almost no yaw angle.

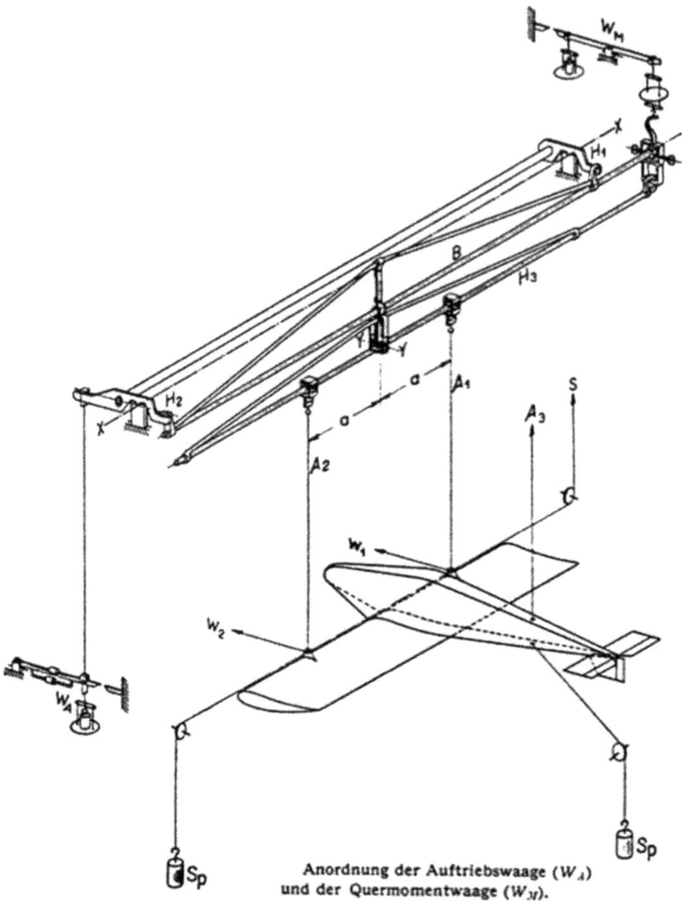

Fig. 4.5 Wire supported model on overhead external balance

4.3 Correction of Mounting and Balance Elasticity

As discussed in the introduction to this chapter, the correct angle of attack must be known within an uncertainty of a few hundredths of a degree, otherwise the high precision of a balance cannot be fully exploited. Several methods are used in the different wind tunnels for accounting for changes in angle of attack and these are now discussed.

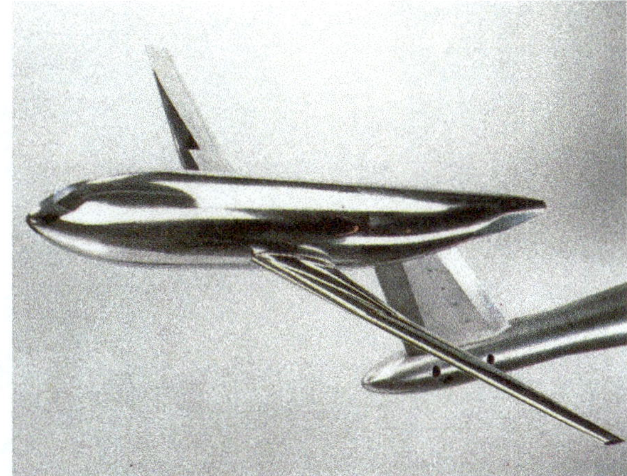

Fig. 4.6 Tail sting with fin attachment on lower side

Fig. 4.7 Tail sting through engine nozzle

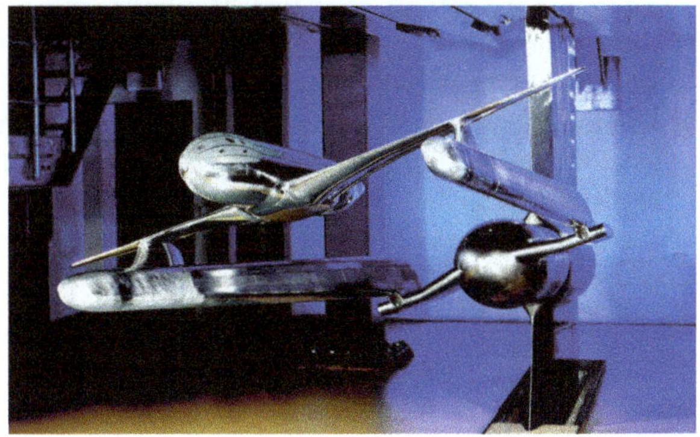

Fig. 4.8 ETW twin sting rig

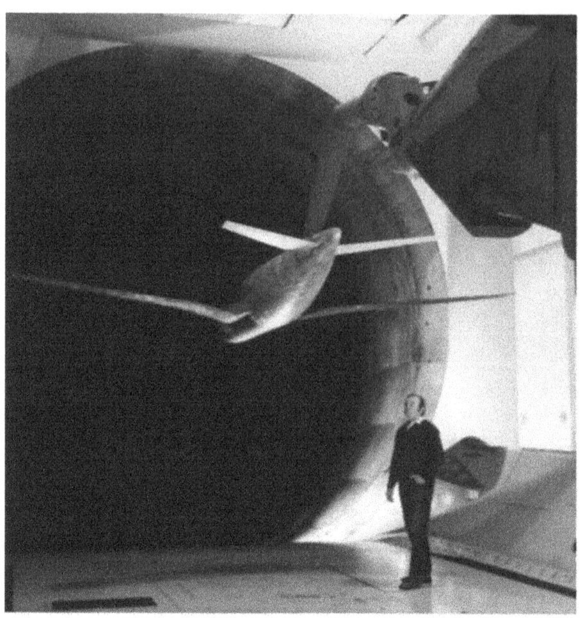

Fig. 4.9 Airbus model with fin sting mounting

4.3.1 Correction by Sting Deformation Measurement

There are several ways to make corrections of the sting deformation:

- Measure the sting deformation on the sting itself. In this case strain gauges have to be applied onto the sting like on a balance and the strain gauge signals must be calibrated for deformation. Well-known loads must be applied and the deformations at the sting end have to be measured. The relation between the loads and the signal yield the deformation correction of the sting during testing.
- Use the balance loads to determine the sting deformations. To be able to do this a deformation calibration has to be performed for the sting. Loads must be added at the sting end and the corresponding deformations must be measured. For the correction the balance loads can be used to determine loads at the sting interface and with these loads the sting deformations can be calculated. Of course the balance deformations must be added.

In both of these cases the balance deformations must be added to the sting deformation. To be able to do this the deformation sensitivities of the balance must be known. To know the balance deformations, a separate calibration of balance signals versus balance deformation has to be performed, because the deformation measurement normally causes small side loads that are too large to have the deformation measurement system in use during force calibration.

4.3.2 Correction by Model Position Measurement

The attitude of the model related to the tunnel axes must be measured inside the model or by a system outside the tunnel. Both methods are used. Most predominate is the use of an inclinometer inside the model, delivering the correct angle of attack. To obtain the correct angle in a rolled position, two inclinometers must be used. Obtaining the correct yaw position is a little more difficult and often optical systems are used.

4.3.3 Correction of Vibration by Active Sting Damping

Considering the combined model and balance mass and the sting elasticity, the potential for the system to oscillate must be examined. Due to unsteady flow or flow separation, model vibrations can be excited and the signals on the balance can become unstable. In the worst scenario, the excitation frequency is equal to the natural frequency of the system and the entire system goes into resonance, damaging the balance or even causing damage to the wind tunnel fan or other systems. Resonance must be avoided in all cases and an active sting damping system is one possibility to reduce unsteady movements.

Such a system can be very simple, like an oil damper outside the tunnel that is connected by a wire to the sting. This system was very successfully used in the low speed wind tunnel of the Technical University of Darmstadt (TUDa). The system does not influence the force measurement data and has only very little impact on the flow.

Another simple method is the implementation of a mass damper into the model, although this is sometimes only a limited option because there is rarely enough space in the model for such a system and the effect is restricted to a small band of frequencies.

A much more sophisticated system was developed at the European Transonic Wind-tunnel (ETW, Fig. 4.10). To reduce the vibrations between sting and balance an active damping system was integrated. The active device consists of piezoelectric actuators that counteract the accelerations measured by the balance signals [2, 3].

4.4 Internal Balance Joints

Most internal balances are separate instruments and must be connected to the model on one side and to the support on the other side. There is a broad variety of balance and sting joints in use throughout the wind tunnel community. Nearly every possible mechanical connection has been used in the past and nearly every wind tunnel has its own special types of balance interfaces. The disadvantage of this situation is that nearly no interchange of balances between the wind tunnels is possible.

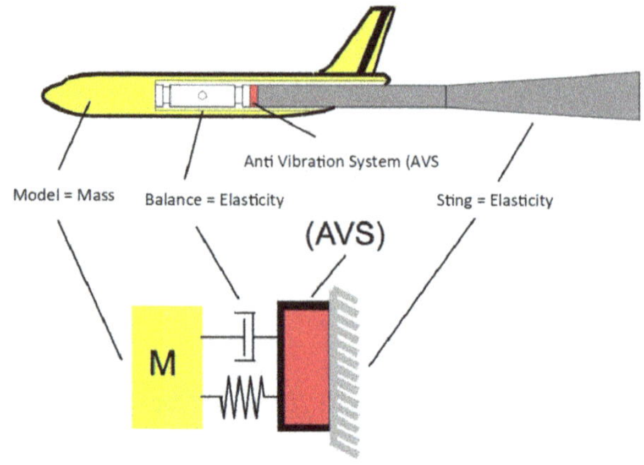

Fig. 4.10 Anti-vibration system of ETW built by the ERAS Company, (Göttingen)

The discussion about the balance joint is a never-ending story. Most balance joints that are used are not optimal with respect to positioning repeatability or load transfer from model to balance or model to sting. The pros and cons of the most widely used types, in light of the requirements placed on the balance/model joints, will be discussed. On balances for existing tunnels there is usually no choice for the interfaces of the balance, because they are already equipped with stings. Other hardware must be purchased when the interfaces are changed.

Often the joint to the model side and the sting side are the same, but this is not necessary. Especially on the model side the joint can be adapted to a new model design to optimize the positioning repeatability. The magnitude of the moments and forces that act on the model are responsible for the dimensioning and shape and for determining the acting stresses at the interface. For a symmetrical balance they are equal on both ends.

If the attitude of the model is measured within the model, good repeatability of the model position is only necessary for the model-balance interface. On the other hand, the primary requirement for the balance-sting interface is a low stress transformation of the loads with a minimum diameter, because this minimizes the aerodynamic sting-fuselage interaction.

If the attitude is not measured in the model, the requirements for both interfaces are the same, but the available space at the model-balance side is normally different to that of the balance-sting area. So in this situation two different joints might better fulfill the requirements.

The strength requirements between model joint and sting joint do not differ very much as long as the balance reference point is in the middle between these joints. In the case of long balances, where most of the sting is part of the balance, the loads on the sting joint are much higher than on the model joint because of the distance to the balance reference point. Such configurations are used when a balance-sting joint cannot be integrated into the sting, because the complete diameter is needed to keep the stress at reasonable levels.

The requirements for the joints are:

- Exact and repeatable positioning relative to the model in all directions and around every axis.
- Reliable transport of all loads from model to the support.
- No contact or friction between model and support beside the balance.
- No play or subsidence.
- Easy mounting and dismounting.

The first ultimate requirement for the model-balance joint is an excellent positioning repeatability. During calibration the calibration model or the orientation of the balance in the machine defines the load directions and accordingly, the axis system of the balance. In the wind tunnel test the loads on the model are exactly measured in these directions and the position of the model related to balance must be perfectly known, otherwise a systematic error will arise from the disorientation.

What are the requirements for the orientation repeatability? Two examples will demonstrate this:

1. The effect of a difference in the roll angle on side force:
   Normally the side force load range of a balance for aircraft testing is 2–7 times smaller than the lift force range. Assuming the side force range is two times smaller, which is the best case, and the side force should be measured with an uncertainty less than 0.1%, the allowed roll angle misalignment is 0.03°.
2. The effect of a misalignment in axial direction on the pitching moment in a flight performance test:
   If the pitching moment should be measured with an uncertainty better than 0.1%, the allowed difference in axial positioning is about 0.05 mm. This means either the re-positioning of the balance in the model related to the position during calibration must be better than 0.05 mm or the axial position of the balance in the model must be measured with an uncertainty less than 0.05 mm.

Note that misalignment is only one source of error. If the test results should have a required uncertainty better than 0.1%, which is a typical requirement in an aircraft wind tunnel test, the positioning accuracy must be even better. Knowing this, the tolerances for the re-positioning or the determination of misalignment of a balance in the model for angular tolerances are about 0.02° and for lateral tolerances about 0.02 mm.

Depending on the load ranges and the test requirements, the tolerance can be a little bit higher or must be even tighter. In any case, positioning errors can cause significant measurement errors and these errors can be caused by the design of the interfaces.

The second requirement is the transfer of load without stress peaks or without too much elasticity in the interface. Stress peaks restrict the overload capability and safety. The elasticity causes orientation errors, if the model-balance joint elasticity is different between calibration and test. For example for a cone joint this is the case if the model conical sleeve has different outer diameters in the calibration setup and the test setup.

There are only two ways to distribute the load over a joint. One is radial, like in a flange, or axial, like in a cone. Normally there is not much available space in the radial direction and this is the reason why cones are more popular as balance joints. In the following, the pros and cons for the different joint types are discussed in terms of the most commonly used joints: cones, cylinders, blocs and flanges.

4.4.1 Cone

The most popular balance interfaces are cones and although there are no cone interfaces without problems, there is no real alternative for a cone of given diameter to maintain the stress level within given limits. The cone connection is very attractive with respect to the transfer of all loads except roll, when only a limited diameter is available.

The major difference of cones is characterized by the taper angle or the taper ratio. The taper ratio of a cone, C, is the ratio between the difference of the maximum diameter D_{max} and the minimum diameter D_{min} of the cone and the distance l between them.

$$C = \frac{D_{\text{max}} - D_{\text{min}}}{l} \tag{4.1}$$

There are very flat cones with a higher part of friction for the transport of loads from one end to the other and the quick release cones that need some elements with form fit, because friction is not high enough to transport the loads from one side to the other. Both types have advantages and disadvantages, which is the reason that a large variety of tapers with different cone angles exist. Some of the most commonly used cones are collected in Table 4.1.

This table shows that even over a small number of balances a large variety of different taper angles have been realized. Although some "regional" preferences can be distinguished, it is more likely that wind tunnel operators simply use what has been in long-time use in that tunnel, insuring continuity in the tunnel operation. A systematic error may be present, but it then remains constant.

- **Slender Cones**

Slender cones with a taper ratio greater than 1:5 or an angle of taper less than 5° and are self-locking on one side when the friction coefficient μ in axial direction is $\mu = 0.1$. For a friction coefficient of $\mu = 0.05$ the taper ratio for self-locking must be 1:9. So self-locking depends heavily on the friction coefficient between the surfaces, and cones with angles of taper lower than 3° can be considered as self-locking. Therefore, only balances with a taper ratio over 1:10 should be assumed as self-locking.

Table 4.1 Example cones used in different balances

Balance	Taper notation	Taper angle (one side)	Taper ratio
ARA 1 in.	–	2, 8624°	1:10
TASK balances	0,625 in./ft	1.492°	1:19.2
NTF 104	Taper 1:12 ; in./ft	2.382°	1:12
W 64	Morse-Taper 5 (DIN 228)	1.5072°	1:19
W 605	Taper 1:3,429 (7:24)	8.297°	1:3.43
W 606	Morse-Taper 5 (DIN 228)	1.5072°	1:19
W 607	Morse-Taper 5 (DIN 228)	1.5072°	1:19
W 609	Morse-Taper 5 (DIN 228)	1.5072°	1:19
W 611	Morse-Taper 2 (DIN 228)	1.43056°	1:20
ONERA S1 Balance 125 ϕ	Taper 7:24	8.297°	1:3.43
AVA-balance 20ϕ	Taper 1:20	1.43222°	1:20
ETW balance predesign (ARA)	–	8°	1:3.56
ETW balance predesign (TUDa)	–	2.75°	1:10.41
ETW W 621 taper	7:24	8,297 8.297°	1:3.43
AEDC ϕ53mm	0.625inch/ft	1.492°	1:19.2
AEDC ϕ15mm	0.625inch/ft	1.492°	1:19.2

The self-locking condition for a taper is:

$$\mu > \frac{C}{2} \tag{4.2}$$

Self-locking cones with a taper ratio of 1:20 are widely used in the tool making industry. These cones have a good load transfer characteristic because they are relatively long and the transfer of moments is distributed over this long distance. Consequently, the differences in pressure between the front end and rear ends are small. They have no radial positioning ability, so additional keys must be provided. Such devices can reduce the friction contact area and this must be taken into account during design.

Positioning

The repeatability of the axial position depends on the mounting force, the conical sleeve wall thickness, their manufacture tolerances and the surface conditions that affect the friction during mounting. Also the people performing the mounting can

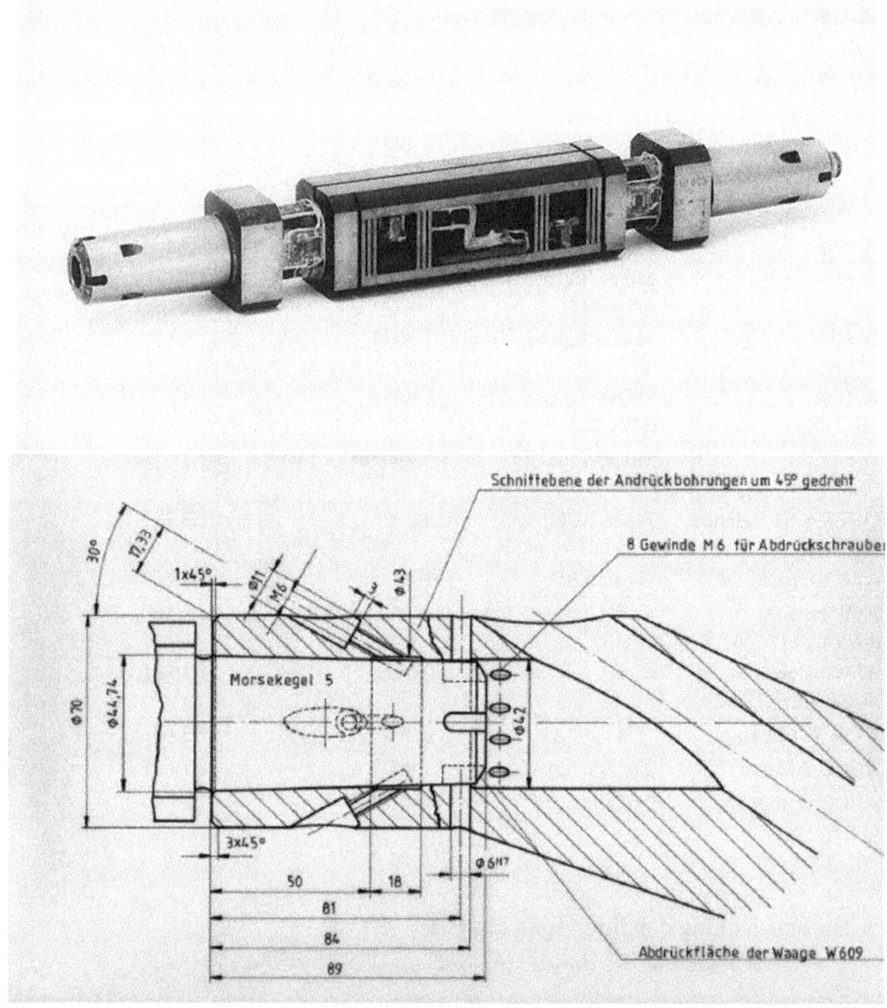

Fig. 4.11 Drawing of sting for balance W609 with cavities for pull-on screws and roll pin

make a difference. All of these factors can influence the axial position within a millimeter.

A careful measurement must be performed to define the exact position of the model related to the balance reference point; otherwise a systematic error occurs in the measurement of the pitch and yaw moments. The slender cone needs a mechanism to maintain it absolutely in place during the test and to prevent it from loosening. For a slender cone, this device can be simple (screws) and must not carry much load.

To transfer the rolling moment and to position the cone in the roll direction, additional keys or pins must be provided (Fig. 4.11).

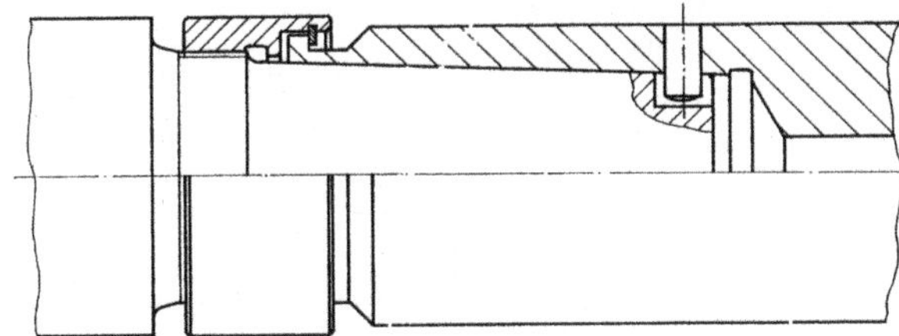

Fig. 4.12 Cone connection with coupling nut and pin

A contact of the sleeve face with the balance must be avoided, because this inhibits the control over the contact pressure. The manufacture of the cone and sleeve must be very precise to ensure a contact area over 80%; otherwise the connection can move under dynamic load. Slender cones must be precisely manufactured and ground to ensure the best adhesive strength. Even if the contact pattern is controlled by die spotting to achieve the best bearing strength, the hysteresis in the measurement in most cases is higher with cones than with flanges or other joints with form closure.

Subsidence is also a problem that can be reduced by static and dynamic pre-loading and re-tightening of the connection after the pre-loading. For the correct roll positioning a pin in the sleeve and a notch at the cone end is used.

- Another problem of the slender self-locking cones is the dismounting of the connection. The forces that are necessary to disconnect such a connection are much higher than the forces that are needed for the mounting and they even can be higher than the working loads. This has to be taken into account during design and the force-away mechanism must be designed much stronger than the press-in mechanism. For example, one experience with a cone was that it was not possible to dismount the calibration sleeve after the calibration with the normal pull-off tool. The mechanism was set to the maximum allowed pull force. Then the sleeve was heated and it jumped off after a certain temperature difference was reached. The elastic energy stored in the joint was strong enough to eject the sleeve away several meters.
- A possible pull-off mechanism are pull off screws in the opposite direction to mounting and fixing screws, but the number of these screws must be doubled (see Fig. 4.11).
  Another mechanism is shown in Fig. 4.12. In this case there is one thread inside the sleeve nut and a snap ring fixes the other end of the nut. A sliding pin at the end of the cone insures roll fixation.

In this case, the nut is at the end of the sting sleeve and so the balance becomes a little longer, but the whole diameter is available for the cone.

The next example is a cone connection with a sleeve nut with a left-hand thread on one side and a right-hand thread on the other side, as shown in Fig. 4.13. By turning the nut clockwise, the cone is pulled into the sting sleeve. The task of the key is to prevent the balance from rotating and to transfer the rolling moment during the test. By turning the sleeve-nut counterclockwise, the balance and the sting are separated.

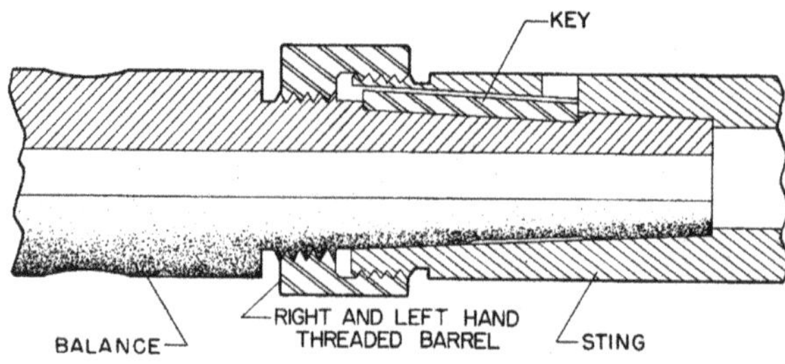

Fig. 4.13 Cone connection with sleeve nut (threaded barrel, NASA)

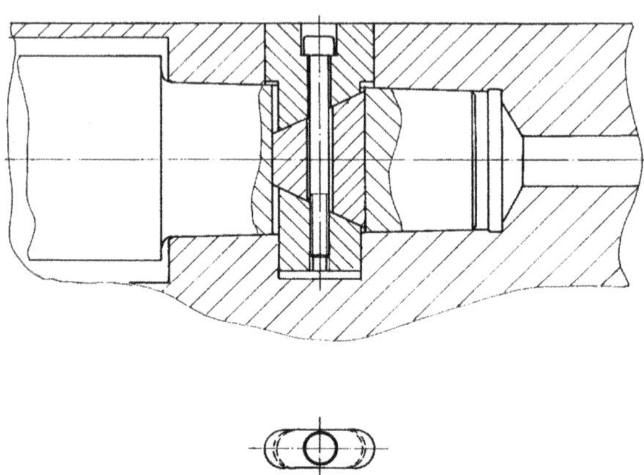

Fig. 4.14 Slender cone with clamping keys

For the nut, some space is needed and therefore the complete diameter cannot be used for the cone to carry the loads.

An example for a mechanism that fulfills all requirements for a cone connection is shown in Fig. 4.14. The drawing shows the key arrangement for pulling in the cone by driving the screw. The keys in this position also fix the cone inside the sting and additionally they also function as roll keys. To push off the cone the keys must be turned 180° disconnecting the screw cone and sleeve. The disadvantage of this construction is that the keys require the complete diameter of the cone. Therefore, there is no room for a cable duct and it can be only used at the model side of a balance.

A very attractive cone connection has been proposed by NASA Langley Research Center (Fig. 4.15). This connection was especially designed for easy mounting of the model. Even the nose of a model must only have a hole for the screw wrench to enable turning of the drive gear shaft. With this drive-gear, the expander is moved up and down the cone for clamping or releasing the connection. A dowel pin transfers the rolling moment and axial centering. This joint requires a certain additional length and is relatively complicated, but the model adapter is very simple and easy to manufacture. Normally, more models are tested on the same balance so the effort for a relatively complicated model joint might pay off over time.

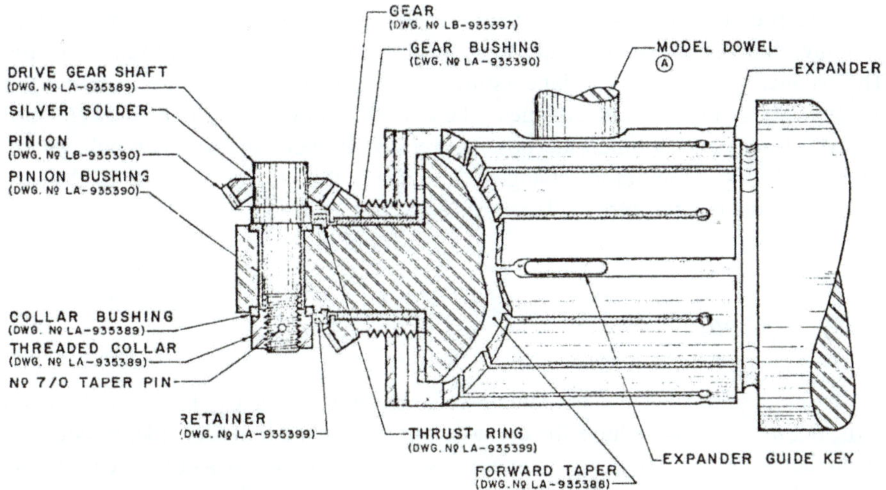

Fig. 4.15 Slender cone with expander clamping by NASA Langley Research Center

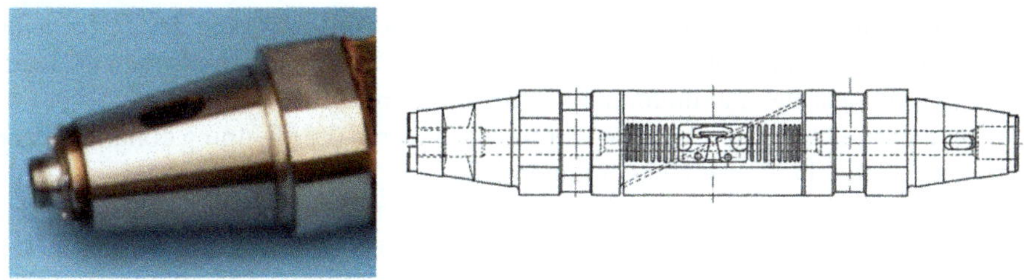

Fig. 4.16 Steep cone with conical center key lock

• Steep Cone

Cones with tapers less than 1:5 are easier to remove, but they require an additional mechanism to keep the sleeve in place under load. This mechanism must also be capable to carry the rolling moment, because the friction is not high enough to transfer the rolling moment. The possibility that the connection becomes loose under dynamic loads is very high and steep cones require a careful re-tightening process to work reliable under test conditions.

The problem of dismounting the balances with slender cones can be overcome by using cones with a taper angle greater than 7°–8°. Depending on the friction coefficient for steel-steel surfaces, it should be greater than 8° to definitely release without any frictional force. Figure 4.16 shows a balance with such a steep cone. In this balance, there is one key from both sides to allow enough space in the center of the balance for the cable duct. By turning the keys by 180°, they function either as pull-in or as push-off keys.

A steep cone like this is in not self-locking, therefore dynamic loads and temperature changes cause problems with such steep cones during the tunnel testing. The balance-model or the balance-sting interface tends to get loose, especially in the presence of temperature changes. The keys must have enough pre-stress to carry all axial forces and compensate possible relative movements between balance and adapters during temperature changes. To achieve this, a readjustment of the keys after a pre-test temperature cycle is necessary.

4.4.2 Cylinder

The cylinder joint is easy to manufacture. Some balances have cylinder joints at both ends, but most balances have the cylinder joint only on the model side. Related to a given diameter, the cylinder joint cannot carry as much load as a cone because some part of the diameter is needed for the model adapter. The fixation is realized by a clamping device, like the one is shown in Fig. 4.18. In this case, additional to the clamping device, a central pin (see Fig. 4.17) is used to achieve the alignment in the axial direction and roll. For good repeatability in mounting, some device for axial centering and the fixing of roll in the cylinder joint is always required when the loads attain a level near the maximum transferable friction loads.

Another possibility to fix the model on a cylinder is the clamping collar, that can either be custom built or bought commercially.

The force balance also uses another common cylinder connection, where the outside of the balance is a cylinder shell that is placed into a cylindrical model adapter (Fig. 4.19). The loads are transferred by the cylinder and a pin that fixes the balance inside the model transfers the rolling moment from model to balance.

To enable the use of a moment balance with models that have a cylinder for a force balance, an adapter for a moment type balance with cones at both ends is conceivable (Fig. 4.20).

Fig. 4.17 Balance W614 with cylinder as model interface

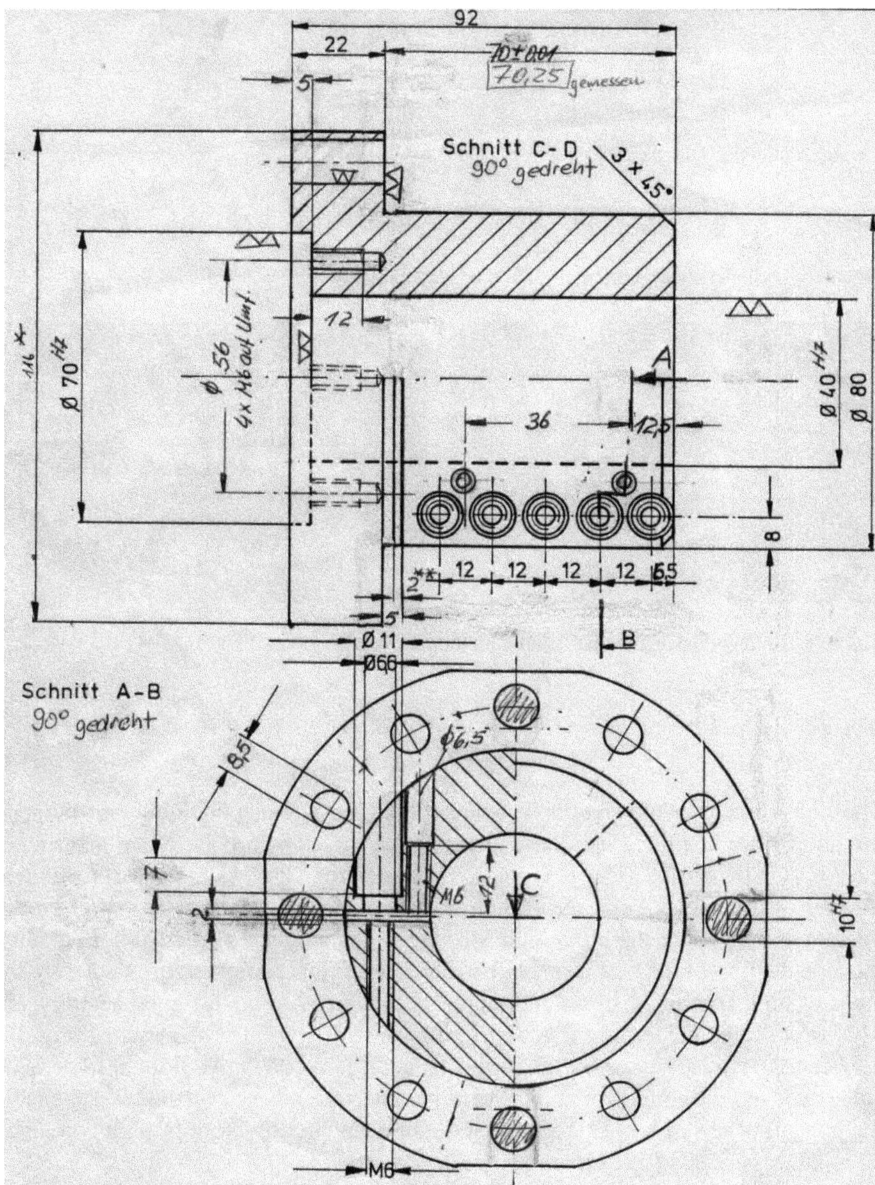

Fig. 4.18 Model adapter for balance W614 with central hole for roll adjustment

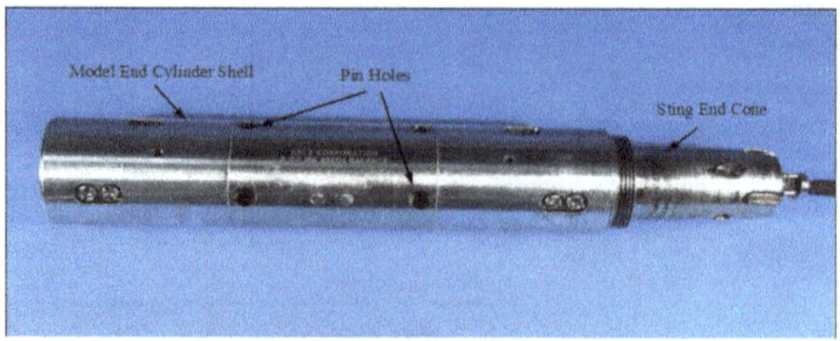

Fig. 4.19 Cylinder shell mounting of a force balance

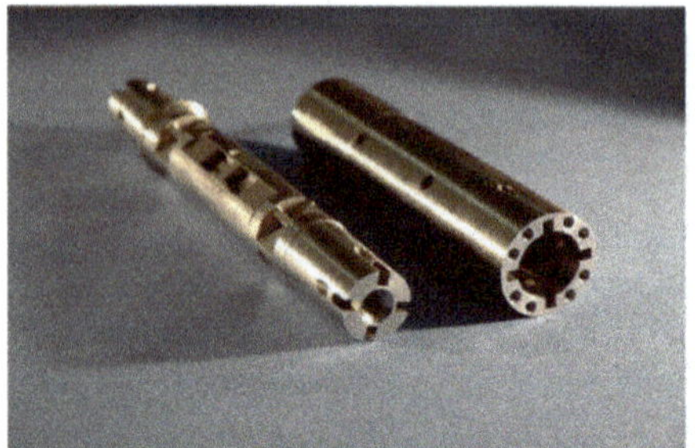

Fig. 4.20 Moment balance without and with mounted shell adapter

4.4.3 Bloc

The bloc joint is a rather simple connection. The lateral positioning in the side direction, normal direction and roll direction is enabled by precise manufacturing of the bloc and the sleeve hole. This is also necessary for a large contact area for a good transfer of the loads. Axial positioning can be achieved with pins or dowel screws that are needed to fix the bloc in the sleeve. If the manufacturing is of high quality, the repeatability of the bloc connection is excellent. The corners of the sleeve are the areas of high combined stress and therefore they need to be designed accordingly. This is possible inside the model but not in the sting. That is the reason why this connection is mostly used for the model balance connection. Additional to this, the cable duct and the connector at the end of the interface must be on one side to enable the holes for the screws. This weakens the bloc and the problem of high combined stress in this area becomes more severe.

The bearing surface of the sting for the balance joint in Fig. 4.21 is a prism. The axial positioning is achieved by the contact of the vertical end of the balance to the sting or by diagonal screws through the sting and balance Fig. 4.22.

Front Mounting of Connector MIKO 55S

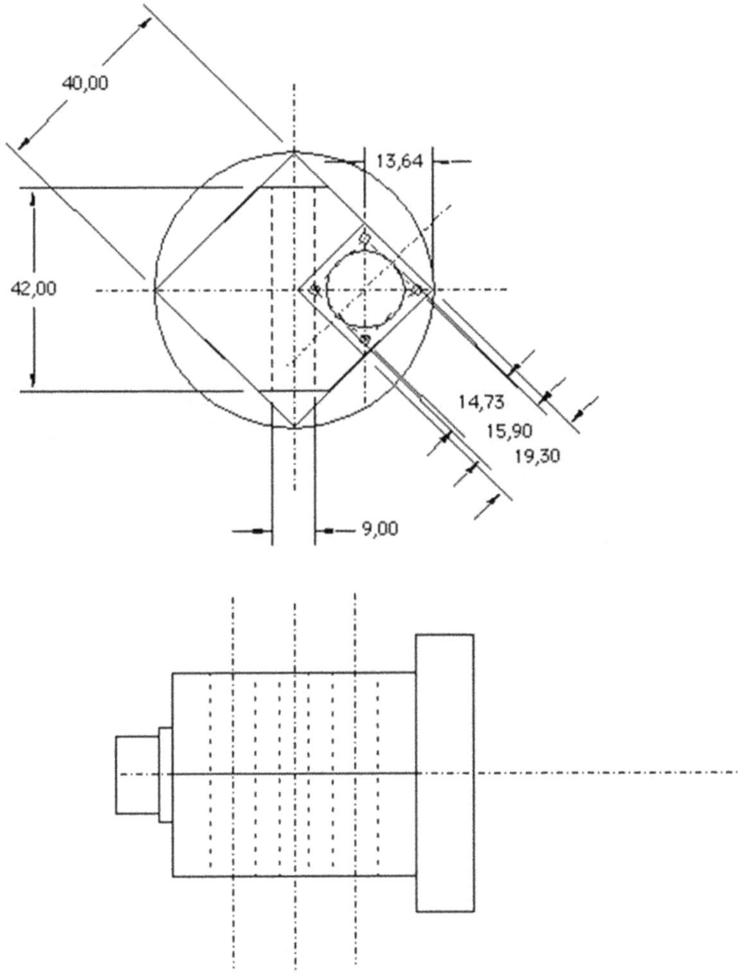

Fig. 4.21 Bloc-sting adapter of a balance with connector for prismatic mounting

4.4.4 Flange

While in the USA cone connections are favored, in Europe the flange connection has become more predominant. Some of the most important wind tunnels like the DNW-LLF or the cryogenic wind tunnels ETW and KKK have now a majority of flange balances (Fig. 4.23). The reason for this is that the flange has the best performance in repeatability and can be easily mounted and dismounted. It can be measured

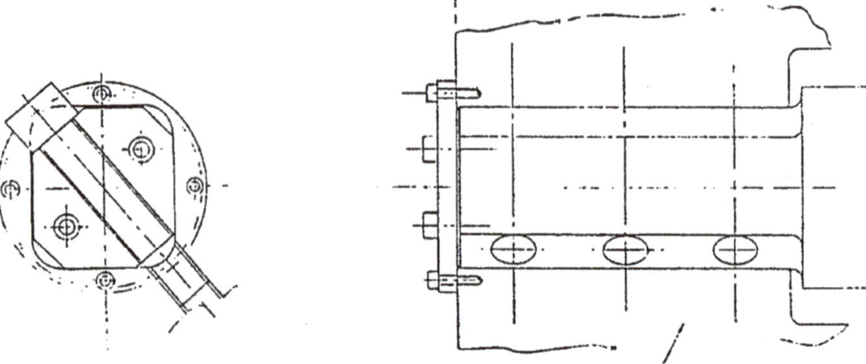

Fig. 4.22 Bloc-model adapter with diagonal screws for mounting

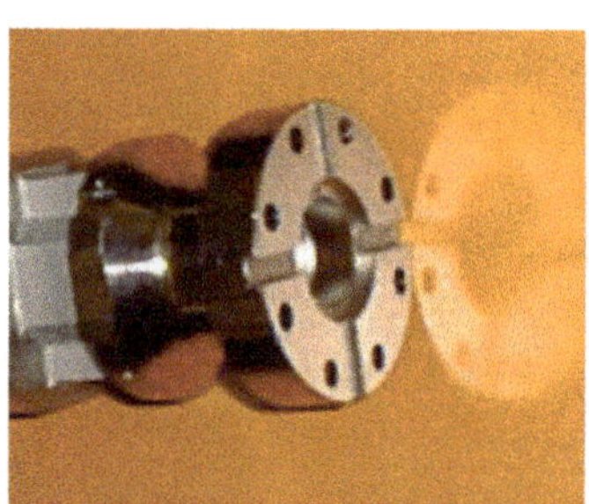

Fig. 4.23 Flanges with keys for roll and lateral positioning (left), with centering for lateral and bolts for roll positioning (right), rotatable by 90° (right side)

separately and these measurements yield the exact position. Even after dismounting the model, the flange remains at exactly the same place for a new test. Most tests in a wind tunnel are comparative tests. To avoid errors caused by misalignment, this is a very useful feature. The disadvantage is the relatively large radial area required by a flange, because the loads in the interface are distributed in the radial direction. The challenging task in the design of a flange is to build it as small as possible.

The flange fixes the axial position and the four keys in radial direction fix all lateral and the roll positions. The flange on the right side allows also a mounting of the balance at a 90° rolled position, while the bolt circle of the flange on the left side allows only the upright or upside down mounting. Instead of four crosswise arranged keys, some balances have flanges that use two dowel pins instead of keys. The repeatability of the two dowel pins is also excellent, but in case of large temperature changes and different flange materials the dowel pins are damaged by the difference in heat expansion.

Another possibility to design a flange with highly repeatable positioning is to use a serration (*Hirth Tooth Coupling*) instead of four radial keys. Rings with the

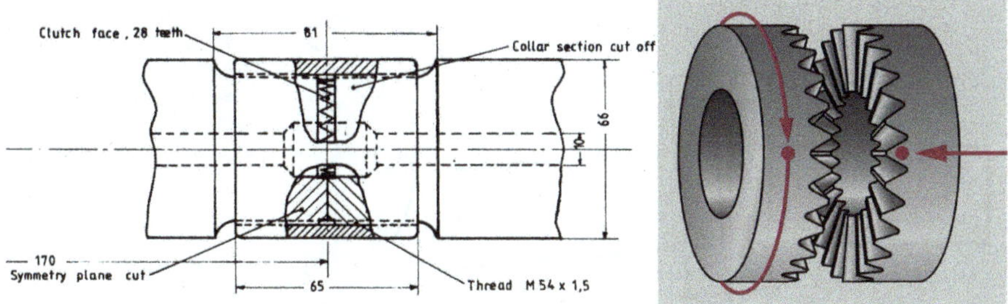

Fig. 4.24 Balance interface with coupling nut and serration (*Hirth Tooth System*)

serration can be commercially bought and screwed onto the flange. Once aligned and fixed, balance and model can be mounted and dismounted without any difference in position. A variation of the rotational position with the tooth pitch is possible.

Figure 4.24 shows a balance joint using a coupling nut with a left-hand thread on one side and a right-hand thread on the other side. By turning the nut, both ends are tightened together and the positioning is achieved by the serration. This design needs very little space and is useful whenever a low diameter connection is required. The problem of this connection is that the thread and the nut cross-section have to carry all the loads and thus, the load capacity of such a connection is smaller than that of a cone. Serrations are very useful for external balances because they can be commercially purchased in standard dimensions. For the use on an internal balance flange, the serrations must be manufactured directly on the balance. That is relatively expensive and is therefore not recommended.

For the calculation of a flange connection, standard equations can be used to determine the bolt loads as long as the diameter is large enough. Ewald in [1] proposed an equation for the load transfer characteristic of a flange, S_{FL}:

$$S_{FL} = \frac{M_B}{{D_{FL}}^3} = \frac{\sqrt{\left(Z \cdot \frac{l}{2} + M_y\right)^2 + \left(Y \cdot \frac{l}{2} + M_z\right)^2}}{D_{FL}^3} \tag{4.3}$$

The load transfer characteristic is the ratio between the acting bending moments on the flange M_B divided by the diameter D_{FL} raised to the power of three.

If the load transfer characteristic S_{FL} is lower than 9 MPa the flange can be assumed to be a standard flange and standard equations can be used. The assumption for a standard flange is that the bending moments act around the base point of flange diameter D (see Fig. 4.25).

The maximum screw force then acts in the screw with the largest distance to the contact point. This is a conservative assumption that allows a quick estimation of the maximum screw force and an estimation of the feasibility of the flange design. It must be taken into account that for the different moment directions the screw with the maximum force is found at different positions. The resulting moment (M_B) is therefore the absolute value of the vector sum of the moments.

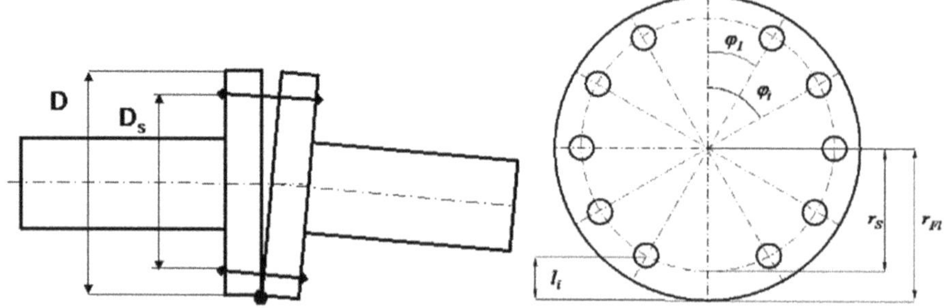

Fig. 4.25 Assumption of flange deformation to calculate maximum screw force

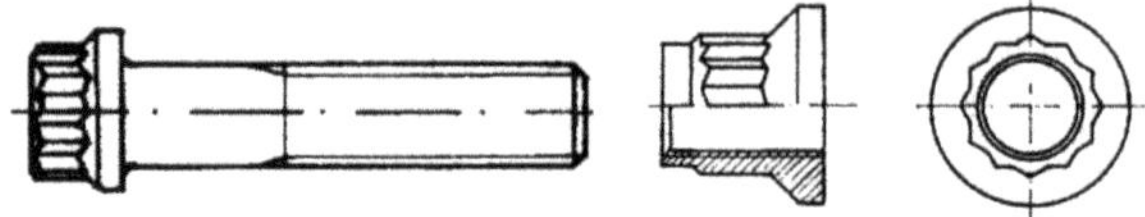

Fig. 4.26 Double hexagon high strength screws

$$M_B = \sqrt{(Z \cdot \frac{l}{2} + M_y)^2 + (Y \cdot \frac{l}{2} + M_z)^2} \tag{4.4}$$

In this equation *Z*, *Y* are the normal and side forces acting at the reference point of the balance and M_y, M_z are the moments acting there.

The maximum screw force can be calculated by using the following equation:

$$F_{\max} = \frac{M_B \cdot (r_{Fl} + r_s \cdot \cos \varphi_1)}{\sum_1^n (r_{Fl} - r_s \cdot \cos \varphi_i)^2} + \frac{F_x}{n} \tag{4.5}$$

Additional to the screw force by the flange moment, part of the axial force must be added. The number of screws is n and φ_1 is the initial angle, if the pitch is not symmetric, $n \cdot \varphi_1 \neq 360°$. For a symmetric distribution of the screws $n \cdot \varphi_1 \neq 360°$, $\varphi_1 = 0$.

The disadvantage of the flange is that there is always space required in the radial direction for the screw heads. Therefore, the neck behind the flange has the smallest diameter and this diameter limits the maximum transferable loads. To optimize this situation, the screws must be placed as far as possible away from the center, the screw heads or the nut diameters must be as small as possible, and the strength of the material used must be as high as possible. This requires special screws and nuts. Screws and nuts that fulfill this requirement are commercially available for aerospace applications. An example of such a high strength screw is shown in Fig. 4.26.

Fig. 4.27 High capacity flange with center keys for positioning

The required tool diameter is exactly the maximum diameter of the screw so the remaining diameter for the balance neck is also a maximum and the peak stresses in the changeover from flange to balance can be minimized.

If the available diameter is smaller than the maximum standard diameter, a detailed Finite Element analysis is necessary to evaluate the maximum local stresses. By doing this a specific load parameter of 13 MPa instead of 9 MPa is possible. Such a high loaded flange is shown in Fig. 4.27.

To check the screw stability, standard codes must be used, for example VDI-Guideline 2230. However, even with a low specific load parameter the maximum screw load is very high and high strength material must be used for the screws. In some cases special screws were manufactured of maraging steel and a special ultrasonic tightening tool is used to pre-stress the screws exactly to 90% of yield stress.

4.4.5 Summary Balance Joints

In Table 4.2 an overview of the major joint features is given. This is just a preliminary qualification that might help in initial decisions about which joint design to pursue. The given attributes for each joint do not have an absolute character; they should be taken as relative to the joint with the best features (Table 4.3).

Table 4.2 Major characterising features of balance joints

Load capacity	The level of transferable load for a given diameter related to a section without any interface
Positioning	Evaluates the capability to obtain an exact pre-defined position during mounting
Repeatability	Gives an indication how precise a position can be repeated from one test to the next
Load transfer quality	Informs about possible problems for the transfer of the loads
Subsidence	Informs about the possibility that the position and the load transfer quality can change within a test
Mounting	Evaluates the amount of work that is necessary for precision mounting
Dismounting	Informs about the risk that problems occur during dismounting

Table 4.3 Balance interface characteristics

	Slender cone	Steep cone	Cylinder + pin or equiv.	Bloc	Flange
Load capacity	High	High	Middle	Middle	Low
Positioning	Poor	Middle	Good	Good	Excellent
Repeatability	Poor	Middle	Good	Good	Excellent
Load transfer qual.	Hysteresis	Hysteresis	Hysteresis	Good	Excellent
Subsidence	Possible	Possible	Possible	Possible	None
Mounting	Problematic	Good	Problematic/good	Easy	Easy
Dismounting	Problematic	Good	Easy	Easy	Easy

4.4.6 Adapters for the Use of Internal Balances as Box Balances and Vice Versa

Normally a sting balance will only be used with a back sting support and a box balance will be used with some kind of a center strut. If there is enough space for a simple adapter, a sting balance can be used with a center strut and a box balance can be used with a rear sting.

The basic idea is to put a box around the balance that is cut along its diagonal into two pieces. One piece is connected to the sting mount and the other is connected to the model mount of the balance. Thus the internal balance intersection plane is changed from parallel to the normal direction of the X axis, and for the box balance, from normal to parallel to the axial axis.. The principle is shown in Figs. 4.28 and 4.29.

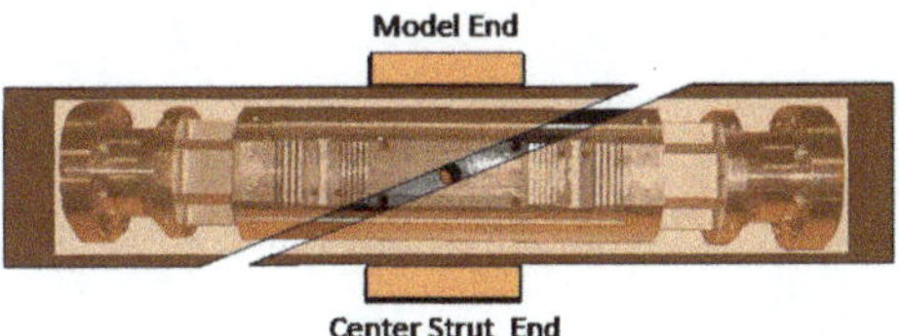

Fig. 4.28 Conversion of sting balance from rear sting support to center strut support

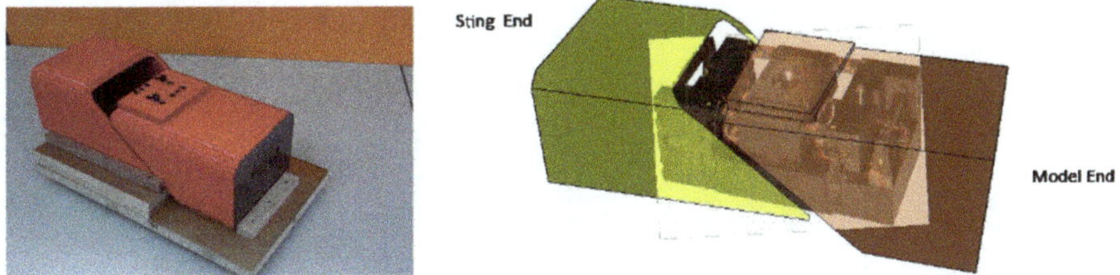

Fig. 4.29 Conversion of box balance to rear sting mounting

Though it requires a lot of space, it is also possible to convert a box balance to a rear sting mounting. In Fig. 4.29 such an application is shown. The reason for this application was not the use in a wind tunnel; the mounting was used to calibrate the balance in an automatic calibration machine that was built for sting balances.

References

1. Ewald, B.: Interne Windkanalwaagen. Tech. Rep. Report A38/99, TU Darmstadt (2000)
2. Fehren, H., Gnauert, U., Wimmel, R.: Validation testing with the active damping system in the European Transonic Wind Tunnel. Tech. Rep. AIAA-2001/0610, Reno (2001)
3. Fehren, H., Gnauert, U., Wimmel, R., Hefer, G., Schimanski, D.: Aktive Schwingungskompensation für den Europäischen Transsonischen Windkanal. Adaptronic Congress, Apr, Berlin (2001)
4. N.N.: VDI Guide Line 2230, Systematic Calculation of Highly Stressed Bolted Joints or Joints with One Cylindrical Bolt. VDI (2015)

Chapter 5
Specification

The normal design procedure for a balance, be it internal or external, is to first collect the design requirements, like the number of components, the range of each component, the maximum space available for the balance, sensitivities, resolutions and accuracy. All these requirements form the specification of a balance and must be taken into account from the beginning. These specifications define not only type and shape of the balance, they also determines the effort in design and manufacturing, the quality and eventually the price of a balance. Specifications can vary from "best effort", meaning do it as best as your means allow, to a large number of documents filled with rules and regulations for every step, starting with the requirement for a special letterhead over the report on technical performance to regulating business conditions. These two extremes define the variation between nearly no specifications and an over-regulation. Experience shows that neither fulfill the basic purpose of a specification sheet to adequately describe customer needs and wishes to the balance manufacturer. An in-depth discussion with the user is always an essential step to refine the initial specifications. Indeed, designing and building to initial specifications can often end in unsatisfactory performance. This occurs because specifications are often generated mixing demands from many different test setups, which are seldom possible to fulfill. For example, a well-designed balance to measure the cruise flight performance of an aircraft does not need a sensitive side force and yawing moment measurement. The side forces and yawing moments are very low at cruise conditions and they are of minor interest. More important for such a balance is a high stiffness in side force direction to secure good dynamic stability. In this case the sensitivity is a requirement for lift, pitch and drag and stiffness while a reasonable signal is the requirement for side force and yaw. This is one reason why an inventory of balances with load ranges and sensitivities adapted to the tasks they usually have is often a better strategy than attempting to have one balance for all applications.

K. Hufnagel, *Wind Tunnel Balances*, Experimental Fluid Mechanics,
https://doi.org/10.1007/978-3-030-97766-5_5

For very complex tunnels with high construction and running costs, like cryogenic wind tunnels, the relative cost of the balance is low so that more expense can be devoted to balances. For these tunnels the time required to design and build a custom balance is more of a problem.

Small tunnels and most of the subsonic wind tunnels cannot generate enough funds from their testing; hence, the cost of a balance is relatively high compared to the wind tunnel operating costs. In this case large investments for a large balance inventory are not possible. Therefore the specifications of balances for these tunnels often aim to cover all the testing possibilities of the tunnel. This generally means that the largest model defines the loads and the smallest model defines the space for the balance. The result is a balance that is too flexible for the large models and has a poor sensitivity for the small models.

To target a good compromise in such cases, the data taken in the tunnel during former tests must be analyzed carefully and the focus of the specifications must be aligned with the loads occurring most frequently. Higher loads than the specified maximum combined loads can be taken into account by a certain overload capacity for single loads or a restricted number of load combinations. Higher sensitivity for lower loads can be realized by using side force and yaw for lift and pitch by turning the balance by 90° around the X axes.

There is no general rule to find the best compromise for a certain tunnel, but there is a recommendation for the tunnel staff to look very closely and detailed into their own data before starting to write specifications for a balance.

By defining loads, sensitivity and dimensions for a balance, the shape of the balance is completely determined and there is no room left for design optimization. After the first set of specifications has been made, some iterations of the specifications in the pre-design phase can be helpful to find a better compromise.

Therefore, the first phase of balance design is the specification phase, the second phase is the pre-design and the last phase is the final design.

Most of the work on the specifications has to be carried out by the wind tunnel staff/operator, while the work in the pre-design phase is a cooperative effort between the wind tunnel operator and the balance manufacturer. The final design is performed solely by the balance manufacturer.

The tools that can be used for meeting a set of specifications are more or less a statistical analysis and an analysis of the current situation.

For all types of balances the contents of the balance specification are more or less the same and this is the most important document that defines all characteristics of the balance. So normally it is up to the balance user to write the specifications. It should be the result of a discussion between balance manufacturer and balance user to find an optimum solution for the needs of the user.

In the first phase of a balance purchase a basic specification is needed for the tenders of all possible suppliers. Usually in this phase there are some iterations, since a request for quotation inevitably generates questions. The reason for this is that most customers do not write specifications for a balance very often and for fear of making mistakes limits are set too stringent and some are even not formulated correctly or are inherently contradictory. In this sense, this chapter is directed more towards the balance customer/user than to the balance designer. The closer the specifications fits the real needs of the user, the easier it is for the manufacturer to make a realistic proposal. Specifying limits that are too stringent results in a high price, and limits not stringent enough will not satisfy the quality requirements.

5.1 Definition of Load

In this chapter the word "load" will be used for forces and moments. The task of a balance is to measure the aerodynamic loads, which act on the model or the components of the model. On a wind tunnel model six different components of aerodynamic loads can act, three forces in direction of the axes and the moments around the axes.

These components are measured in a certain axis system that can be either fixed to the model or to the wind tunnel. For an internal balance the loads are always measured in the model axis system and have to be transferred into the wind axis system. For an external balance the loads can be either measured in the wind axis system (most full model cases) or they can be measured in the model axis system, for example with the half model setup. For external balances the mechanism for setting angle of attack, yaw and roll determines in which axis system the loads are measured.

For the measurement of loads on model parts like rudder, flaps and missiles, usually less than six components are required (Fig. 5.1).

5.2 Specification of Balance Load Ranges

Before a balance can be designed the major specifications are the definition of the load ranges and the available space for the balance. This is the most complicated process prior to the design of a balance, because cost and accuracy considerations must be made long before the first tests are performed.

The maximum combined loads specify the load ranges of the balance for the design. For the maximum design loads of a balance there are different definitions. They can differ between

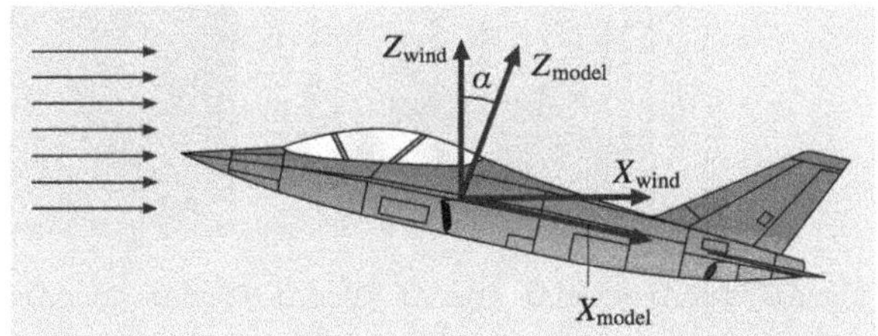

Fig. 5.1 Wind axis system/model axis system

- The loads are acting all act together at the same time. They are called maximum combined loads.
- If the maximum load is acting alone, it is defined as maximum single load.

Usually no single loads appear in wind tunnel tests and combined loads stress the balance in a much more complicated manner. The stress analysis of the balance has to take into account this situation. With the combination of only two loads, the balance is usually able to carry higher loads. To estimate which load can be carried in a given situation the balance manufacturer provides loading diagrams, which allow the test engineer to decide whether the load combination of the planned test is within the limits of the balance.

For an external balance the available space is mostly not a problem. An external balance is used over several decades in a wind tunnel, therefore specifications of the design load ranges have to be orientated more on the capabilities of the wind tunnel itself and the different possible model setups in the tunnel, like half model and full model testing, or the possibility of aircraft, car and building testing. For the design of the external balance, all load ranges of the principal balance configuration must be first known. Two different options are possible:

1. The turntable is mounted inside the weighbridge. In this case the balance always remains in the wind axis system and therefore the balance always measures the wind loads. The disadvantage of this option is that the entire turntable mechanism has to be on the metric side of the balance and so the balance is pre-loaded by the weight of this mechanism.
2. The entire balance is mounted on a turntable. In this case some components stay in the wind axis system and some stay in the model axis system, so a calculation of the aerodynamic loads from the balance loads is necessary.

For example, for a full model setup as shown in Fig. 5.2b, the balance always measures the aerodynamic loads while the angle of attack changes by moving the afterbody sting up and down and the balance stays in place. In a half model setup, as shown in Fig. 5.2a, the balance moves with the model when the angle of attack changes and now it measures the load in the model axis system.

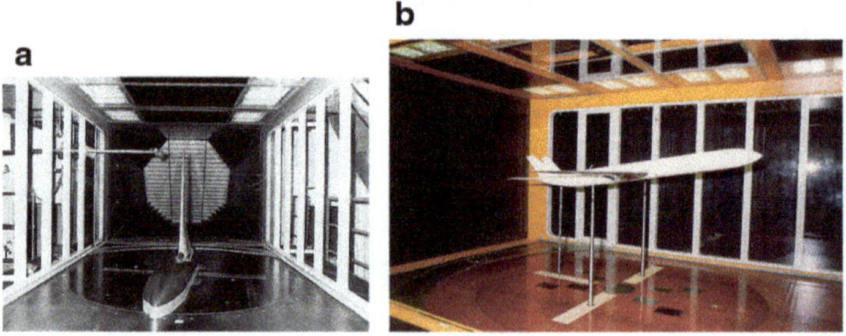

Fig. 5.2 **a** Half model on external balance. **b** Full model on external balance

This makes the determination of the balance load ranges rather difficult. Therefore, it is useful to fill out a table where first the maximum loads for the different test setups are calculated in the wind axis system. For this some assumptions of the aerodynamic coefficients and the model size must be made. Subsequently, by using the assumptions for the maximum angle of attack or yaw angle, the maximum loads in the model axis system can be calculated.

From Table 5.1 the maximum of each component can be taken as the maximum load for the balance. Naturally this leads to a balance with rather high load ranges and so for some cases the load range could be too high to measure with high resolution. However, if a certain test requires a high resolution and this kind of test is frequently performed in the tunnel, it is better to take the lower load range to insure higher resolution and higher accuracy. These considerations must be made for all components to insure a balance with the best fit of the load range for the normal operation of the tunnel.

For an internal balance the available space turns out to become increasingly more of a problem. The available space for balances is restricted by the diameter of the fuselage. In times where transport aircraft become larger, the scale of the models becomes smaller, since the cross-sections of the wind tunnels do not enlarge to the same extent as the aircraft. As a consequence the available diameter for the balance becomes smaller.

For combat aircraft the loads related to the available restricted space inside the model are very high because wind tunnel tests with this type of aircraft are mostly performed in pressurized wind tunnels in order to achieve the correct Mach number. The high static pressure leads to high air density in the tunnel and this leads to high loads on the model. On the other hand the space for the engine air ducts require space inside the model and this reduces the available space for the balance. These two effects lead to higher specific loads on the balance and this makes it much more complicated to develop a balance with high accuracy. Therefore the definition of the load ranges and the definition of the available space must be performed very carefully to enable a balance design for best usability. The specification for an internal balance should therefore be made related to the model and the loads on this model during the test and not for the tunnel capabilities. If this is not done, it will always lead to internal balances with less sensitivity and less accuracy for the test.

Unlike an external balance, where some of the components are always fixed to the wind axis system, the internal balance always measures the loads in the model axis system; thus, lift and drag are for example always a combination of axial force and normal force. Therefore, it has to be taken into account at which angle of attack the maximum loads occur. For fighter aircraft maximum loads can act at an angle of attack up to 40° whereas the maximum forces for a clean transport aircraft may occur at 15° angle of attack.

Because the balance is mounted inside the model and does not change the orientation relative to the model, no maximum single loads occur. For different test setups (transporter, fighter, high lift, cruise condition, etc.), different maximum combined loads occur. Again, it is useful to use a table such as shown in Table 5.2 to determine the maximum combined loads for the balance.

Table 5.1 Maximum combined loads for external balance

			Wind axis system							Model axis system					
Test type	q	A	C_x	C_y	C	Drag	Side force	F, M	max α, β, γ	Fx	Fy	Fz	Mx	My	Mz
	Pa	m^2				N	N	N, Nm	deg	N	N	N	Nm	Nm	Nm
Full model															
Half model															
etc.															

Table 5.2 Determination of maximum combined load for an internal balance

			Wind axis system							Model axis system					
Test type	q	A	C_x	C_y	C	Drag	Side Force	F, M	max α, β, γ	Fx	Fy	Fz	Mx	My	Mz
	Pa	m^2				N	N	N, Nm	deg	N	N	N	Nm	Nm	Nm
Transporter															
Landing															
Cruise															
Fighter															
etc.															

5.3 Dynamic Loads

It is very difficult to specify a dynamic load spectrum, because it would imply that the wind tunnel operator knows what kind of test he is going to perform in the future and which dynamic loads will occur during these tests. Normally these data are not available and only few customers provide specifications on dynamic loads.

In special cases like a rotating balance, where the rotor dynamic is known by the desired RPM, some forecast on the dynamic load spectra can be made.

The dynamic behavior of a model strongly depends on the model and on the tunnel. Therefore only some assumptions can be made that are based on the experience of past tests in the tunnel. To obtain the necessary information a dynamic data analysis of past tests must be performed. To be able to do this, data of the load monitoring system can be used. Some tunnels have such a system, not necessarily intended for a dynamic load forecast, but to be able to perform a dynamic lifetime prediction of the balance and sting. For lifetime prediction the same problem of not knowing what happens in the future occurs, but the knowledge of the dynamic past can be of help.

If data of a load monitoring system are available, data analysis must be performed in the following manner. Frequency spectrum can be calculated by a Fourier transformation. To count the number and the magnitude of loads, a counting algorithm must be used. For dynamic load prediction the "Rainflow Algorithm" is used to count loads and magnitude of loads [1, 11].

If no dynamic data is available, other assumptions can be made. An example for a transonic tunnel is the dynamic load specification of ETW. The requirement is $\pm 20\%$ of combined static load for the amplitude at frequencies typical from 5 to 80 Hz. For a normal design these dynamic loads are not critical for the lifetime of a balance because the normal gap between acting stresses and yield stress is about 2–4 times the stresses of static loads.

5.4 Maximum Combined Load; Maximum Single Load

The definition of the maximum loads can differ. If the loads are all acting together at the same time they are called maximum combined loads. If the maximum load is acting alone it is defined as maximum single load. The maximum single loads form a load trapeze, which does not naturally cover the test requirements (see Fig. 5.3). The maximum combined loads specify the load ranges of the balance for the design. Usually no single loads appear in wind tunnel tests and combined loads stress the balance in a much more complicated way, so the stress analysis of the balance has to take this situation into account. With the combination of only two loads instead of all six components, the balance usually can carry higher loads than defined by the maximum combined loads. To estimate which load can be carried in a given situation the balance manufacturer provides loading diagrams (load rhombus) which allow the

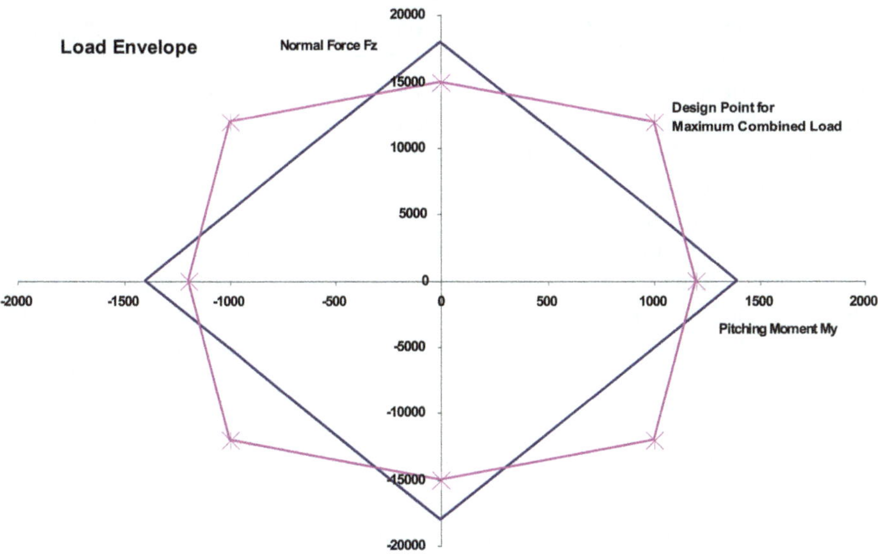

Fig. 5.3 Load rhombus, load trapeze

test engineer to determine whether the load combination of the planned test is within the limits of the balance.

Which one of the above-mentioned loads are used as "Full Scale Loads" depends on the balance manufacture's philosophy. A customer should specify the definition of the full scale load, because using this definition the relative uncertainty of the balance can vary by the factor of 2 without any difference in the absolute uncertainty.

If the loads are specified for an existing tunnel, an analysis of the measured data can provide sufficient information about the load ranges needed. The envelope around the measured data is one possibility, but this will lead to a very robust definition of load ranges. If the definition of the load ranges is set to the values with the most frequent occurrence, then there are higher loads or load combinations that do not occur very often during a test. During the detailed design phase it should be checked whether the balance can carry these loads without damage. By doing this the balance is adapted much better to most of the load situations inside the tunnel.

5.5 Safety Factors

The definition of safety factors strongly depends on the balance type. A normal internal or external balance that is intended for long-time use in the tunnel must be designed for infinite lifetime and good repeatability and therefore low hysteresis. These measurement requirements imply a large safety margin between yield stress and the existing stresses in the balance. The ratio between yield stress and stress under the strain gauge should be 3–5. For stress concentrations away from the gauge areas this ratio should be higher in the ideal case. Normally this cannot be achieved due to space restrictions and safety factors around 3 are common for stress calculations with Finite Element Methods (FEM). Especially in the areas of the interfaces of the balances, lower safety factors of 1.5–2 are sometimes accepted. Stresses in these areas do not necessarily affect the quality of the strain measurement, but usually this cannot be estimated by a FEM calculation because the resolution and the accuracy of a calculation are not sufficient to deliver these results. Effects of less than 0.01% must be predicted and FEM calculations are far removed from providing results with this precision. To check whether the strain in the high stressed areas does not affect measurements, experimental tests with dummies can be performed.

5.6 Deflections

Deflections and sensitivity cannot be specified co-existent because these requirements are contradictory. To obtain a particular sensitivity, a particular deformation is needed. In a well-designed balance these deformations are caused by the strains of the measuring sections and so they are needed for the measurement. A request for a higher stiffness will consequently reduce the sensitivity. So if a high stiffness is an ultimate requirement, the sensitivity must be achieved by changing the strain sensor, for example by using semiconductor gauges or switching to piezoelectric sensors. In this case other restrictions on signal stability or temperature sensitivity must be taken into account.

5.7 Constraints Due to Model Design (Space and Position)

The considerations above do not take into account that the internal balance must fit into the model, rather they are based on the assumption that the balance reference point is close to the center of gravity of the real vehicle. The specified moments are related to the center of gravity of the real vehicle, not to the center of gravity of the model, because the mass distribution inside the model does not necessarily match the mass distribution of the real vehicle.

In some cases the balance load specifications fit to the test load envelope, but the balance cannot be mounted near the center of gravity. Then the moment load ranges of the balance will not fit anymore to the test moment ranges. Or if the balance must be designed especially for such a case, high moment load ranges make it very complicated to design a balance with defined sensitivities for the force and the moment measurement. In some cases it is even impossible to design a balance with nearly the same sensitivity for forces and moments. In this case, usually a lower sensitivity for the forces must be accepted, or the position of the balance inside the model must be changed.

In preparing for a test, it must be determined whether the design of a balance is suitable for the given geometry and setup position. To do this the specific load parameter and the principle design equation discussed in the next Sects. 5.8 and 5.9 can be used.

5.8 Specific Load Parameter

The feasibility of a new balance can only be judged after defining the maximum loads, the maximum combined load and the available space. To determine whether the given diameter inside the model allows designing an appropriate balance, the specific load parameter can be used.

Before designing the bending section, it is essential to determine whether the balance is highly loaded or not. High loading means that the ratio between the loads and the available volume for the balance is high or not. This ratio will express the magnitude of the stress level inside the balance before starting with the calculation. This ratio is defined as "*Specific Load Parameter*" [3] by the following equations:

$$S_{round} = \frac{N + L/2 + M}{D^3}; \quad S_{rectangular} \frac{N + L/2 + M}{1.7 \cdot B \cdot H} \; \mathrm{N/cm^2} \tag{5.1}$$

where L [cm] is the active length of the balance without interfaces, D [cm] is the diameter, and B, H [cm] are the maximum available width and height of the balance. In these equations N [N] is a force and M [Ncm] a moment (Fig. 5.4).

The first equation is used when the main cross-section of the balance is circular and the second equation is used when the main cross-section is rectangular. Experience shows that highly loaded balances have a specific load parameter greater than

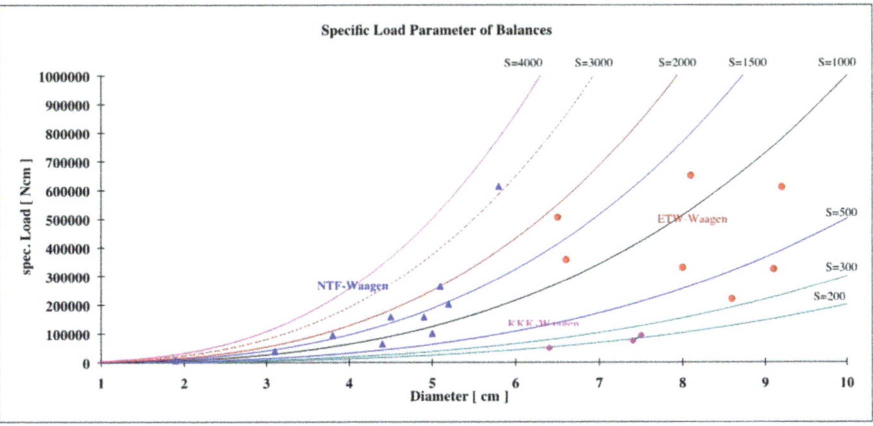

Fig. 5.4 Specific load parameter of some balances

$1000\,\mathrm{N/cm^2}$. In this case the highly stressed areas occur not only in the area where strain gauges are applied; a complex deformation of the balance can produce stress concentrations in areas where they are not expected.

Balances with a specific load factor of less than $1000\,\mathrm{N/cm^2}$ can be calculated with standard handbook formulas that are applied to the gauging areas and some other obviously high stressed areas, like the flexures and the interfaces. Experience shows that in these cases the calculated signals match well with measured signals and that finite element calculations do not necessarily produce better results. So the possible diameter or maximal height and width are the first dimensions that are given as balance dimensions.

5.9 Principle Design Equations (Feasibility)

After defining the possible cross-section, the length of the balance has to be estimated. All internal balances have in common that a moment and a force are measured in two different sections along the X axis. The distance between the two sections defines the separation of the signal between force and moment. In the ideal case half of the signal should be proportional to the force and the other half should be proportional to the moment (see Sect. 6.1); [4].

Moment and Force Separation

To obtain the same output for the force and the moment, the distance $l = l_1 + l_2$ between the two sections has to be calculated. For a moment type balance these relations can be described by the following equations (Fig. 5.5):

$$\sigma_1 = \frac{M_{B1}}{W_1} = \frac{M + N \cdot l_1}{W_1}; \sigma_2 = \frac{M_{B2}}{W_2} = \frac{M - N \cdot l_2}{W_2}; W_1 = W_2; l_1 = l_2 \quad (5.2)$$

where σ_1 and σ_2 are the stresses in section one and two, caused by the moment (M) and the force (N). (W_1) and (W_2) are the section modulus. The sum and the difference of these two stresses are:

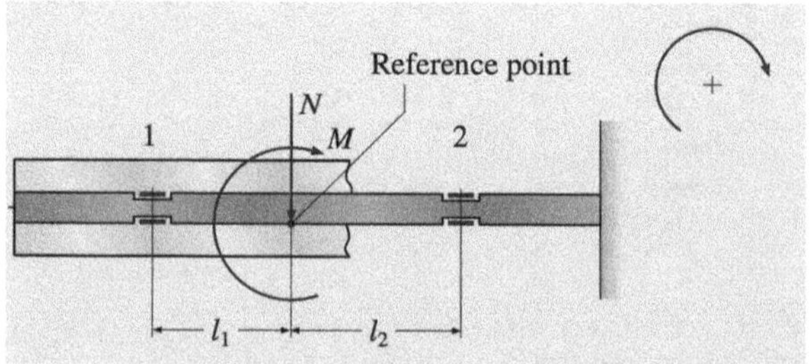

Fig. 5.5 Force and moment acting on a sleeve over the reference point

$$\sigma_1 + \sigma_2 = M(\frac{1}{W_1} + \frac{1}{W_2}) + N(\frac{l_1}{W_1} - \frac{l_2}{W_2}) = \frac{2M}{W} \tag{5.3}$$

$$\sigma_1 - \sigma_2 = M(\frac{1}{W_1} - \frac{1}{W_2}) + N(\frac{l_1}{W_1} + \frac{l_2}{W_2}) = \frac{N \cdot l}{W} \tag{5.4}$$

In Eq. (5.3) it can be seen that the sum of the stresses is only proportional to the moment (M) and in Eq. (5.4) it can be seen that the difference of the stresses is only proportional to the force (N). As a consequence, the sum of the signals of bridge (S_1) and bridge (S_2) are proportional to the moment (M) and the difference of the signals of bridge (S_1) and bridge (S_2) are proportional to the force (N)

$$\begin{aligned} \sigma_1 + \sigma_2 &\approx \Delta U_M \\ \sigma_1 - \sigma_2 &\approx \Delta U_N \end{aligned} \tag{5.5}$$

Now the ratio between the sum and the difference of the signals is calculated leading to the following equation:

$$\frac{\Delta U_M}{\Delta U_N} = \frac{2 \cdot M}{N \cdot l} \tag{5.6}$$

With a given moment (M) and a given force (N) and the design aim that the ratio of the signal should be one, the distance (l) between the measuring sections can be calculated:

$$l = \frac{2 \cdot M}{N} \tag{5.7}$$

This equation is also valid for a force type balance. The equation reveals that to achieve the same output for the force and the moment, there is one optimum distance l between the measuring sections. Thus, by the definition of the load components the length of the balance will be determined. If the required load combination results in a length that does not fit into the model, the distance between the measuring sections must be a compromise and either the signals for the force or the signals for the moment will be smaller than the other.

Another problem can appear when the distance for lift and pitch in the X-Z plane of the balance is different from the distance for side force and yaw in the Y-X plane of the balance. A solution for this can be two different measuring sections, but this will enlarge the total length of the balance, which is seldom acceptable.

For a single test setup, normally the balance length can be easily optimized for the model, but balance load range definition is often a compromise of requirements for different test setups, where the maximum loads for all tests form the envelope of the combined load range specification. In this case a good compromise for the balance dimensions can seldom be found. The ideal distribution between force and moment is a ratio of 50–50%. Lower values than 40–60% or vice versa should only be accepted if the component with the lower sensitivity is not that important. By determining the distance l between the bending sections in a moment type balance or the tension and compression sensors in a force balance, the total length of the balance can be estimated, and now the overall dimensions of the balance are more or less fixed.

5.10 Specification of Resolution, Repeatability and Sensitivities

5.10.1 Resolution

The specification of the resolution of a strain gauge based transducer is not really necessary (see Chap. 2). The resolution of a strain gauge is limited by the thermal noise inside the wire that effects the resistance measurement. Based on the usual resistance of a strain gauge, a resolution is 1:2,000,000 is common, which is equal to 0.5 ppm, or based on full scale output of 1 mV/V, the resolution is 0.5 nV/V. More information about the resolution of a strain gauge is given in subsection 9.1.9.

For a given balance system the measurement equipment limits the resolution. Modern electronic systems are capable of resolutions of about 1:300,000 up to 1:1,000,000, That is close to the strain gauge capabilities. To check whether there is sufficient resolution for the force measurement, it is only necessary to check the resolution of the signal processing system.

5.10.2 Repeatability

In the specifications the repeatability can be limited to the performance of the balance itself or to the overall repeatability, taking into account the uncertainties of the environmental conditions and the uncertainties by the calibration (see Chap. 2). Which values should be specified here depends on the scope of the order.

If only a balance with calibration is ordered, the repeatability at constant environmental conditions is equal to the resolution of the balance. However, practically a repeatability of about 0.05–0.005% of full-scale output is usually required. Repeatability can be checked in this case by short-term repetition of loads applied at the absolute same position (even then there is some uncertainty by the loading). By loading and unloading the test weights, the balance signals must remain constant within the given limits when the environmental conditions change over the desired range..

If a complete balance system is ordered, the repeatability of the signal processing system could be the limiting factor and this has to be checked separately.

5.10.3 Sensitivity

The specification of sensitivity depends on the number of components that strain one measurement section. In an internal balance there are usually one or two components that load one strain gauge bridge. One limit depends on the signal processing system. To determine this limit the maximum possible input signal of the system must be divided by the number of components that stress the bridge. For an internal balance this usually means an output between 1 and 1.5 mV/V.

The other limit is the stress under the strain gauge that should not exceed a corresponding signal of 2–3 mV/V. The lower value is preferred because this guarantees low creep, low hysteresis and a higher stiffness.

Modern signal processing systems have excellent performance, so the 1 mV/V signal level for the combined output of a strain gauge bridge is reasonable if not more than two loads affect the measurement. This is a good compromise between sufficient signal and high stiffness.

In other types of balances, like half model balances or box balances, one strain gauge bridge might be affected by more than two loads, so the maximum output for a single load should not be higher than 0.5–1 mV/V to remain below the 2 mV/V level for the combined loading.

To specify only a high sensitivity does not necessarily result in a good balance. To enable some potential for the balance designer to find a good comprise between output, good stability of the signal and stiffness of the balance, a practical range for the sensitivity should be specified.

5.11 Specification of Uncertainty

The specification of the allowed uncertainty for a calibrated balance is a rather complicated problem and there is even no common definition for what is the uncertainty of a balance. The problem of such a definition is that even if the definition is restricted to the load measurement at a stable temperature, there are six measurements influencing each other and thus affecting the uncertainty of one component. If temperature variations are allowed, the problem becomes even complex.

To solve this problem and to come to a common understanding *Ewald* and *Graewe* [7] formulated an equation that accounted for interactions on the uncertainty. These equations and the factors used where first agreed between *Airbus Bremen* and the *DNW*. Airbus Bremen built the first three internal balances for the *DNW-LLF* wind tunnel and they agreed on this description for the allowed uncertainty of the balance. Subsequently, this equation was extended with terms that take the effect of temperature into account. These equations were used in the specifications of the balances for *ETW*.

After the delivery of the balance it has to be demonstrated that the uncertainty of any measurement does in fact lie within the designated limits. However, the equations do not express the actual quality and the real uncertainty of the delivered balance. The discussion how to calculate the actual uncertainty of a combined force measurement out of the calibration data is still ongoing. What is required is a commonly accepted procedure so that the quality of different balances can be compared. The procedure to use the equations for the specifications to set a minimum required quality has been proven to be feasible. For a conventional balance at nearly stable temperature conditions the equations are:

$$\delta \leq [a \cdot F_{i,max} + b \cdot F_i] \cdot [c_i + d_i \cdot \sum_{n=1,n\neq i}^{6} |\frac{F_n}{F_{n,max}}|] \tag{5.8}$$

The allowed uncertainty ∂_i should be smaller than the product of uncertainties produced by the measured force F_i multiplied by the uncertainties produced by the sum of the relative interactions $F_n/F_{n,max}$. The factors a are global uncertainty factors and the factors c_i is an individual uncertainty factor for the component i. d_i is the uncertainty factor for the interactions.

To consider the effect of a global change of temperature and temperature gradients inside the balance Eq. (5.8) is expanded with two terms.

$$\delta \leq [a \cdot F_{i,max} + b \cdot F_i] \cdot [c_i^2 + (d_i \cdot \sum_{n=1,n\neq i}^{6} |\frac{F_i}{F_{i,max}}|)^2 + (e_i \cdot \Delta T_1)^2 + (f_i \cdot \Delta T_2)^2]^{\frac{1}{2}} \tag{5.9}$$

The factor e_i is the uncertainty factor for change of temperature ΔT_1 on the balance between wind on and wind off. The factor f_i is the uncertainty factor for temperature difference ΔT_2 between sting end and model end (Table 5.3).

Equations (5.8) and (5.9) differ a little, because according to the error propagation, the errors do not sum up directly. To obtain the resultant uncertainty the square root of the sum of the squared individual uncertainties must be taken. The desired factors a to f_i must be defined by the balance customer.

According to the different influences of the interactions on the measured component, every factor for the interaction must be defined individually. The quantities c_i and d_i are vectors with up to 6 elements (Table 5.4).

5.12 Specification of Thermal Characteristics

The specification of the thermal characteristics of a balance has an strong influence on the cost of a balance. Thus, it is most important for the customer of a balance to specify their needs as exactly as possible. Stricter specifications than needed will raise the cost in a highly nonlinear fashion.

Normally the operation temperature range is defined by the operating conditions of the tunnel, but the larger the dimensions of a balance, the more temperature

Table 5.3 Nomenclature "Allowed Uncertainty"

Param	Definition	Units
∂_i	Allowed uncertainty for the component i	N; Nm
a	Global uncertainty factor related to full scale	–
b	Global uncertainty factor related to measured value	–
$F_{i,max}$	Full scale of component i	N; Nm
F_i	Measured value of component i	N; Nm
c_i	Uncertainty factor for component i	–
d_i	Uncertainty factor for the interactions on component i	–
F_n	Measured interaction n	N ;Nm
$F_{n,max}$	Full scale of measured interaction n	N; Nm
e_i	Uncertainty factor for temperature change	N/K; Nm/K
f_i	Uncertainty factor for temperature gradients	N/K; Nm/K
ΔT_1	Temperature change on balance between two wind offs	K
ΔT_2	Temperature difference between sting end and model end	K

Table 5.4 Typical value ranges for the uncertainty factors

Factor	Value	Units
a	0.0005…0.001	
b	0.0001…0.0005	
c_i	0.4…2.0	
d_i	0.2…1.0	
e_i	0.02	N/K; Nm/K
f_i	0.01…0.1	N/K; Nm/K

gradients become the limiting problem of thermal stability. So specifying the thermal characteristics is not only a function of the tunnel temperature it is also a function of the dimensions of a balance.

5.12.1 Operating Temperature Range

Normally, conventional wind tunnel operators know the gas temperature of their tunnel very well and so it is easy to define exactly this gas temperature range as the operating temperature range for the balance. In tunnels where the temperature is stabilized by a cooler, the total range should be limited to the stabilized range and best results can be obtained by pre-conditioning of the model and the balance to these temperatures.

In tunnels without a cooler the temperature range depends on the variation of tunnel temperature between summer and winter. For good temperature conditioning some very large conventional tunnels stabilize the tunnel temperature and balance temperature before starting a measurement. With such a procedure the specified temperature range can be less stringent and at the same time the balance zero output can be easily stabilized at very low values, which improves the quality of a test without generating high cost for compensation measures.

5.12.2 Zero Drift

Zero drift is defined and explained in Chaps. 2 and 8 in detail. Without going into more detail here, the zero drift can be defined as the change of a balance signal between two repeated situations. This can be two wind-off situations with exactly the same model position. So the definition of the minimum required zero drift depends not only on the tunnel temperature range, but also depends on the tunnel operating procedure. If more wind-off situations are taken, the range for the maximum zero drift can be limited to the maximum change of temperature between two wind-off tests. More wind off tests reduce the limits for the specification of the balance and as a consequence, reduce the cost for the balance. However, also the tunnel efficiency is reduced. Increasing the tunnel efficiency by building a totally temperature compensated but expensive balance is therefore not always of interest for a wind tunnel, because this will reduce the occupation time for the tunnel, which is good for the tunnel user but bad for the profit of a tunnel. This is curious, but it demonstrates that increasing the quality does not always increase the financial bottom line. Without knowing the individual tunnel circumstances only a range for the zero drift specification can be given. Depending in part on which tunnel is being considered, the zero drift for the balance should normally be specified within the range of: 0.02 and 0.003%/K related to full scale output.

5.12.3 Sensitivity Drift

Sensitivity drift is also described in detail in Chaps. 2 and 8, thus, only a short definition is given here. For the balance sensitivity drift is the change of the sensitivity to an applied load due to a change of temperature. Uncertainty of a balance measurement caused by the sensitivity drift can only be minimized by keeping it low in the balance itself, stabilizing the tunnel temperature within narrow limits, or having evaluation matrices for different temperatures. Realizing one of these solutions is costly and to achieve the best results every procedure has to be realized. In cryogenic wind tunnels there is no alternative and all measures have to be ecploited.

The specification of sensitivity drift also depends on which tunnel is being considered and the range is generally within 0.001 and 0.0005%/K of nominal sensitivity.

5.12.4 Temperature Gradients

As long as the tunnel temperature changes slowly and as long as the overall change is not large, temperature gradients inside a balance my not be an issue. Nevertheless, even small temperature differences can conceivably affect a measurement seriously. A brief analysis of what can occur when a temperature difference arises is given here.

With the change of tunnel gas temperature the model normally changes its temperature relatively quickly because of the large surface exposed to the flow. All other components like the balance and the sting and support that have lower surface-to-volume ratio, follow slower. The result is a constant heat flow over the model that acts as a heat exchanger. This heat flow generates the temperature gradients. The higher the surface-to-volume ratio, the higher the sensitivity is to temperature gradients. Additionally, on objects with large dimensions it is more likely that temperature differences arise than on small objects, so these are the two main reasons why large balances are naturally more sensitive to temperature gradients.

The measurement effect of a temperature gradient is that the temperature differences cause deformation in one part of the structure and strain gauge bridges measure these deformations. In summary, temperature gradients generate a mechanical effect that is sensed by the balance. This mechanical effect Δl is proportional to the thermal expansion α coefficient and the maximum temperature difference ΔT that act over the characteristic length $\boldsymbol{l}$.

$$\Delta l = \alpha \cdot \Delta T \cdot l \tag{5.10}$$

For example the characteristic length l of an internal balance is the distance between the bending sections. In a balance for cryogenic tunnels the temperature gradients can be up to 20 K during a test. With a typical thermal expansion coefficient of $\alpha = 10 \times 10^{-6}$ 1/K and a typical distance of 300 mm, the deformation of balance is about 0.06 mm. The total deformation of a typical strain gauge sensor is about 0.3 mm. Comparing these two values, the temperature gradient effect can be about 20% of the signal. With such simple considerations the effect of temperature gradients can be estimated.

The manner and magnitude of temperature gradients must be known or measured and documented in the specifications so that the balance designer can take these into consideration.

5.13 Balance Interfaces

The function of the balance interface is to achieve a homogeneous repeatable transfer of the loads from the model to the balance and from the balance to the sting (see Chap. 4).

For the model-balance interface of internal balances it is hard to fulfill the two requirements of transferring the loads homogeneously with a high repeatability.

Mostly the space restrictions do not allow complicated precision interfaces and these issues are the subject of many design decisions. For the definition of the interface of an internal balance the possibility to turn the balance by 90° should be considered, because this allows a variation of the load ranges, for example to use side force bridges for lift.

If there are no stings in the inventory of a tunnel, the definition of the sting to balance interface must be done very carefully. Decisions made here have a long lifetime. Model-sting interfaces can be varied much easier because new models can be designed with different adapters.

For external balances there is normally no problem when defining the interface, because there is enough space that allows flexibility for different adapters that fulfill all the requirements.

5.14 Miscellaneous Specifications

5.14.1 Reference Planes and Reference Point

It already has been mentioned in Chap. 2 the difference between the model reference point and the balance reference point. Due to the importance of these definitions they are discussed once again, but now in more detail.

The desire is to measure the loads related to the model axis system, whose origin is named model reference point and not related to the balance axis system, whose origin is named balance reference system. The problem is that the correct position of the model is not known when a balance is specified and additionally, the model reference point varies from model to model. To overcome this problem a fictive reference point must be specified related to the metric end of the balance, because the metric end of the balance is fixed to the model and so the differences between the deformations of the model to the deformation of the balance are a minimum at this point. The position of this fictive model reference point is identical to the balance reference point only in the unloaded condition. That is the reason why the two definitions are often misunderstood (see Sect. 2.1).

At first the fictive model reference point can be defined at any position, it must not even be defined within the balance length. For practical reasons however, it is better to define it exactly at the midpoint between the bending sections or the sections that are used to determine moments. If a model is placed with its center of gravity over this point, the load ranges of the balance define the maximum loads and moments that are allowed in the wind tunnel test. If these points do not match, the ranges for the forces are the same but the moment ranges change.

In any case, the position of the fictive model reference point related to the center of gravity of the model must be exactly known. "Exactly" in this case means within a few hundreds of a millimeter and a few seconds of a degree. To enable this during the mounting of the model it must be possible to level the balance and to measure the relative position of the model center to the leveled position of the balance. Therefore on the metric end of the balance a horizontal plane and a vertical plane have to be foreseen. The balance customer should define these planes in the specification, because this defines the entire rigging process.

5.14.2 Moisture Protection

Most balance materials are not corrosion resistant. If the tunnel conditions are humid, a moisture protection must be specified for the balance. If the balance material is not corrosive the strain gauges must be protected by some kind of coating, because the signal is influenced by moisture and the contamination of dust and dirt. Suitable coatings are available from strain gauge manufacturers. Using in-house coatings require tests before applying them to a balance.

References

1. ASTM, E.: 1049-85. Standard Practices for Cycle Counting in Fatigue Analysis (2005)
2. Cahill, D.: Balance calibration uncertainty. In: 6th Symposium on Strain Gauge Balances, First Panel Session (2008)
3. Curry, T.: Force Measurement Devices and Techniques. The Boeing Company, Paper 1963 ERA/ISA Exposition at Seattle, Washington (1963)
4. Ewald, B.: The status of internal strain gage balance development for conventional and for cryogenic wind tunnels. In: AGARD Conference Proceedings. AGARD (1998)
5. Ewald, B.: The uncertainty of internal wind tunnel balances. Definition and verification. In: 3rd International Symposium on Strain-Gauge Balances, Symposium Papers, Darmstadt, pp. 13–16 (2002)
6. Ewald, B., Krenz, G.: The accuracy problem of airplane development force testing in cryogenic wind tunnels. In: 14th Aerodynamic Testing Conference, West Palm Beach, p. 776 (1986)
7. Ewald, B., Graewe, E.: ETW Balance Predesign Study. Tech. Rep. Report T.H. Darmstadt A 65/89 (MBB TE 2-1749), T.H. Darmstadt (1989)
8. Heyser, A.: Entwicklung von Windkanaleinbauwaagen mit Dehnungsmessstreifensystemen. DVL-Bericht Nr.261 (1963)
9. Hufnagel, K.: The truth about true loads.In: 5th Symposium on Strain Gauge Balances, Modane (2006)
10. Kammeyer, M.E., Tunnel, H.W.: Uncertainty analysis for force testing in production wind tunnels. In: NASA Conference Publications, pp. 221–242 (1999)
11. Köhler, M., Jenne, S., Pötter, K., Zenner, H.: Zählverfahren und Lastannahme in der Betriebsfestigkeit. Springer (2012)
12. Roberts, P.: Determination of bias and precision uncertainties of a strain gage balance by replicated machine calibration. In: 2nd Symposium on Strain Gauge Balances (1999)
13. Wright, F.L.: Experiences relative to the interaction between the balance engineer and the project engineer with regard to measurement uncertainty. In: NASA Conference Publications, pp. 243–278 (1999)

Chapter 6
Design of Balances

For the pre-design stage of a balance, codes are needed to provide a first description of the main dimensions of the balance, length of the active part, bending sections, the global dimensions of the axial force system and the calculated direct sensitivities.

The relations in this chapter will describe how stresses and signals in the major part of a balance can be calculated. A code that uses these standard relations can provide an easy and rapid first design. Such codes use linear mechanics and are relatively accurate in the prediction of the signals as long as the stresses are low. For balances with higher stress levels the predictions are less reliable, thus the dimensions obtained can only be used as a starting point for a subsequent optimization process using finite element codes.

The main design philosophy is that the highest stress should occur in the area where the strain gauges are applied. The second aim is to design the balance with a maximum stiffness, although high stiffness is counterproductive for a high sensitivity, since an increase in deformation is required for a higher sensitivity.

6.1 Internal Balances

Before describing detailed calculations, some general guidelines for the design of an internal balance will be discussed. To obtain a first estimate of the dimensions of an internal balance, the load ranges and the available space must be known from the specifications. Although this sounds like a trivial statement, many subsequent problems can be traced back to omitting this step. Either too many or too few features are often specified. Thus, the specification step demands a very close dialog with the end user in developing the specifications.

K. Hufnagel, *Wind Tunnel Balances*, Experimental Fluid Mechanics,
https://doi.org/10.1007/978-3-030-97766-5_6

1. Assuming that loads and available space are well discussed and known, then a first basic calculation using the "Basic Design Equations" is performed to determine the length of the balance. This calculation quickly reveals whether a proper separation of load and moment measurement is possible within the given length.
2. The "Specific Load Parameter" must be calculated, which indicates whether reasonable stress levels can be maintained within the given cross-section of a balance. If these two conditions can be met, then in principle a balance can be designed to meet the specifications.
3. The next step is to characterize the dimensions of the bending sections. To obtain a rough estimate of how to adjust the bending section to the different requirements for the longitudinal motion (F_z, M_y) and the yaw motion (F_y, M_z), the calculation of the stresses on the surfaces of a bending section with a rectangle shape must be performed. The width and the height of the rectangle should be set such that the signals for F_y, M_z and F_z, M_y reach a certain optimum within the given limits. This cross-section is the starting point for every subsequent optimization. If the dimension of this rectangle is very small compared with the available cross-section, a cage shape bending section is recommended. This leads at least to the same sensitivities, while achieving a much stiffer balance.
4. All these calculations define roughly the dimensions of the active part of the balance, except the axial force section. To obtain an estimate for the shape of the axial force section, possible dimensions of the flexure must be determined in a sketch of the balance overall cross-section. With these estimations a first calculation of the balance parallelogram system must be performed using a simple rectangular axial force measuring beam to estimate the output for the axial force.

After going through this process once, all important dimensions of the balance are roughly known and the optimization of the design can begin. To optimize the structure, two major iterations must be performed. One iterative process optimizes the bending section and the other must be made to optimize the parallelogram section. In rare cases, if stresses inside the balance exceed the allowed stress level and can only be reduced by changing the length of the parallelogram section, a re-dimensioning of the bending section is necessary and the entire process must start again from the beginning.

6.1.1 Moment and Force Separation Basic Design Equation

All internal balances have in common that a moment and a force are measured in two different sections along the X axis. The distance between the two sections defines the separation of the signal between force and moment. In the ideal case, half of the signal should be proportional to the force and the other half should be proportional to the moment.

To obtain approximately the same output for the force and the moment, the distance $l = l_1 + l_2$ between the two sections has to be calculated. For a moment type balance these relations can be described by the following equations (Figs. 6.1 and 6.2):

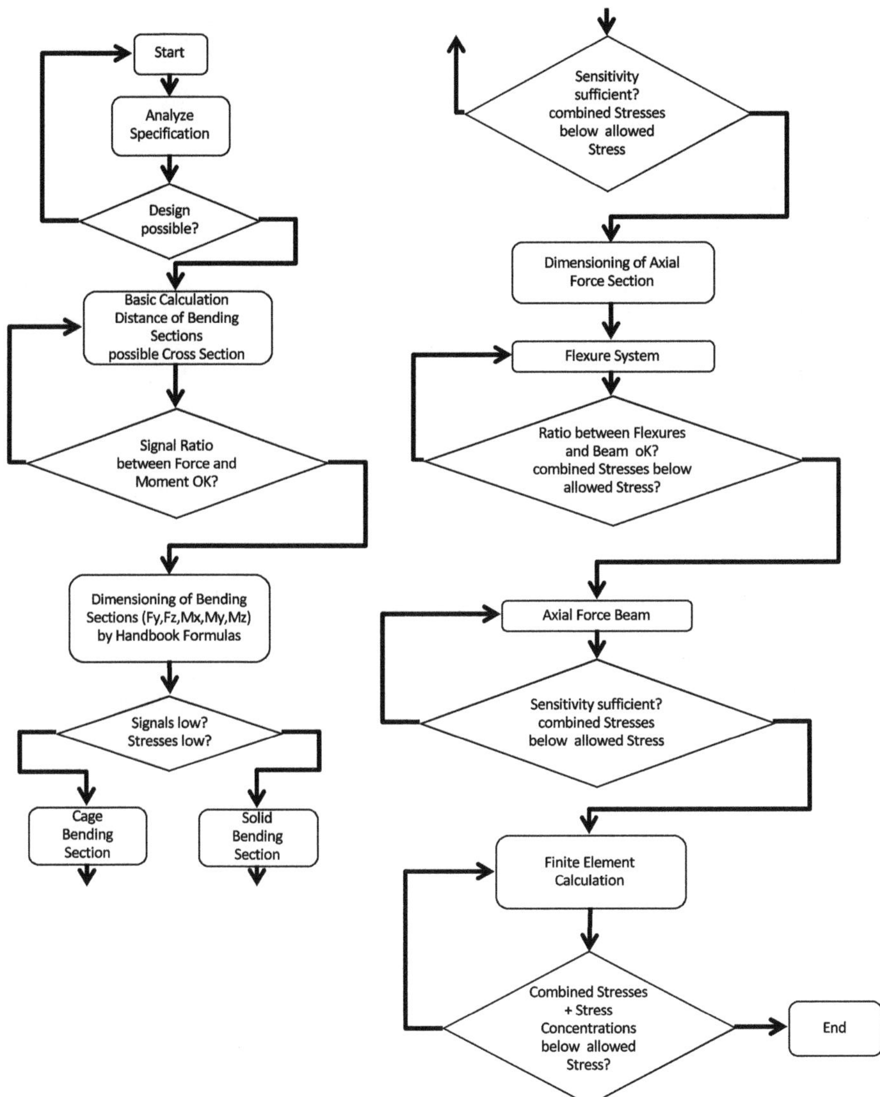

Fig. 6.1 Flowchart for balance calculation

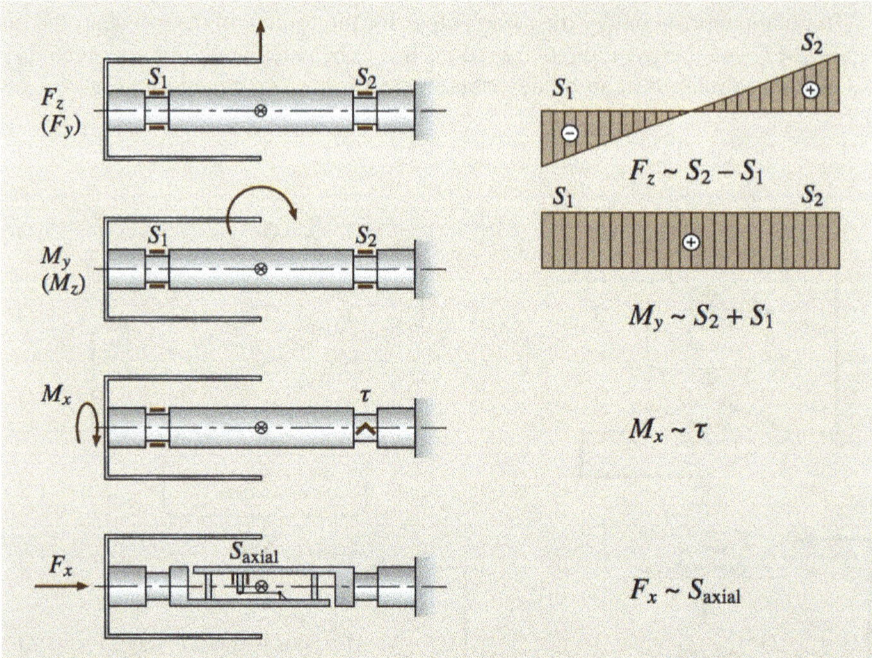

Fig. 6.2 Measurement sections for forces and moments of a six-component balance

$$\sigma_1 = \frac{M_{B1}}{W_1} = \frac{M + N \cdot l_1}{W_1}; \quad \sigma_2 = \frac{M_{B2}}{W_2} = \frac{M - N \cdot l_2}{W_2}; \quad W_1 = W_2; \quad l_1 = l_2 \tag{6.1}$$

σ_1 and σ_2 are the stresses in section one and two, caused by the moment (M) and the force (N). (W_1) and (W_2) are the section moduli. The sum and the difference of these two stresses are:

$$\sigma_1 + \sigma_2 = M(\frac{1}{W_1} + \frac{1}{W_2}) + N(\frac{l_1}{W_1} - \frac{l_2}{W_2}) = \frac{2M}{W} \tag{6.2}$$

$$\sigma_1 - \sigma_2 = M(\frac{1}{W_1} - \frac{1}{W_2}) + N(\frac{l_1}{W_1} + \frac{l_2}{W_2}) = \frac{N \cdot l}{W} \tag{6.3}$$

In Eq. (6.2) it can be seen that the sum of the stresses is proportional only to the moment M and in Eq. (6.3) it can be seen that the difference of the stresses is proportional only to the force N. Because of that, the sum of the signals of bridge S_1 and bridge S_2 is proportional to the moment M and the difference of the signal of bridge S_1 and bridge S_2 is proportional to the force N.

$$\begin{aligned} \sigma_1 + \sigma_2 &\approx \Delta U_M \\ \sigma_1 - \sigma_2 &\approx \Delta U_N \end{aligned} \tag{6.4}$$

Now the ratio between the sum and the difference of the signals obtaining the following equation is calculated:

$$\frac{\Delta U_M}{\Delta U_N} = \frac{2 \cdot M}{N \cdot l} \tag{6.5}$$

With a given moment M and a given force N and the design aim that the ratio of the signal should be unity, the distance l between the measuring sections can be calculated:

$$l = \frac{2 \cdot M}{N} \tag{6.6}$$

This equation is also valid for a force type balance. The equation reveals, that to achieve the same output for the force and the moment, there is one optimum distance between the measuring sections. Thus, by the definition of the load components, the length of the balance will be determined. If the required load combination results in a length that does not fit into the model, the distance between the measuring sections must be a compromise and either the signals for the forces or the signals for the moments are smaller than the others.

Another problem can appear when the distance for lift and pitch in the X-Z plane of the balance is different than the distance for side force and yaw in the Y-X plane of the balance.

A solution for this can be two different measuring sections, but this will enlarge the total length of the balance, which in most cases is not desirable.

For a single test setup, normally the balance length can be easily optimized for the model, but balance load range definition is mostly a mix of the requirements for different test setups, where the maximum loads for all tests form the envelope of the combined load range specification. In this case, very seldom can a good compromise for the balance dimensions be found.

The basic assumption for this kind of force and moment separation is that the load reference point is exactly in the middle between the two bending sections and the section moduli W_1 and W_2 are identical. So after the balance is ready, the length of the balance must be measured and the exact distance of middle from the model interface end defines the reference point of the balance.

If the load reference point is not defined exactly in the middle between the bending sections, a force and moment separation is not always possible. Examining Eqs. (6.2) and (6.3) and assuming that the distances l_1, l_2 and the section moduli W_1 ,W_2 are not identical, it can be easily seen that the sum and the difference of the stresses σ_1, σ_2 are a function of force N and moment M. Some special cases where the load reference center is not exactly in the middle are now discussed.

For the first case it is assumed that the load reference center is somewhere between the bending sections (see Fig. 6.3). If the section modulus of section 1 is identical to the section modulus of section 2, Eq. (6.3) can be transformed to:

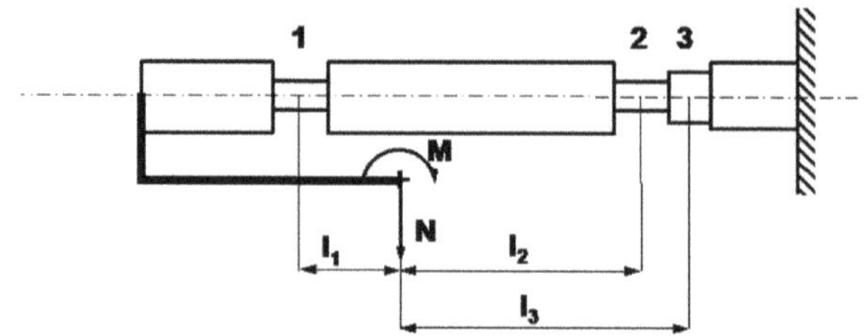

Fig. 6.3 Load reference point between section 1 and 2

$$\sigma_1 - \sigma_2 = N(\frac{l_1}{W_1} + \frac{l_2}{W_2}); \quad W_1 = W_2 \tag{6.7}$$

Again the difference of the bending stresses of section 1 and 2 are proportional to the force, but stress in Eq. (6.2) is only proportional to the moment if the ratios l_1/W_1 and l_2/W_2 are identical.

$$\sigma_1 + \sigma_2 = M(\frac{1}{W_1} + \frac{1}{W_2}) + N(\frac{l_1}{W_1} - \frac{l_2}{W_2}); \tag{6.8}$$

This is not possible because this is not consistent with the requirement of Eq. (6.8). To obtain a signal that is only proportional to the moment, an additional bending section (see Fig. 6.3) must be foreseen and dimensioned in such a way that the relation in Eq. (6.9) is fulfilled.

$$\sigma_1 + \sigma_3 = M(\frac{1}{W_1} + \frac{1}{W_3}); \quad \frac{l_1}{W_1} = \frac{l_3}{W_3} \tag{6.9}$$

Again the signal generated by the sum of the bending stresses σ_1, σ_3 is only proportional to the moment M.

For the second case it is assumed that the load reference center is in the position of the model end bending section 1.

If the section moduli W_1 and W_2 are identical, then again the result of Eq. (6.3) is that the

$$\sigma_1 - \sigma_2 = N \cdot \frac{l_2}{W_2} \tag{6.10}$$

signal for the force is proportional to the difference of the two bending stresses, but the sum of the bending stresses depends on both force and moment (Fig. 6.4).

$$\sigma_1 + \sigma_2 = M(\frac{1}{W_1} + \frac{1}{W_2}) - N\frac{l_2}{W_2} \tag{6.11}$$

In this special case the signal for the moment M is proportional to the bending stress σ_1 of Sect. 6.1.

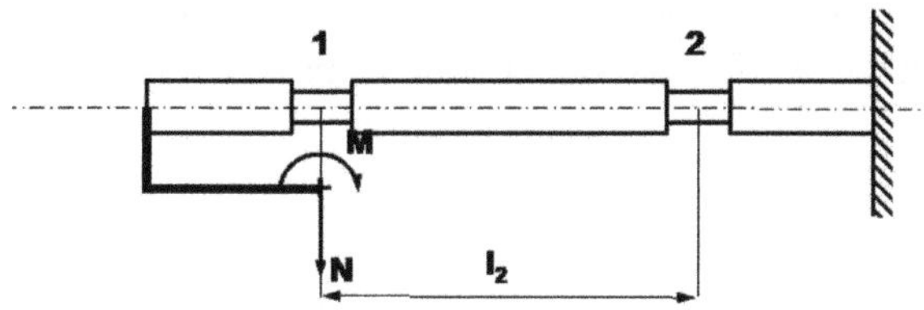

Fig. 6.4 Load reference center placed at bending section 1

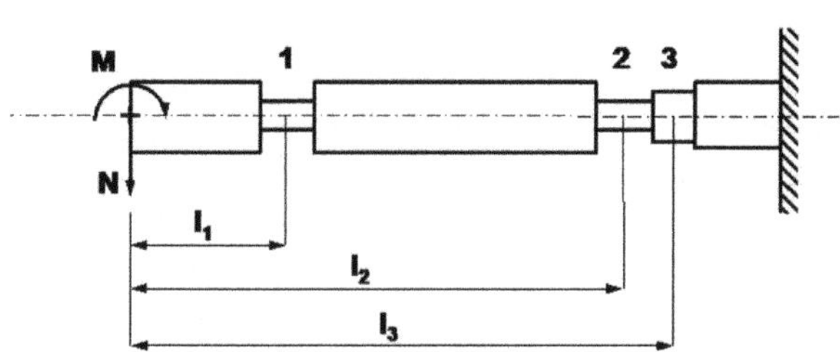

Fig. 6.5 Load reference center on model end of the balance

Finally, for the third case it is assumed that the load reference center is placed at the distance l_1 away from bending section 1 on the model end of the balance (see Fig. 6.5).

For this configuration Eqs. (6.2) and (6.3) change into the following versions caused by the change of the moment directions.

$$\sigma_1 + \sigma_2 = M(\frac{1}{W_1} + \frac{1}{W_2}) - N(\frac{l_1}{W_1} + \frac{l_2}{W_2}) \tag{6.12}$$

$$\sigma_1 - \sigma_2 = M(\frac{1}{W_1} - \frac{1}{W_2}) - N(\frac{l_1}{W_1} - \frac{l_2}{W_2}) \tag{6.13}$$

If the section modulus of section 1 is identical to that of section 2, again the difference [Eq .(6.13)] of the bending stresses is only proportional to the force.

$$\sigma_1 - \sigma_2 = N(\frac{l_2}{W_2} - \frac{l_1}{W_1}) \quad \text{for} \quad W_1 = W_2 \tag{6.14}$$

If the relation $l_1/W_1 = l_2/W_2$ is fulfilled, Eq. (6.13) is only proportional to the moment

$$\sigma_1 - \sigma_2 = M(\frac{1}{W_1} - \frac{1}{W_2}) \quad \text{for} \quad \frac{l_1}{W_1} = \frac{l_2}{W_2} \tag{6.15}$$

The two bending stress sections cannot be designed such that they fulfill both requirements. If the bending section 1 and 2 are geometrical identical and a third bending section is added which fulfills the requirement $l_1/W_1 = l_3/W_3$, a separation for force and moment measurement is possible. Then the difference of the bending stresses from section 1 and 2 is proportional to the force N and the difference of the bending stresses 1 and 3 is only proportional to the moment M. These considerations show that it is possible to separate force and moment measurement even when the load reference center is not in the middle between the bending sections; however when the load reference center is in the middle it is much easier to design a symmetrical balance.

6.1.2 Specific Load Parameter

Before designing the bending sections it is good practice to first get an impression whether the balance is highly loaded or not, before starting with further calculations. High loading means that the ratio between the loads and the available volume for the balance is high. This ratio indicates the magnitude of the stress level inside the balance. This ratio is defined as "*Specific Load Parameter*" by the following equations:

$$S_{\text{round}} = \frac{N + L/2 + M}{D^3}; \quad S_{\text{rectangular}} \frac{N + L/2 + M}{1.7 \cdot B \cdot H} \text{ N/cm}^2 \tag{6.16}$$

L [cm] is the active length of the balance without interfaces, D [cm] is the diameter, and B, H [cm] are the maximum available width and height of the balance. Force N [N] is the normal force and moment M [Ncm] is the pitching moment. The first equation is used when the main cross-section of the balance is circular and the second equation is used when the main cross-section is rectangular. Experience shows that highly loaded balances have a specific load parameter greater than 1000 N/cm^2. In this case the highly stressed areas occur not in the area where strain gauges are applied, but more likely on the flexures of the axial force elements.

Balances with a low specific load parameter can be designed by using only handbook formulas, because the main stressed areas can be located easily and for these locations the stresses and the signals can be predicted with sufficient accuracy. Balances with a high load parameter need to be analysed using a finite element calculation to find the critical areas and to redesign the balance to decrease stresses in these areas. It is not recommended to build balances with specific load parameters greater than $S = 2000\,\text{Ncm}$ (Fig. 6.6).

6.1.3 Routine Design Methods for Internal Balances

6.1.3.1 Basic Formulas for Stress Calculations

Modern Finite Element (FE) programs offer the possibility to calculate the stress inside a balance very precisely and with moderate computational effort, i.e. on a personal computer. The main problem with this approach is that the computation must be repeated for every minor design change. The parameters that can be varied

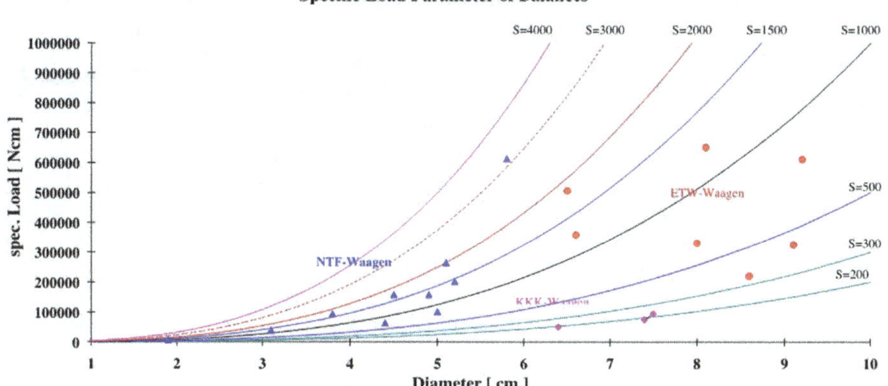

Fig. 6.6 Specific load parameter of some balances

for the optimization of a design are so numerous that this represents a significant effort before arriving at the best compromise for minimum stress level and best signal distribution. Thus, codes that use handbook formulas are still a convenient tool to rapidly approach an optimum design. Subsequently, the fine-tuning can be performed using a FE computation. That is also the justification for now discussing the following basic relations for stress calculations.

If these equations are implemented in programs or spreadsheets, a wide variation of parameters can be rapidly explored to find the most appropriate starting point for more detailed calculations. After the determination of the distance between the bending sections, the main forces and moments that are acting in these sections can be determined.

To start, a rough sketch of the balance with the major allowed dimensions and the location of the balance center are required. In the following discussion, the balance center is always set in the middle between the bending sections.

To determine the stresses in the bending sections, three fundamental beam models are used:

1. A beam fixed at one end and loaded with a moment acting on the other end
2. A beam fixed at one end and guided at the other end in the normal direction and loaded with a force on the guided end
3. A beam fixed at one end and torqued at the other end.

These three simplified mechanical models represent the loading of an internal balance under all components except axial force, which will be discussed in Sect. 6.1.6.

Model 1 represents the loading situation by a pitching or yawing moment at the center of the balance.

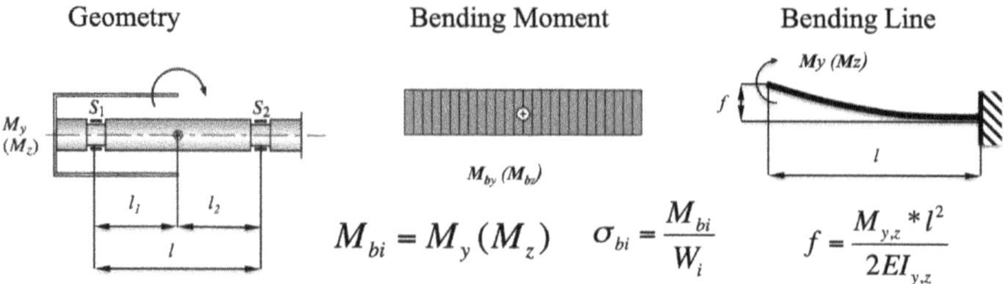

Fig. 6.7 Moment distribution, stress and strain formulas

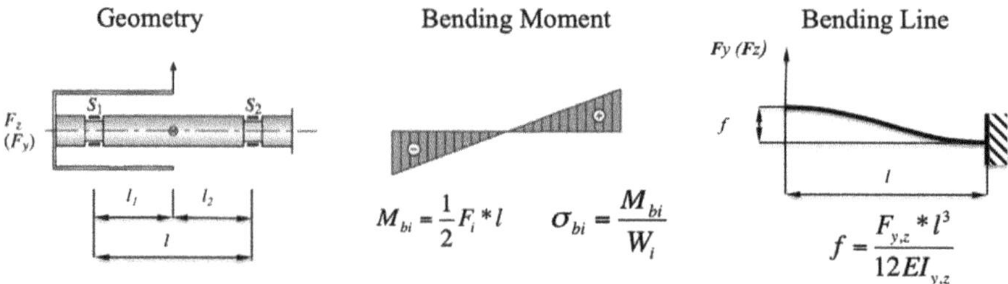

Fig. 6.8 Bending moment distribution by force, stress and strain formulas

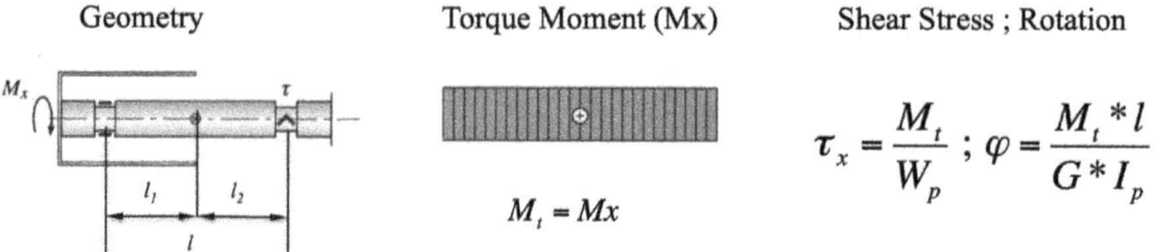

Fig. 6.9 Torque moment distribution, stress and strain formulas

To calculate the bending stress σ_{bi} [N/mm^2] in the measuring section the moments acting M_{bi} [Nm] are the specified moments M_y;Mz [Nm]. Unknown are the section moduli W_i [mm^3] for the cross-section. Section moduli for some frequently used bending sections are described in the Sects. 6.1.3.3–6.1.3.9 (Fig. 6.7).

Model 2 represents the loading situation by lift or side force F_z; F_y [N] at the center of the balance.

To calculate the bending stress σ_{bi} [N/mm^2] related to the forces, the bending moment M_{bi} [Nm] can be calculated using the equations given in Fig. 6.8, where the force F_i [N] is either the specified lift or side force F_z; F_y [N]. Here also, only the section moduli W_i [mm^3] are unknown and have to be calculated from the chosen cross-section geometry.

Model 3

Represents the loading of rolling moment at the center of the balance (Fig. 6.9).

The torque moment is constant over the length of the balance and therefore the specified torque moment M_x can be used to calculate the corresponding shear stress in every section. What is unknown here is the polar section modulus and examples for typical cross-sections will be calculated in separate Sects. 6.1.3.3–6.1.3.9

6.1.3.2 Section Moduli for Different Cross-Sections

Once the forces and moments acting in a certain plane of the balance have been calculated, the section moduli for given cross-sections can be derived to yield the stress on the surfaces of the bending sections where the stresses and signals for the desired loads must be calculated. The examples chosen will cover most applications, without being exhaustive. Nevertheless, for special applications alternative designs may deliver better results in distributing the stresses to achieve a better signal efficiency or a stiffer design.

The general nomenclature for this section is given as follows:

Dimension	Name	Unit
Area of cross-section	A	mm^2
Area moment of inertia Y axis	I_y	mm^4
Section modulus Y axis	W_y	mm^3
Area moment of Inertia Z-Axis	I_z	mm^4
Section modulus Z axis	W_z	mm^3
Area moment of inertia X axis	I_p	mm^4
Section modulus X axis	W_p	mm^3

6.1.3.3 Section Moduli for Rectangular Cross-Section

A rectangular cross-section with the dimensions shown in Fig. 6.10 can be used for a massive bending section as well as in the middle of a cage design with several other beams arranged around the central beam.

The section moduli for bending are relatively simple. The center hole is optional.

$$A = B \cdot H - \frac{\pi D_i{}^2}{4}; \quad I_y = \frac{B \cdot H^3}{12} - \frac{\pi \cdot D_i{}^4}{64}; \quad I_z = \frac{H \cdot B^3}{12} - \frac{\pi \cdot D_i{}^4}{64}$$
$$W_y = \frac{B \cdot H^2}{6} - \frac{\pi \cdot D_i{}^4}{32H}; \quad W_z = \frac{H \cdot B^2}{6} - \frac{\pi \cdot D_i{}^4}{32H} \tag{6.17}$$

For a torque (M_x) moment around the axial axis, the calculation of the moduli is somewhat more complicated and can be computed using the following equations:

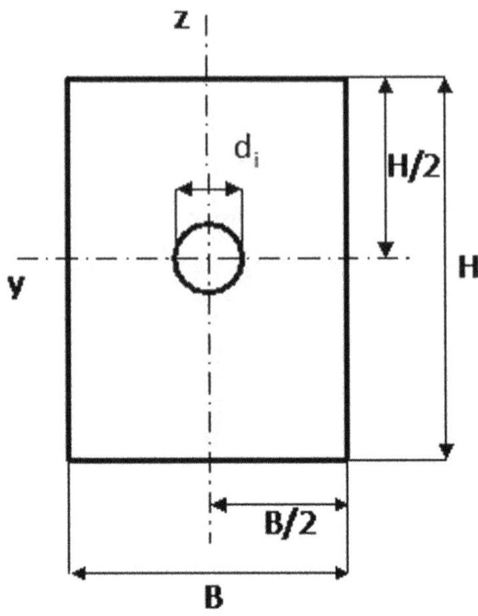

Fig. 6.10 Dimensions of a rectangular cross-section

$$I_p = H \cdot B^3 \cdot [\frac{1}{3} - 0.21 \cdot \frac{B}{H}(1 - \frac{B^4}{12 \cdot H^4})] - \frac{\pi \cdot D_i{}^4}{32}$$

$$W_p = \frac{H \cdot B^2 \cdot [\frac{1}{3} - 0.21 \cdot \frac{B}{H}(1 - \frac{B^4}{12 \cdot H^4})] - \frac{\pi \cdot D_i{}^4}{32}}{1 - \frac{0.65}{1+(\frac{H}{B})^3}} \quad for \quad H \geq B \tag{6.18}$$

For $D_i = 0$ the equation reduces to a cross-section without a center hole.

6.1.3.4 Section Moduli for Octagonal Cross-Section

The octagonal cross-section can also be used for a massive bending sections using a center hole. This is an ideal cross-section for a very highly loaded balance, where there is not much room for a cage design and a cylindrical shape of the balance for the bending section is almost unavoidable.

Expressions for the cross-sectional area and the moduli for bending are given as follows (Fig. 6.11):

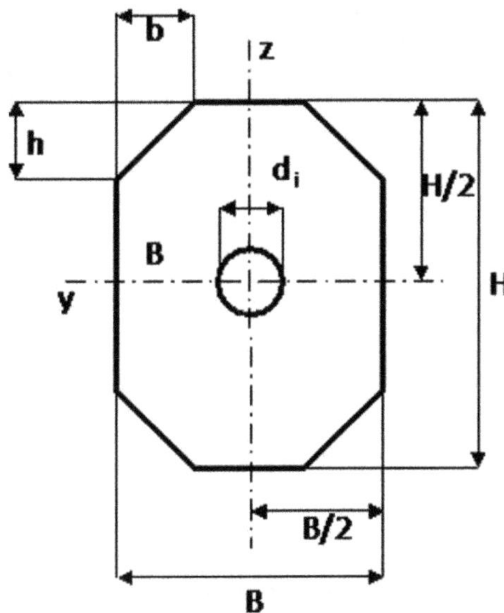

Fig. 6.11 Dimensions for the octagonal cross-section

$$A = H \cdot B - 2 \cdot b \cdot h - \frac{\pi \cdot {d_i}^2}{4} \tag{6.19}$$

$$I_y = \frac{B \cdot H^3}{12} - [\frac{b \cdot h^3}{9} + \frac{b \cdot h}{18} \cdot (3H - 2h)^2 + \frac{\pi \cdot {d_i}^4}{64}] \tag{6.20}$$

$$W_y = \frac{2 \cdot I_y}{H} \tag{6.21}$$

$$I_z = \frac{H \cdot B^3}{12} - [\frac{h \cdot b^3}{9} + \frac{b \cdot h}{18} \cdot (3B - 2b)^2 + \frac{\pi \cdot {d_i}^4}{64}] \tag{6.22}$$

$$W_z = \frac{2 \cdot I_z}{B} \tag{6.23}$$

and for torque:

$$I_p = \frac{A^4}{40 \cdot (I_y + I_z)} - \frac{\pi \cdot {d_i}^4}{32} \tag{6.24}$$

$$W_p = \frac{I_p}{C \cdot B}; \qquad C = \frac{1 + 0.15 \cdot (\frac{\pi \cdot B^2}{2 \cdot A})^2}{1 + (\frac{\pi \cdot B^2}{2 \cdot A})^2} \tag{6.25}$$

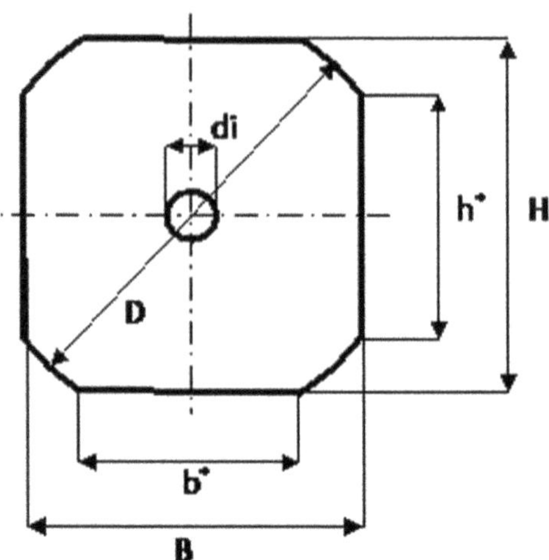

Fig. 6.12 Dimensions of a circular cross-section with flattenings

6.1.3.5 Section Moduli for Circular Cross-Section with Flattenings

This cross-section is also very useful when the loads related to the maximum possible area are relative high. With this cross-section most of the available area can be used for the bending section.

The corresponding relations are given by (Fig. 6.12):

$$\begin{aligned} A &= \frac{\pi}{4} \cdot (D^2 - d_i^2) - \frac{\pi}{180} \cdot (\arcsin \frac{b^*}{D} + \arccos \frac{h^*}{D}) \\ &+ \frac{1}{D} \cdot (b^* \cdot \sqrt{1 - \frac{b^{*2}}{d^2}} + h^* \cdot \sqrt{1 - \frac{h^{*2}}{D^2}}) \\ h^* &= \sqrt{D^2 - B^2}; \qquad b^* = \sqrt{D^2 - H^2} \end{aligned} \tag{6.26}$$

$$\begin{aligned} I_z &= \frac{\pi}{64} \cdot (D^4 - d_i^4) - \frac{D^4}{64} \cdot (2\beta - \frac{\sin 4\beta}{2} + 2\alpha - \frac{4}{3} \sin 2\alpha + \frac{1}{6} \sin 4\alpha) \\ I_y &= \frac{\pi}{64} \cdot (D^4 - d_i^4) - \frac{D^4}{64} \cdot (2\alpha - \frac{\sin 4\alpha}{2} + 2\beta - \frac{4}{3} \sin 2\beta + \frac{1}{6} \sin 4\beta) \end{aligned} \tag{6.27}$$

$$\alpha = \arcsin \frac{b^*}{D}; \qquad \beta = \arcsin \frac{h^*}{D} \tag{6.28}$$

$$W_y = \frac{2I_y}{H}; \qquad W_z = \frac{2I_z}{B} \tag{6.29}$$

$$I_p = \frac{A^4}{40 \cdot (I_y + I_z)} \tag{6.30}$$

$$W_p = \frac{I_y}{C \cdot B}; \qquad C = \frac{1 + 0.15 \cdot (\frac{\pi B^2}{4A})^2}{1 + (\frac{\pi B^2}{4A})^2} \tag{6.31}$$

6.1.3.6 Section Moduli for Cruciform Cross-Section with Fillets

$$A = B \cdot H - 4bh + e^2 - \frac{\pi {d_i}^2}{4} \tag{6.32}$$

$$\begin{aligned} I_y = &\frac{1}{12} \cdot [(B - 2b) \cdot H^3 + 2b \cdot (H - 2h)^3 + \frac{e^4}{3} \\ &+ 12e^2 \cdot (\frac{H}{2} - h + \frac{e}{2\sqrt{3}})^2] - \frac{\pi d_i^4}{4} \end{aligned} \tag{6.33}$$

$$\begin{aligned} I_z = &\frac{1}{12} \cdot [(H - 2h) \cdot B^3 + 2h \cdot (B - 2b)^3 \\ &+ \frac{e^4}{3} + 12e^2 \cdot (\frac{B}{2} - b + \frac{e}{2\sqrt{3}})^2] - \frac{\pi d_i^4}{4} \end{aligned} \tag{6.34}$$

$$W_y = \frac{2I_y}{H} \quad ; \qquad W_z = \frac{2I_z}{B} \tag{6.35}$$

There is no analytic expression for the area moment and the section modulus to calculate the shear stress related to the torque moment M_x, so an approximation has to be made. This approximation is taken from [19], according to Figs. 6.13 and 6.14.

The approximation of the linear fillet is made with a radius $r = 1.09 \cdot e$.

$$I_p = K_1 + K_2 + \alpha \cdot D_{ein} - \frac{\pi \cdot {d_i}^4}{32} \tag{6.36}$$

$$D_{ein} = e + \sqrt{2} \cdot (\frac{B + H}{2} - b - h) \tag{6.37}$$

$$K_1 = B \cdot (H - 2h)^3 \cdot [\frac{1}{3} - 0.21 \cdot \frac{H - 2h}{B} \cdot (1 - \frac{(H - 2h)^4}{12B^4})] \tag{6.38}$$

$$K_1 = 2h \cdot (B - 2b)^3 \cdot [\frac{1}{3} - 0.105 \cdot \frac{B - 2b}{h} \cdot (1 - \frac{(B - 2b)^2}{192h^4})] \tag{6.39}$$

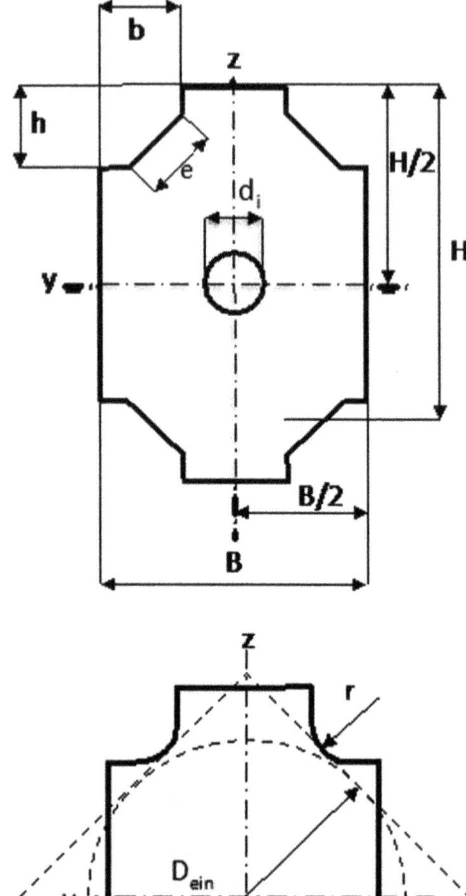

Fig. 6.13 Dimensions of a cruciform cross-section with fillets

Fig. 6.14 Approximation for cruciform cross-section with fillets

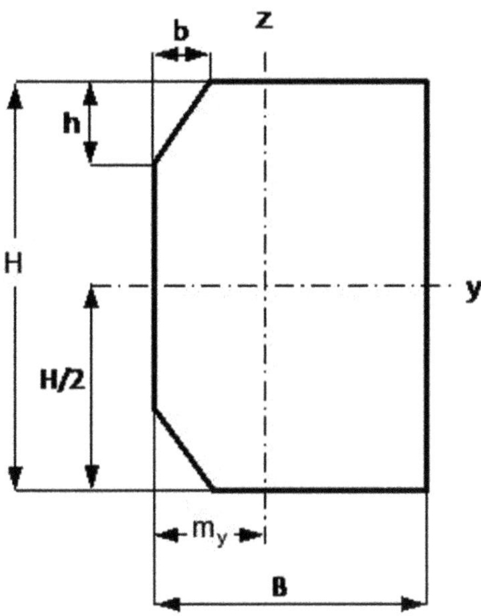

Fig. 6.15 Dimension for hexagonal cross-section on side

$$\begin{aligned}\alpha &= \frac{t}{t_1} \cdot (0.15 + 0.0707 \cdot \frac{e}{H - 2h}) \\ t &= (H - 2h);\ \text{if}(H - 2h) < (B - 2b);\ \text{else}\, t = (B - 2b) \\ t_1 &= (H - 2h);\ \text{if}(H - 2h) > (B - 2b);\ \text{else}\, t_1 = (B - 2b)\end{aligned} \tag{6.40}$$

6.1.3.7 Section Moduli for Hexagonal Cross-Section on Side

The formulas for the following cross-sections are not intended for use as single beams in a bending section. These shapes can be used in a cage arrangement with or without a central beam. The calculation of stresses and signals in a cage section will be discussed in Sect. 6.1.5 (Fig. 6.15).

$$A = B \cdot H - b \cdot h \tag{6.41}$$

$$I_y = \frac{B \cdot H^3}{12} - h \cdot b \cdot [\frac{h^2}{18} + (\frac{H}{2} - \frac{h}{3})^2] \tag{6.42}$$

$$I_z = \frac{H}{3} \cdot ({e_1}^3 + {e_2}^3) - b \cdot h \cdot [\frac{b^2}{18} + (e_2 - \frac{b}{3})^2]$$

$$e_1 = \frac{H \cdot B^2 - 2Bbh + \frac{2}{3}hb^2}{2 \cdot (BH - bh)} \tag{6.43}$$

$$e_2 = B - e_1; \quad m_y = e_2;$$

$$W_y = \frac{2I_y}{H} \quad ; \quad W_z = \frac{I_z}{e_2} \tag{6.44}$$

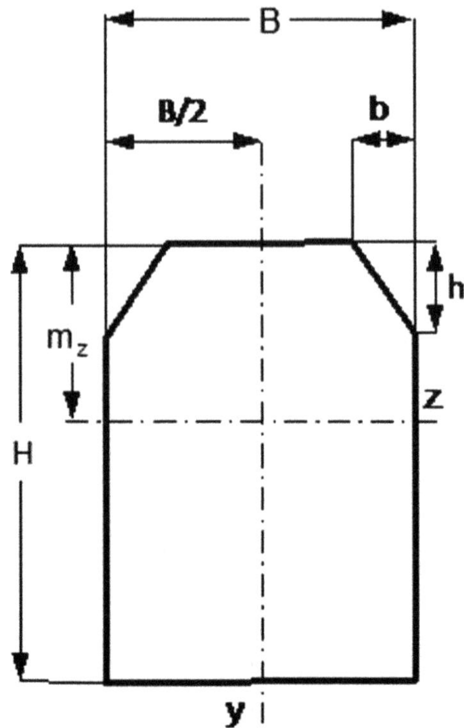

Fig. 6.16 Dimension for hexagonal cross-section on top

$$I_p = \frac{A^4}{40 \cdot (I_y - I_z)} \tag{6.45}$$

$$W_p = \frac{I_p}{C \cdot B}; \quad C = \frac{1 + 0.15 \cdot (\frac{\pi \cdot H^2}{4 \cdot A})^2}{1 + (\frac{\pi \cdot H^2}{4 \cdot A})^2} \tag{6.46}$$

6.1.3.8 Section Moduli for Hexagonal Cross-Section on Top

See Fig. 6.16.

$$A = B \cdot H - b \cdot h \tag{6.47}$$

$$I_z = \frac{H \cdot B^3}{12} - h \cdot b \cdot [\frac{b^2}{18} + (\frac{B}{2} - \frac{b}{3})^2] \tag{6.48}$$

$$I_y = \frac{B}{3} \cdot ({e_1}^3 + {e_2}^3) - b \cdot h \cdot [\frac{h^2}{18} + (e_2 - \frac{h}{3})^2]$$

$$e_1 = \frac{B \cdot H^2 - 2Hbh + \frac{2}{3}bh^2}{2 \cdot (BH - bh)} \tag{6.49}$$

$$e_2 = H - e_1; \quad m_z = e_2;$$

$$W_z = \frac{2I_z}{B} \quad ; \quad W_y = \frac{I_y}{e_2} \tag{6.50}$$

$$I_p = \frac{A^4}{40 \cdot (I_y - I_z)} \tag{6.51}$$

$$W_p = \frac{I_p}{C \cdot B}; \quad C = \frac{1 + 0.15 \cdot (\frac{\pi \cdot B^2}{4 \cdot A})^2}{1 + (\frac{\pi \cdot B^2}{4 \cdot A})^2} \tag{6.52}$$

6.1.3.9 Section Moduli for Rectangles with Rib on the Side

See Fig. 6.17.

$$A = B \cdot H - 2 \cdot b \cdot h \tag{6.53}$$

$$I_y = \frac{(B - b) \cdot H^3 + b \cdot (H - 2h)^3}{12} \tag{6.54}$$

$$I_z = \frac{1}{3}[\cdot H \cdot {e_1}^3 - 2h \cdot (e_1 - B + b)^3) + (H - 2h) \cdot {e_2}^3] \tag{6.55}$$

$$e_1 = \frac{HB^2 - 4Bbh + 2hb^2}{2 \cdot (BH - 2bh)}$$

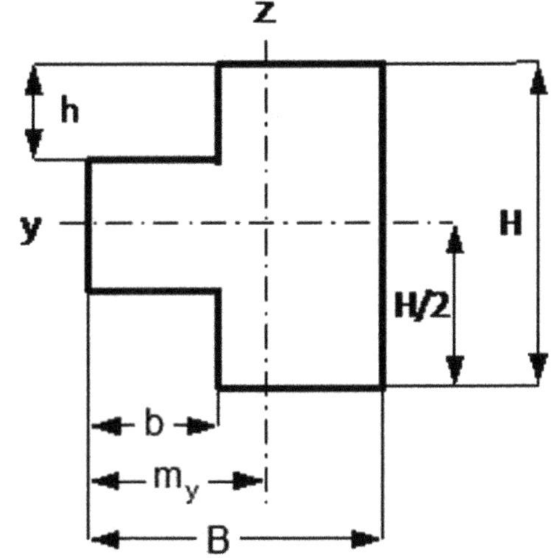

Fig. 6.17 Dimensions for rectangles with rib on the side

$$e_2 = B - e_1; \quad m_y = e_2$$

$$W_y = \frac{2I_y}{H}; \quad W_z = \frac{I_z}{e_2} \tag{6.56}$$

For this kind of shape there is no analytical equation for the section modulus to calculate shear stress due to torque. This is not a disadvantage because this shape is intended for the use in cage design bending sections, where the torsion is transferred into bending of two beams with such a cross section.

6.1.3.10 Section Moduli for Rectangles with Rib on Top

See Fig. 6.18.

$$A = B \cdot H - 2 \cdot b \cdot h \tag{6.57}$$

$$I_z = \frac{(H-h) \cdot B^3 + h \cdot (B-2b)^3}{12} \tag{6.58}$$

$$I_y = \frac{1}{3}[\cdot B \cdot {e_1}^3 - 2b \cdot (e_1 - H + h)^3) + (B - 2b) \cdot {e_2}^3] \tag{6.59}$$

$$e_1 = \frac{H^2 B - 4Hbh + 2h^2 b}{2 \cdot (BH - 2bh)}$$

$$e_2 = H - e_1; \quad m_z = e_2$$

$$W_z = \frac{2I_z}{B}; \quad W_y = \frac{I_y}{e_2} \tag{6.60}$$

For the polar section modulus see the comment above in Sect. 6.1.3.9.

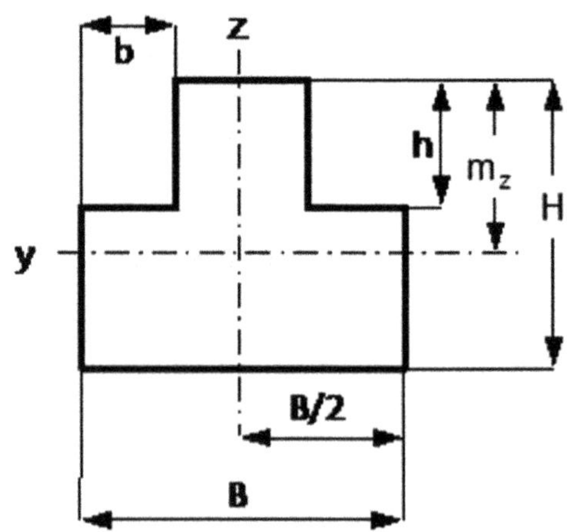

Fig. 6.18 Dimensions for rectangles with rib on top

6.1.4 Design of Solid Bending Section of Internal Balances

With the determination of the distance between the bending sections using the basic design equations of Sect. 6.1.1, the distribution of the bending stress acting in one plane between the corresponding force and moment must been determined. With the design of the cross-sectional shape at the bending measurement location, the separation of the sensitivity for the different planes can be examined. Further to the distribution of load and moment sensitivity, the other major basic design issue of an internal balance is the separation of the signals within the different measurement planes.

The cross-section has to be designed in such a manner that the signals due to the loads acting in the X-Z plane (normal force F_z and the pitching moment M_y) and the signals due to the loads that are acting in the X-Y plane (side force F_y and the yawing moment M_z) are close to the maximum allowed signal amplitude. Usually the allowed maximum signal amplitude is about 2 mV/V for a strain gauge full bridge. If this value can be achieved for the combined output of force and moment, the ideal maximum signal for each component is 1 mV/V. To achieve this value is not always possible, so a good compromise lies between 0.5 mV/V and 1.5 mV/V for each component. With high quality measurement equipment there will then be sufficient resolution for a precise measurement.

To transfer these signal requirements to a stress level, the stresses generated by the loads in the measurement areas depend on the material that is used for the balance and the strain gauge bridge setup. Assuming full bridges are used, the required stress level can be calculated with the following equation:

$$\sigma = \frac{\Delta U}{U} \cdot \frac{E}{k} \qquad [\mathrm{N/mm^2}] \tag{6.61}$$

where E [N/mm^2] is the Young's modulus of the balance material, $\Delta U/U$ [mV/V] is the desired signal output of the strain gauge full bridge, and $k \approx 2$ is the strain gauge factor.

For steel ($E \sim 200{,}000\,\mathrm{N/mm^2}$) the design target should be between $50\,\mathrm{N/mm^2}$ and $150\,\mathrm{N/mm^2}$ for each component. For the combination of force and moment these values should not exceed $300\,\mathrm{N/mm^2}$.

The four loads F_y, F_z, M_y, M_z are measured in the same cross-sections of the balance, but the strain gauges are glued on different side areas and so the sensitivity of the gauges to the strain generated by the loads can be adjusted to the load by the shape of the cross-section. Numerous different shapes are possible and many parameters can be varied to design a cross-section for optimizing the sensitivity for the loads that are acting.

To obtain a good estimate for the necessary cross-section, it is useful to start with a rigid bending section and a simple geometry, like a rectangle or a circle with flattenings. After that, further optimization can be done by adding more parameters, by adding center holes, cutouts or switch to a cage design.

The following tables show the stress relations to calculate the stresses on the different surfaces and the maximum stress acting in the section. The stresses on the surface can be used to calculate the signal related to the load and the maximum stress is needed to check the material utilization level and safety.

6.1.4.1 Rectangular Cross-Section

Stress by:	X-Y Plane	X-Z Plane	
F_x	$\sigma_{z1} = F_x/A$	$\sigma_{y1} = F_x/A$	(6.62)
F_z	$\sigma_{z2} = F_z \cdot l_1/W_y$	$\sigma_{y2} = \sigma_{z2}$	(6.63)
F_y	$\sigma_{z3} = \sigma_{y3}$	$\sigma_{y3} = F_y \cdot l_1/W_z$	(6.64)
M_x	$\tau_z = \tau_y \cdot f(\frac{H}{B})$	$\tau_y = M_x/W_p$	(6.65)
M_y	$\sigma_{z4} = M_y/W_y$	$\sigma_{y4} = \sigma_{z4}$	(6.66)
M_z	$\sigma_{z5} = \sigma_{y5}$	$\sigma_{y5} = M_z/W_z$	(6.67)
Ma.:	$\sigma_{zmax} = \sum_{i=1}^{5} \sigma_{zi}$	$\sigma_{ymax} = \sum_{i=1}^{5} \sigma_{yi}$	(6.68)
Comb	$\sigma_v = \sqrt{\sigma_{max}^2 + 3\tau^2}$(Mises)	$\sigma_v = \frac{1}{2}(\sigma_{max} + \sqrt{\sigma_{max}^2 + 4\tau^2})$(Rankine)	(6.69)

The maximum stress exists in the corner of the rectangle where only tension and compression stresses occur. The maximum shear stress is in the middle of the plane with the longest dimension.

Normally $H > B$, so the maximum shear stress exists in the X-Y plane. For $B > H$ the formulas for the shear stress must be exchanged and the maximum stress occurs in the X-Z plane.

For the combined stress either the von Mises stress for the comparison with the yield stress of the material or the Rankine stress can be taken for the comparison with the ultimate stress of the material.

$$f(\frac{H}{B}) = 1.6868 - 1.089\frac{H}{B} + 0.504\frac{H^2}{B^2} - 0.1145\frac{H^3}{B^3} + 0.0126\frac{H^4}{B^4} - 5.31 \times 10^{-4}\frac{H^5}{B^5} \quad (6.70)$$

6.1.4.2 Octagonal Cross-Section

Stress by:	X-Y Plane	X-Z Plane	
F_x	$\sigma_{z1} = F_x/A$	$\sigma_{y1} = F_x/A$	(6.71)
F_z	$\sigma_{z2} = F_z \cdot l_1/W_y$	$\sigma_{y2} = \sigma_{z2} \cdot (H - 2h)/H$	(6.72)
F_y	$\sigma_{z3} = \sigma_{y3} \cdot (B - 2b)/B$	$\sigma_{y3} = F_y \cdot l_1/W_z$	(6.73)
M_x	$\tau_z = M_x/W_p$	$\tau_y = M_x/W_p$	(6.74)
M_y	$\sigma_{z4} = M_y/W_y$	$\sigma_{y4} = \sigma_{z4} \cdot (H - 2h)/H$	(6.75)
M_z	$\sigma_{z5} = \sigma_{y5} \cdot (B - 2b)/B$	$\sigma_{y5} = M_z/W_z$	(6.76)
Max:	$\sigma_{zmax} = \sum_{i=1}^{5} \sigma_{zi}$	$\sigma_{ymax} = \sum_{i=1}^{5} \sigma_{yi}$	(6.77)
Comb	$\sigma_v = \sqrt{\sigma_{max}^2 + 3\tau^2}$(Mises)	$\sigma_v = \frac{1}{2}(\sigma_{max} + \sqrt{\sigma_{max}^2 + 4\tau^2})$(Rankine)	(6.78)

6.1.4.3 Circular Cross-Section with Flattenings

Stress by:	X-Y Plane	X-Z Plane	
F_x	$\sigma_{z1} = F_x/A$	$\sigma_{y1} = F_x/A$	(6.79)
F_z	$\sigma_{z2} = F_z \cdot l_1/W_y$	$\sigma_{y2} = \sigma_{z2} \cdot (h/(D \cdot cos\alpha))$	(6.80)
F_y	$\sigma_{z3} = \sigma_{y3} \cdot (b/(D \cdot cos\beta))$	$\sigma_{y3} = F_y \cdot l_1/W_z$	(6.81)
M_x	$\tau_z = M_x/W_p$	$\tau_y = M_x/W_p$	(6.82)
M_y	$\sigma_{z4} = M_y/W_y$	$\sigma_{y4} = \sigma_{z4} \cdot (h/(D \cdot cos\alpha))$	(6.83)
M_z	$\sigma_{z5} = \sigma_{y5} \cdot (b/(D \cdot cos\beta))$	$\sigma_{y5} = M_z/W_z$	(6.84)
Max:	$\sigma_{zmax} = \sum_{i=1}^{5} \sigma_{zi}$	$\sigma_{ymax} = \sum_{i=1}^{5} \sigma_{yi}$	(6.85)
Comb	$\sigma_v = \sqrt{\sigma_{max}^2 + 3\tau^2}$(Mises)	$\sigma_v = \frac{1}{2}(\sigma_{max} + \sqrt{\sigma_{max}^2 + 4\tau^2})$(Rankine)	(6.86)

6.1.4.4 Cruciform Cross-Section with Fillets

Stress by:	X-Y Plane	X-Z Plane	
F_x	$\sigma_{z1} = F_x/A$	$\sigma_{y1} = F_x/A$	(6.87)
F_z	$\sigma_{z2} = F_z \cdot l_1/W_y$	$\sigma_{y2} = \sigma_{z2} \cdot (H - 2h)/H$	(6.88)
F_y	$\sigma_{z3} = \sigma_{y3} \cdot (B - 2b)/B$	$\sigma_{y3} = F_y \cdot l_1/W_z$	(6.89)
M_x	$\tau_z = M_x/W_p$	$\tau_y = M_x/W_p$	(6.90)
M_y	$\sigma_{z4} = M_y/W_y$	$\sigma_{y4} = \sigma_{z4} \cdot (H - 2h)/H$	(6.91)
M_z	$\sigma_{z5} = \sigma_{y5} \cdot (B - 2b)/B$	$\sigma_{y5} = M_z/W_z$	(6.92)
Max:	$\sigma_{zmax} = \sum_{i=1}^{5} \sigma_{zi}$	$\sigma_{ymax} = \sum_{i=1}^{5} \sigma_{yi}$	(6.93)
Comb	$\sigma_v = \sqrt{\sigma_{max}^2 + 3\tau^2}$(Mises)	$\sigma_v = \frac{1}{2}(\sigma_{max} + \sqrt{\sigma_{max}^2 + 4\tau^2})$(Rankine)	(6.94)

$$C = \frac{D_{ein}}{1 + (\frac{\pi \cdot D^2}{4A})^2} \cdot [1 + (0.118 \cdot In(1 + \frac{D_{ein}}{2r}) + 0.238\frac{D_{ein}}{2r}) \cdot 0.762] \qquad (6.95)$$

6.1.5 Design of Cage Bending Section

Following the basic calculation of the rudimentary cross-section for the bending beam, it can be decided whether the further design optimization is done on a rigid beam or if it is possible to switch to a cage design. The advantage of the cage design is a stiffer balance with the same sensitivity, because the additional numbers of parameters provide more options for optimization. The separation of the stresses related to the loads to be measured can be much better achieved; hence, the interactions can be reduced.

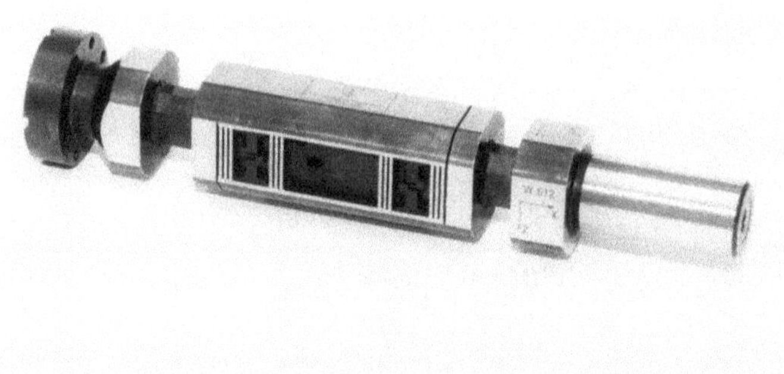

Fig. 6.19 Cryogenic balance W612 for Cologne cryogenic wind tunnel

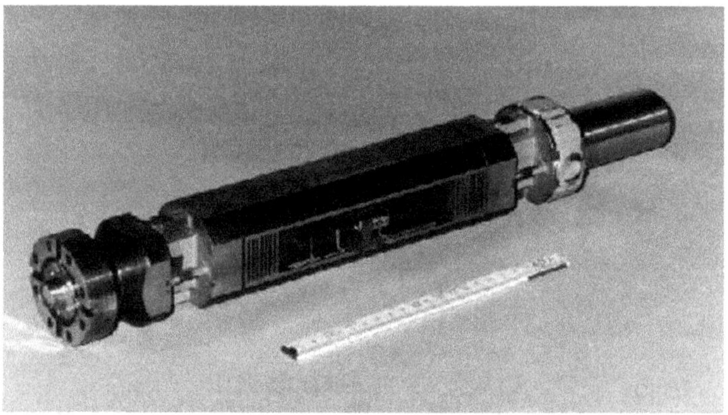

Fig. 6.20 Balance W610 for the TU Darmstadt low speed wind tunnel

There is no general criteria dictating when to switch from a solid bending beam to a cage. If the solid bending cross-section related to the overall cross-section is small, it is generally better to use a cage. This criterion is based on considerations related to pure mechanical stresses. In case of large temperature changes or in the presence of temperature gradients, it may be better to use a small solid section because the concentrated assembly of the strain gauges is much less sensitive to temperature and temperature gradients. As an example, the bending section of balance W612 is shown in Fig. 6.19.

In Fig. 6.19 it can be seen that the solid bending section is of the octagonal type and the cutouts are large so that the appearance of the bending section is small related to the rest of the balance. Normally in this case a cage design is recommended, but this balance was designed for the cryogenic wind tunnel of DLR in Cologne. To minimize the effect of temperature to the strain gauge bridges, the compact design of the solid section was preferred.

The balance W610 for the low speed wind tunnel of the Technical University of Darmstadt (Fig. 6.20) with similar load ranges and the same outer dimension was designed with a cage bending section instead, because only minor temperature effects occur.

Beside a solid rectangular central beam, four small rectangular beams are arranged around this central beam and almost the entire cross-section of the balance is used for the cage. On the surfaces of the small beams, normal force, pitching moment, side force and yawing moment are measured. The shear stress on the center beam is used to measure rolling moment.

To check whether there is sufficient space for a cage design, the specific load parameter given in Eq. 6.16 can be of assistance. If the specific load parameter exceeds a value of 500 Ncm it will be difficult to design a cage with moderate stresses ($<300\,\mathrm{N/mm^2}$).

The trend in aerodynamic development is that the available space in the model fuselage continually decreases due to increased instrumentation. So the specific load of the balances increases and only in rare cases it is possible to design a cage.

The number of parameters increase with the number of beams that are used for the cage and for each beam every cross-section discussed in subsection 6.1.3.2 can be used. The variety of combinations is nearly infinite, so it is not practical to describe each one in detail. However, one basic configuration (five beams $n = 5$) will be discussed to illustrate the methodology of the calculation. This basic configuration is shown in Fig. 6.21.

This configuration also includes the four-beam configuration ($n = 4$) if the center beam is eliminated. To use other shapes, the section modulus in the equations must be exchanged. The assumption is made that beams are always placed symmetrical to one of the axes ($q = 2$) or symmetrical to both axes ($q = 4$). The center beam has a symmetry number of $q = 1$. Asymmetrical balances are possible, but are not discussed here.

For rectangular beams the maximum stress occurs right at the edge of the beam. The six loads on the balance contribute more or less all to the stress on the surface of each beam and in total nine different stresses have to be calculated for each surface of a single beam.

One normal stress caused by the axial force (F_x). One transverse and one bending stress is generated by the normal force (F_z) and by the side force (F_y). Pitching (M_y) and yawing moments (M_z) create one bending stress and the rolling moment (M_x) creates one bending stress and one shear stress.

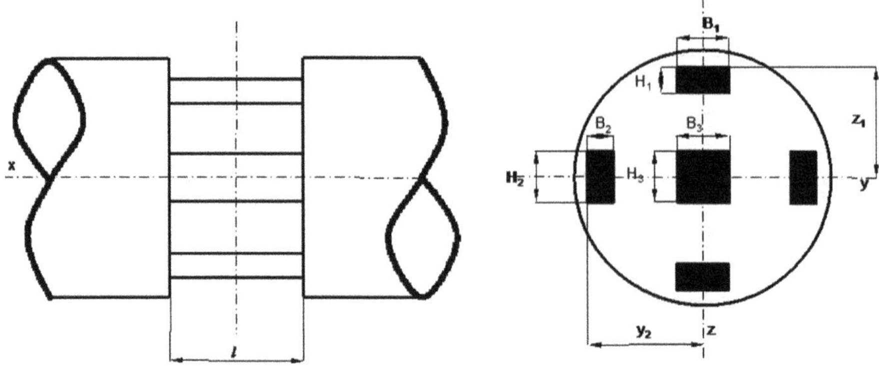

Fig. 6.21 Cage bending section with five rectangular beams

6.1.5.1 Stress Due to Axial Force

The stress created by the axial force on each beam (F_{xi}) can be calculated relatively easily. The total axial load (F_x) is divided into the individual load (F_{xi}), related to ratio of the total longitudinal stiffness $(C_{x_{tot}})$ to the individual stiffness (C_{xi}).

$$Cx_i = \frac{E \cdot A_i}{l_i}; \quad A_i = B_i \cdot H_i \tag{6.96}$$

$$Cx_{tot} = \frac{E \cdot A_{tot}}{l_i}; \quad A_{tot} = \sum_1^n B_i \cdot H_i \tag{6.97}$$

$$Fx_i = Fx_{tot} \cdot \frac{Cx_i}{Cx_{tot}} = Fx_{tot} \cdot \frac{A_i}{A_{tot}} \tag{6.98}$$

Because Young's modulus (E) and the length $(l = l_i)$ is equal for all beams, the load (F_x) is distributed according to the ratio of the cross-section (A_i/A_{tot}). Then the normal stress (σ_{xi}) in each beam can be calculated as:

$$\sigma_{xi} = \frac{F_{x_i}}{A_i} \tag{6.99}$$

6.1.5.2 Stress on the Beam Related to Lateral Forces

Lateral forces $(F_z\ ,\ F_y)$ deform the cage in the z or y direction as it is shown in Fig. 6.22.

For the deformation (δ) either in x or z direction, the stiffness of each beam with the number (i) can be expressed as:

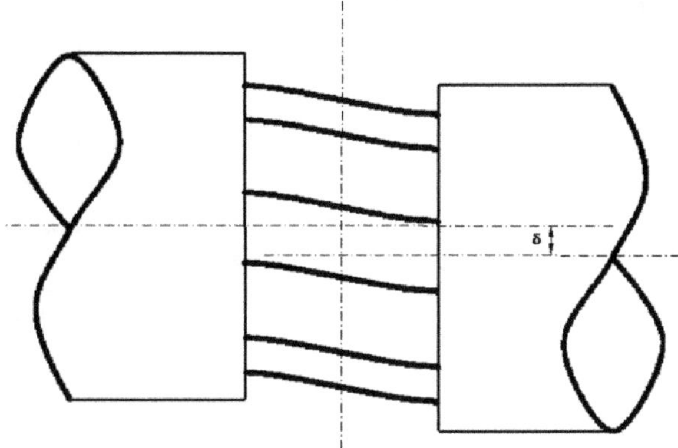

Fig. 6.22 Deformation of cage by lateral forces

$$\frac{F_{zi}}{\delta_{zi}} = \frac{1}{\frac{l_i^3}{12 \cdot E \cdot I y_i} + \frac{1.2 \cdot l_i}{A_i \cdot G}} = C z_i; \quad \frac{F_{yi}}{\delta_{yi}} = \frac{1}{\frac{l_i^3}{12 \cdot E \cdot I z_i} + \frac{1.2 \cdot l_i}{A_i \cdot G}} = C y_i; \quad A_i = B_i \cdot H_i \tag{6.100}$$

For the cage, the stiffness ($C_{z_{ges}}$, $C_{y_{ges}}$.) is the sum of all individual stiffness values (C_{z_i}, C_{y_i}).

$$\frac{F z_{tot}}{\delta_z} = \sum_1^n q_i \frac{F z_i}{\delta_z} = C z_{tot}; \quad \frac{F y_{tot}}{\delta_y} = \sum_1^n q_i \frac{F y_i}{\sigma_y} = C y_{tot} \tag{6.101}$$

The load (F_z, F_y) on the individual beams is distributed related to the ratio of the total stiffness of the cage (C_{tot}) to the individual stiffness of the beam (C_i).

$$F z_i = F_{tot} \cdot \frac{C z_i}{C z_{tot}}; \quad F y_i = F_{tot} \cdot \frac{C y_i}{C y_{tot}} \tag{6.102}$$

Now the partial lateral load (F_{z_i}, F_{y_i}) of each beam is known and with the section moduli (W_i) calculated in Sect. 6.1.3.2, the stresses on the individual beams can be calculated on the different surfaces of each beam.

The transverse stress (σ_{Ti}) due to the lateral forces is calculated as follows:

$$\sigma_{T z_i} = \frac{F z_i}{A_i}; \quad \sigma_{T y_i} = \frac{F y_i}{A_i} \tag{6.103}$$

and the bending stress (σ_{Bi}) due to lateral forces:

$$\sigma_{B z_i} = \frac{F z_i \cdot l_i}{2 \cdot W y_i}; \quad \sigma_{B y_i} = \frac{F y_i \cdot l_i}{2 \cdot W z_i} \tag{6.104}$$

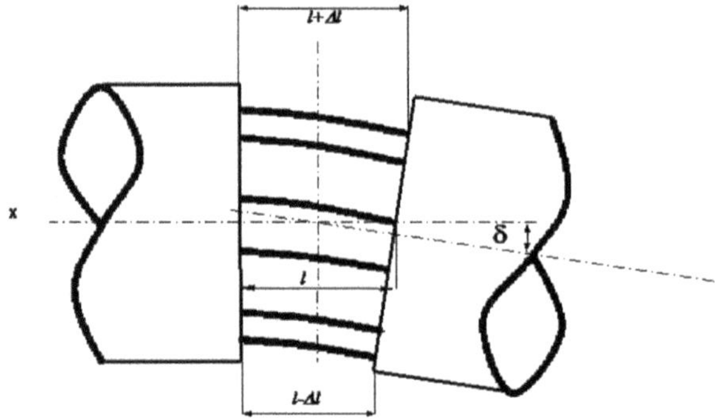

Fig. 6.23 Deformation of cage by M_y or M_z

6.1.5.3 Stresses by Pitching and Yawing Moment

The deformation of the cage by a pitching moment or a yawing moment is shown in Fig. 6.23.

For the calculation of the bending stresses due to pitching (M_y) and yawing (M_z) moment the section moduli of the whole cage (Iy_{tot}, Iz_{tot}) must be first calculated. This is done using the individual section moduli (Iy_i, Iz_i) of subsection 6.1.3.2 and combining them using Steiner's rule. In the following equations this rule is given for rectangular beams and a five-beam cage.

$$Iz_{tot} = q_i \cdot \sum_1^n [Iz_i + A_i(z_i - \frac{H_i}{2})^2]; \quad Iy_{tot} = q_i \cdot \sum_1^n [Iy_i + A_i(y_i - \frac{B_i}{2})^2] \tag{6.105}$$

The moment acting on one beam (M_i) is calculated by multiplying the total moment (M_{tot}) by the ratio of beam stiffness (Cm_i) to the total stiffness (Cm_{tot}).

$$Mz_i = Mz_{tot} \cdot \frac{Cmz_i}{Cmz_{tot}}; \quad My_i = My_{tot} \cdot \frac{Cmy_i}{Cmy_{tot}} \tag{6.106}$$

The deformation of the individual beams by the total bending moment (M_i) can be separated into a tension and compression deformation and a bending deformation. Therefore, the individual stiffness (Cm_i) of a beam can be calculated as the sum of the stiffness of the beam in axial direction and the bending stiffness.

$$\begin{aligned} Cmz_i &= \frac{Mz}{\sigma} = Cx_{beam} \cdot z_i^2 + Cm_{beam}; \\ Cmy_i &= \frac{My}{\sigma} = Cx_{beam} \cdot y_i^2 + Cm_{beam} \end{aligned} \tag{6.107}$$

$$Cmz_i = \frac{E}{l_i} \cdot [Iz_i + A_i \cdot (z_i - \frac{H_i}{2})^2];$$
$$Cmy_i = \frac{E}{l_i} \cdot [Iy_i + A_i \cdot (y_i - \frac{B_i}{2})^2] \quad (6.108)$$

The total stiffness (Cm_{tot}) is:

$$Cmz_{tot} = \sum_1^n q_i \cdot Cmz_i; \quad Cmy_{tot} = \sum_1^n q_i \cdot Cmy_i \quad (6.109)$$

After the individual moment for every beam is calculated, the bending stress on the outer surface of the beam related to this moment is:

$$\sigma_{B_{mzi}} = \frac{M_{zi}}{W_{yi}}; \quad \sigma_{B_{myi}} = \frac{M_{yi}}{W_{zi}} \quad (6.110)$$

6.1.5.4 Stresses by Rolling Moment

The deformation of the cage by the rolling moment is a combined deformation of torque and bending. The whole cage is torqued around the axial axis (Φ). As a result, the central beam is also torqued around the center axis with the torque angle (Φ). The main deformations of the outer beams are a lateral deformation in y direction (Δy) and a lateral deformation in z direction (Δz) and a torque. Under this assumption the deformation for the cage is shown in Fig. 6.24.

The deformation of each individual beam is dependent on the beam location and the twist angle. If all is taken into account, the equations are relatively complicated. To describe the deformations in an easier way, an approximation is made that yields the values without a large error. These assumptions are:

$$\Phi; \quad \delta z_i = \Phi \cdot (y_i - \frac{B_i}{2}); \quad \delta y_i = \Phi \cdot (z_i - \frac{H_i}{2}) \quad (6.111)$$

The relative error for the deformation is below 1% except in the case when one of the coordinates (y_i, or z_i) gets close to zero. Then the relative error is high. In this situation the absolute value for the movement in one direction is small and can be neglected, because the movement in the other direction dominates the deformation.

The energy (A_i) for the deformation of one beam is the sum of the energies that is needed to deform the beam in two lateral directions (σz_i; σy_i) and one torque deformation (Φ). The partial energies (Az_i, Ay_i, Ar_i) can be obtained by multiplying the stiffness with the square of the deformation.

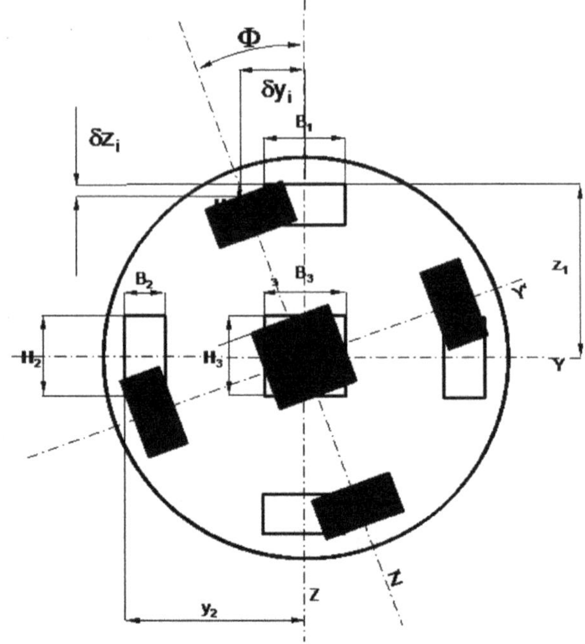

Fig. 6.24 Deformation of the cage by rolling moment M_x

$$
\begin{aligned}
A_i &= Az_i + Ay_i + Ar_i \\
A_i &= \frac{1}{2} \cdot Ct_i \cdot \Phi^2 = \frac{1}{2}(Cz_i \cdot \delta z_i{}^2 + Cy_i \cdot \delta y_i{}^2 + Cr_i \cdot \Phi^2)
\end{aligned} \tag{6.112}
$$

Substituting the deformations (δz_i, δy_i) by the assumptions made in Eq. (6.111), the torsional stiffness (Ct_i) of one beam can be expressed as:

$$
\begin{aligned}
\frac{M_i}{\Phi} &= Ct_i = Cz_i \cdot (y_i - \frac{B_i}{2})^2 + Cy_i \cdot (z_i - \frac{H_i}{2})^2 + Cr_i \\
Cr_i &= \frac{Ip_i \cdot G}{l_i}; \quad Cz_i = \frac{12E \cdot Iy_i}{l^3}; \quad Cy_i = \frac{12E \cdot Iz_i}{l^3}
\end{aligned} \tag{6.113}
$$

The total torsional stiffness of the cage is equal to the sum of all individual stiffness values.

$$
\frac{M_x}{\Phi} = \sum_1^n \frac{M_i}{\Phi}; \quad Ct_{tot} = \sum_1^5 Ct_i \tag{6.114}
$$

After the total torsional stiffness (Ct_{tot}) of the cage is known, the individual loads and the torque moment can be distributed according to the ratio of the individual stiffness (Ct_i) to the total stiffness. The forces (Fz_i, Fy_i) and the torque moment (Mr_i) acting on one beam can be calculated with the following equations:

$$\begin{aligned} Fz_i &= \frac{Mx}{Ct_{tot}} Cz_i \cdot (y_i - \frac{B_i}{2}) \\ Fy_i &= \frac{Mx}{Ct_{tot}} Cy_i \cdot (z_i - \frac{H_i}{2}) \\ Mr_i &= \frac{Mx \cdot Cr_i}{Ct_{tot}} \end{aligned} \tag{6.115}$$

With these loadings on each beam, two bending stresses and one shear stress on the surface of each beam must be calculated using the equations given in subsection 6.1.3.2 to determine the section moduli (Wy_i, Wz_i, Wp_i).

$$\begin{aligned} \sigma_{bz_i} &= \frac{Fz_i \cdot l}{Wy_i} \\ \sigma_{by_i} &= \frac{Fy_i \cdot l}{Wz_i} \\ \tau_{T_i} &= \frac{Mr_i}{Wp_i} \end{aligned} \tag{6.116}$$

6.1.5.5 Combined Stress

To perform the strength calculation, all nine stresses described in the last four subsections must be combined according to the *Rankine* stress theory to a combined stress that can be compared with the yield stress of the material. The *Rankine* rule is used instead of the *v. Mises* rule, because balance materials behave like brittle material. Although the material strength is very high, the difference between yield stress and ultimate stress is very small. For the measurement of the loads, the surface of the beam with maximum stress related to this load must be selected.

6.1.6 Design of Axial Force Section

The axial force section of an internal balance is the section where only one load is measured. The different configurations for the other loads are relatively simple and a balance without the axial force section is relatively easy to machine. The cost for a five component balance (missile balances) without the axial force section can therefore be comparatively low. The problem that makes the axial force system so complicated is that the axial force in aerodynamic measurement of aircraft is

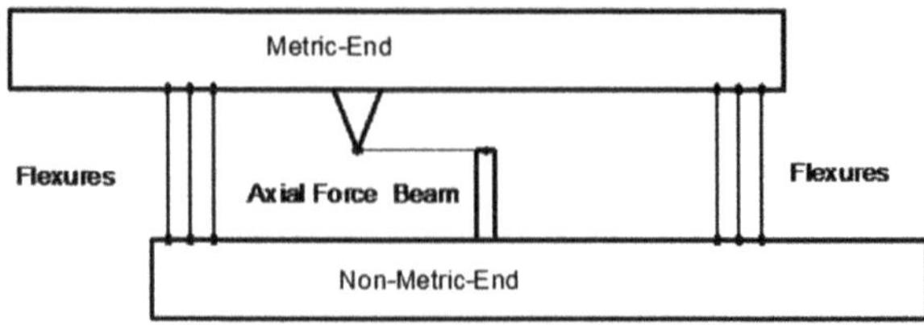

Fig. 6.25 Principle axial force system

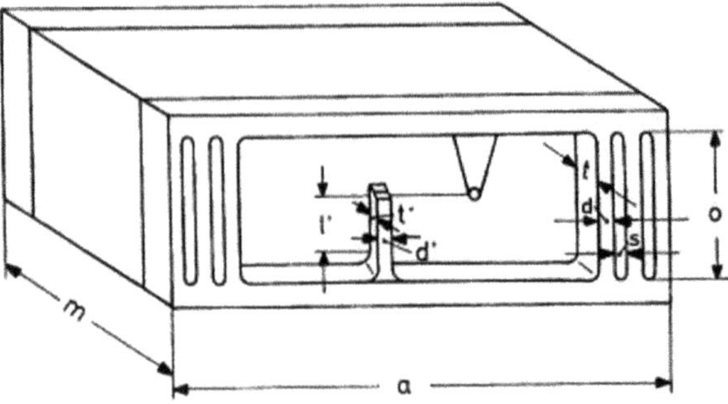

Fig. 6.26 Nomenclature of a basic axial force system

comparatively low and thus, all other loads must bypass a sensitive measurement system without having much influence on the measurement of the axial force. This bypass cannot be achieved without parallel stiffness that takes away some part of the axial force. The common solution for this problem is a combination of a flexure section with a decoupled axial force measurement beam, as shown in Fig. 6.25.

The flexure system has to carry all the forces produced by the normal force (F_z), side force (F_y) pitching moment (M_y), yawing moment (M_z) and rolling moment (M_x), without being too stiff in the axial direction. Usually this problem can be solved such that the axial force distribution between flexures (F_{x_F}) and measurement beam ($F_{x_M B}$) is about 40% to 60%. If the partial axial force on the measurement beam ($F_{x_M F}$) is less than 40%, an acceptable signal level for the axial force is not possible. Nevertheless, the variety of possible configurations is nearly infinite. Therefore only one example for the design of an axial force beam design calculation is given. For all other configurations only the equations for the individual shapes have to be changed. Most of the axial force systems look like the example shown in Fig. 6.26.

In this example it is assumed that the axial force beam in the center of the system is perfectly decoupled from all non-axial forces; hence, carrying no load except this part of the axial force. Under this assumption, all other forces and a part of the axial force must be carried by the flexure system.

6.1.6.1 Stresses in the Flexures

The normal force (F_z) is distributed uniform over all flexures. The total number of the flexures is n so the tension and compression stress (σ_z) in one flexure is:

$$\sigma_z = \frac{F_z}{4 \cdot n \cdot t \cdot d} \tag{6.117}$$

The rolling moment (M_x) generates tension stresses on one side and compression stresses on the other side in the flexures of the axial force system. The magnitude of this stress is:

$$\sigma_{Mx} = \frac{Mx}{2 \cdot n \cdot t \cdot d \cdot (m - t)} \tag{6.118}$$

The axial force is carried by the flexure system (Fx_F) and of course by the axial force measuring beam (Fx_{MB}). The load is distributed between the two systems according to the ratio of their stiffness in axial direction (Cx_{Mb}/Cx_F). This stiffness is:

$$C_{MB} = \frac{Fx_{MB}}{\delta_x} = \frac{2}{\frac{1.2 \cdot (l^* - r^*/2)}{t^* \cdot d^* \cdot G} - \frac{4 \cdot (l^* - r^*/2)^3}{t^* \cdot d^{*3} \cdot E}} \tag{6.119}$$

$$C_F = \frac{Fx_F}{\delta_x} = \frac{4 \cdot n}{\frac{1.2 \cdot o'}{t \cdot d \cdot G} - \frac{4 \cdot o'^3}{t \cdot d^3 \cdot E}}; \quad o' = o - s/2 \tag{6.120}$$

With Eqs. (6.119) and (6.120) the distribution of the forces can be calculated as follows:

$$Fx_{MB} = \frac{Fx \cdot Cx_{MB}}{2 \cdot (Cx_F + Cx_{MB})}; \quad Fx_F = \frac{Fx - 2Fx_{MB}}{4 \cdot n} \tag{6.121}$$

The stresses in the flexures and the measuring beam according to the partial axial loads are bending stresses that occur in the root section of the beams. These bending stresses are:

$$\sigma_{xMB} = \frac{6 \cdot Fx_{MB} \cdot (l^* - r^*)}{t^* \cdot d^{*2}}; \quad \sigma_{x_F} = \frac{3 \cdot Fx_F \cdot o'}{f \cdot d^2} \tag{6.122}$$

Similar to the axial force, the side force also generates a bending stress in the root section of the flexures, but according to the assumptions made, the entire side force (F_y) is carried by the flexures and each flexure carries the same part of the side force. Therefore, the stress related to the side force can be calculated as:

$$\sigma_y = \frac{3 \cdot F_y \cdot o'}{4 \cdot n \cdot t^2 \cdot d} \tag{6.123}$$

It is more complicated to determine the stresses by the other components M_y and M_z. For these loads the different distances of each flexure from the center has to be taken into account. It is assumed that the main beams are very stiff and the deformation by the moments (M_y, M_z) is a pure rotation around the center axis. Under this assumption the load on the outer flexure becomes a maximum. In reality the main beams of the balance are not infinitely stiff and also deform slightly, so that the load on the outer flexure is smaller, while the inner flexures carries more load.

A flexure system where the thickness (d), the width (t) and the length (o) are the same for all flexures, the flexure with the longest distance from the center point carries most of the load generated by the moments. The significant stress for the strength calculation has to be calculated therefore only for this flexure.

The equilibrium between the total moment (M') and the partial flexure forces (F_i) multiplied by their distance to the center of the balance can be written as:

$$M' = F_1 \cdot (\frac{a}{2} - \frac{d}{2}) + F_2 \cdot (\frac{a}{2} - \frac{3d}{2} - s) + \cdots + F_n \cdot (\frac{a}{2} - \frac{(2n-1)d}{2} - (n-1) \cdot s) \tag{6.124}$$

For a pure rotation around the center axis by the moment, the relation between the partial flexure load on the outer flexure (F_1) and any other flexure (F_i) becomes:

$$F_i = F_1 \cdot \frac{[\frac{a}{2} - \frac{(2i-1) \cdot d}{2} - (i-1) \cdot s]}{\frac{1}{2}(a-d)} \tag{6.125}$$

Using this equation and the relation of Eq. (6.124) the total moment can be written as a function of the force on the outer flexure (F_1) and the sum of the force distribution multiplied by the partial length:

$$M' = F_1 \cdot \frac{\sum_{i=1}^{n} [\frac{a}{2} - \frac{(2i-1) \cdot d}{2} - (i-1) \cdot s]}{\frac{1}{2}(a-d)} \tag{6.126}$$

The total moment ($M = M_y, M_z$) for all four flexure packages is now

$$M = F_1 \cdot 8 \cdot \frac{\sum_{i=1}^{n} [\frac{a}{2} - \frac{(2i-1) \cdot d}{2} - (i-1) \cdot s]^2}{(a-d)} \tag{6.127}$$

A parallelogram factor γ can be defined:

$$\gamma = 8 \cdot \frac{\sum_{i=1}^{n} [\frac{a}{2} - \frac{(2i-1) \cdot d}{2} - (i-1) \cdot s]^2}{\frac{1}{2}(a-d)} \tag{6.128}$$

With this factor the tension or compression stress by the pitching moment on the outer flexure becomes:

$$\sigma_{My} = \frac{My}{\gamma \cdot t \cdot d} \tag{6.129}$$

The bending stress by the yawing moment can be calculated using:

$$\sigma_{Mz} = \frac{3 \cdot Mz \cdot o'}{\gamma \cdot t^2 \cdot d}; \quad o' = o - \frac{s}{2} \tag{6.130}$$

Now the maximum combined stress ($\sigma_{F_{tot}}$) in the outer flexure is given as a sum of the stresses by tension or compression and the bending stresses:

$$\sigma_{F_{tot}} = \sigma_z + \sigma_{Mx} + \sigma_{X_F} + \sigma_Y + \sigma_{My} + \sigma_{Mz} \tag{6.131}$$

This combined stress is used for the comparison with the yield stress of the material for the strength analysis of the flexure.

To control the stability of the flexure system, the safety against buckling must be checked. To do this the slenderness ratio (λ) and the buckling stress (σ_K) must be calculated.

$$\sigma_K = \frac{\pi^2 \cdot E}{\lambda^2} = \frac{(\pi \cdot d)^2 \cdot E}{12 \cdot o^2}; \quad \lambda = \frac{o \cdot \sqrt{12}}{d} \tag{6.132}$$

For $\lambda < 80$ or $\sigma_K > 2 * (\sigma_Z + \sigma_{Mx} + \sigma_{My})$ there is no risk of buckling.

6.1.6.2 Stresses in the Measuring Beam and Axial Force Signal

The load on the measuring beam by the axial force and the stress related to this partial load has already been determined using Eqs. (6.121) and (6.122) as a byproduct of the calculation of the partial axial load for the flexures. Equation (6.122) gives the bending stress on the surface of the measuring beam, where the uniform rectangular cross-section ends and the changeover with the radius starts. Normally the stress there is small because the stress for the strain measurement does not exceed $300\,\text{N/mm}^2$. Of more interest is the signal that can be expected. To obtain this, the stress at the position where the strain gauge is installed needs to be calculated. Because the strain gauge integrates the strain over the entire grid area, a section needs to be identified where the stress represents the mean stress of the grid area. In case of a cantilever beam with a constant cross-section, the mean stress is right in the middle of the gauge ($l_{DMS}/2$). To achieve high sensitivity, the gauge grid should be installed as near as possible to the changeover area. Due to practical installation procedures, a distance to the beginning of the radius of 0.15 mm is used. The mean stress now becomes:

$$\sigma_{DMS} = \frac{6 \cdot Fx_{MB} \cdot (l^* - r^* - 0.15 - \frac{l_{DMS}}{2})}{t^* \cdot d^{*2}} \tag{6.133}$$

The signal of an active full bridge then will be:

$$S_x = \frac{\Delta U_x}{U} = k \cdot \frac{\sigma_{DMS}}{E} \cdot 1000 \quad [\frac{mV}{V}]; \quad k = 2 \tag{6.134}$$

6.1.6.3 Types of Flexure Systems

The above described flexure system can be altered in various ways. The equations then must be modified for the calculation of the stresses and signal, but the principal approach remains as described above. There are two major design variations. The first possible variation is a change in the flexure system. Instead a number of uniform springs of thickness d_i or length o of each flexure can be different. The only mandatory requirement is that the changes in the four packages are symmetrical to the center.

This measure is used to change the distribution of the loads on the flexures. In the above described example, the outer flexure carries most of the load generated by the pitching moment (M_y) and yawing moment (M_z). If one of these moments is responsible for high stresses in the outer flexure, the stiffness of the others must be increased so that they carry more of the load, thus decreasing the high stress in the outer flexure.

An example for a balance with a non-uniform distribution of the flexure thickness is balance W621, shown in Fig. 6.27. In Fig. 6.27 it can be seen that the inner flexures are much thicker than the outer ones. This balance has high load ranges for M_y and M_z, and with a uniform distribution of the flexure thickness the outer flexure was overloaded. By increasing the thickness of the inner flexures, this overload problem was solved. The general problem using this trick is that the total stiffness of the flexure system in the axial direction (Cx_{MB}) must remain constant, such that the sensitivity of the axial force measurement is the same or higher.

An example where the length of the flexures is different is shown in the picture of the *NLR* balance (Fig. 6.28). In this case the length and the thickness of the flexure vary the stiffness of each flexure. The big advantage of this design is that the main beam cross-section at the inner flexure is larger and so the stiffness related bending lateral to the axial direction is greater.

The optimum distribution of thickness and length is not easy to determine, because for the six flexures, twelve more parameters can be varied. Such an optimization is very time consuming and labor intensive without a simple calculation program as described above. The results of such an algorithm can be used to minimize the iterations with FE calculations. For high loaded balances, such a design is mandatory because without it stress concentrations cannot be avoided in the outer flexure.

6.1.6.4 Types of Axial Force Measurement Beams

In the method described above to determine the dimensions of the axial force sections, not only is a uniform distribution of the flexure length and thickness assumed, also an axial force beam with rectangular and constant cross-section is assumed. For the first approximate determination of the size of the axial force measuring beam, this assumption is sufficient. The main aim of this step is to determine the stiffness of the flexure system and the stiffness of the axial force beam to obtain flexures to carry all the other loads passing the balance and to obtain enough axial load on the

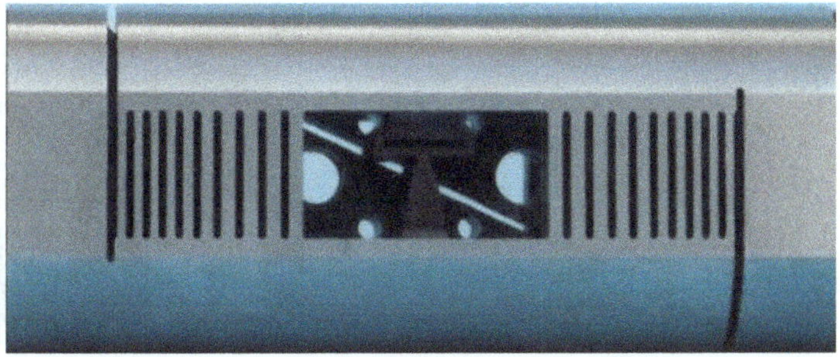

Fig. 6.27 Flexure system of balance W621

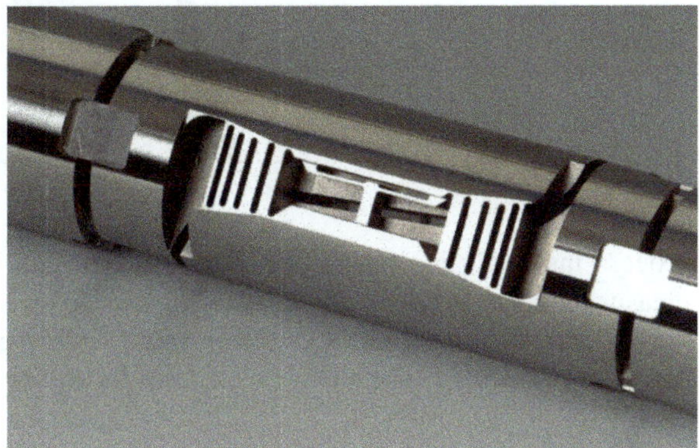

Fig. 6.28 NLR balance with variation in flexure length and thickness

measuring beam for a signal of high amplitude. After such a first characterization of the stiffness, the shape of the axial force measuring beam can be optimized.

Constant Stress Beam:

One target of the optimization is to have a uniform constant stress under the strain gauge. If this is utilized by the beam design, the signal from the axial force is not sensitive to small variations in the strain gauge position. An equal sensitivity of both axial force beams reduce the signal due to the interactions by rolling moment, yawing moment and pitching moment [24], because normally the axial force beams deform in opposite directions related to the moments and so the signals according to these deformations cancel out, but only if the signals have exactly the same value.

On a cantilever beam with constant cross-section, the stress is linearly distributed over the length (maximum at root, zero at tip) and so displacement differences of the gauges cause linear differences in the sensitivity, as illustrated in Fig. 6.26. Results are interactions by the moments on the axial force.

The shape of the beam can be manufactured within tolerances of less then a few hundreds of a millimeter, but the positioning tolerance of the strain gauge is within a few tenths of a millimeter. So the strain gauge application process is the greatest source of inaccuracy in the entire manufacturing process of a balance. To optimize the precision of a balance, it is therefore necessary to make the gauge application

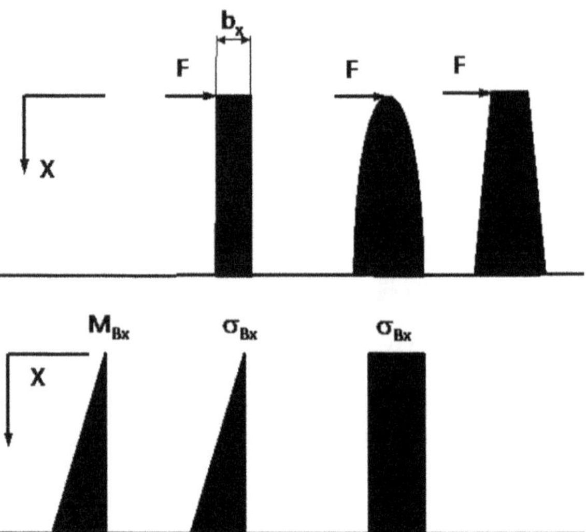

Fig. 6.29 Contours for a constant stress beam

process insensitive to position tolerances. This can be obtained by designing the measuring beams as *constant stress beams*.

To determine the contour of the constant stress beam, the following calculation has to be performed. The stress at a certain location can be calculated as:

$$\sigma_{Bx} = \frac{M_{bx}}{W_{Bx}} = \frac{12 \cdot F \cdot x}{h \cdot {b_x}^2} \quad \rightarrow b_x = \sqrt{\frac{12 \cdot F}{\sigma_{Bx} \cdot h}} \cdot \sqrt{x} \qquad (6.135)$$

Here σ_{Bx} is the stress under the strain gauge that has to be constant. M_{Bx} is the bending moment at the location x and W_{Bx} is the section modulus at the location x. The thickness of the beam h is kept constant and so the width of the beam at the location x can be calculated using Eq. (6.135). If this is done, the shape of the beam is a parabola, as it is shown in Fig. 6.29.

A parabola cannot easily be manufactured, so a good approximation is a trapeze beam, also shown in Fig. 6.29. The best fit of the trapeze to the parabola should be in the area of the strain gauge application, then there will be a rather large area of constant stress there and the desired effect has been achieved.

There are many other different possible solutions to achieve constant stress areas on a cantilever beam. For example if the width b is kept constant, the thickness h_z has to be changed linearly, like it is shown in Fig. 6.30.

In Fig. 6.30 the picture on the left and in the middle show a trapeze beam where the trapeze is an approximation of the parabola shape, and on the right side the trapeze orientation is orthogonal to the force. In both cases the FE calculations show that in the gauge application area the strain is almost constant.

The choice of an optimum design for a beam depends on two different factors. First it must be possible to machine the beam inside the balance body and second it

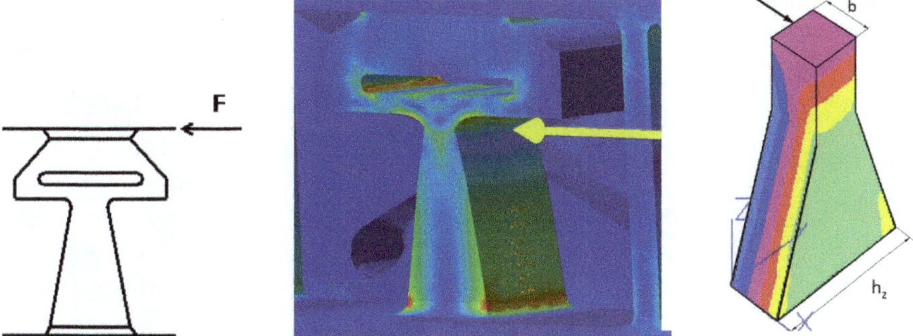

Fig. 6.30 Two trapeze cantilever beams with constant stress zones (normal or orthogonal to the trapeze area)

must be possible to install the gauges on the beam surface. Especially if both surfaces are curved or not parallel to each other, it is more complicated to build a clamp that presses the gauges with the correct pressure at the correct place during curing. If there is some shear stress on the gauge during curing, the chance of misalignment is much higher than for parallel surfaces.

6.1.6.5 Reduction of Interactions

Interactions of other loads on the measuring beam for the load to be measured are another problem that can be influenced by the design of the beam. The usual method to reduce the interaction of other loads, especially on the axial force, is the so-called decoupling system. A cantilever beam is used as an axial force measuring device that is linked to the non-metric side of the balance by flexures, such that only tension and compression forces act on the beam. Figure 6.28 of the NLR balance shows a typical axial force measuring beam with thin links on both sides to decouple the beam as much as possible from deformations by other loads. Further minimization of interactions is achieved by placing the beam as close as possible to the centerline of the balance. Links at both sides have the advantage that the structure is symmetrical and there is much higher stability in the system. As a consequence, the risk of buckling in the compressed linking rod is lower, but in principle one link is enough.

A complete analysis of the interaction on the axial force was given in [24]. Some results from this study show that there is indeed a reduction of interaction for the decoupled system, but for a constant stress beam in a coupled symmetric arrangement (Fig. 6.31) these advantages are relative small. The reason is that the interaction signals in the left and right beam of an axial force system cancel out because they are of opposite sign.

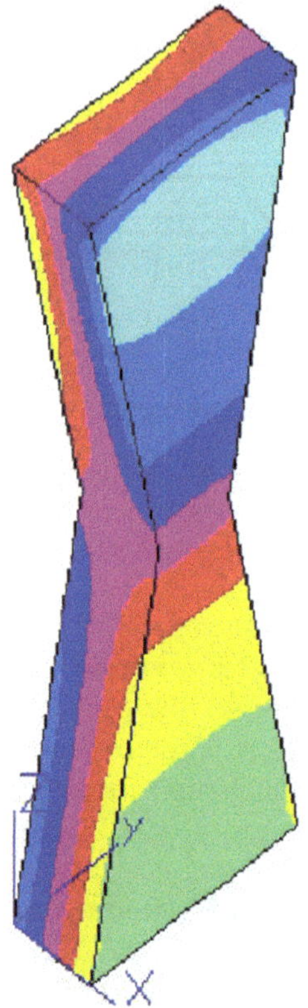

Fig. 6.31 Coupled double trapeze axial force beam

The advantage of such a coupled arrangement is that the heat flux through the element is faster then in the decoupled version and so this kind of element is better in balances where large temperature gradients can occur.

It is not possible to give a general rule whether a decoupled or coupled element is better in terms of interaction. A constant stress beam design in any case minimizes the interactions, because the sensitivity of both beams match each other better.

To determine which shape of axial force element is the optimum for a specified balance is one major task of the design engineer. One recommendation is to calculate a balance with more than one design of axial force beam to reveal the differences.

6.1.6.6 Temperature Gradient Influence

Up to now it has been assumed that the axial force section is always at a constant temperature, or only minor temperature changes occur. In balances for cryogenic or high temperature wind tunnels this is no longer valid. The tunnel operation requires not only the change from ambient to low or high temperatures, also temperature variations up to 40 K between runs are made. Due to cost or run-time capability it is not possible to wait until the total model setup stabilizes at the new temperature, and so temperature gradients inside the balance occur. However, before embarking on design descriptions for overcoming this problem, the difference between measurement errors arising from the absolute temperature or from temperature gradients should be clarified.

For pure temperature changes, the whole balance and model system has overall the same temperature. Structural changes arising from a change in temperature are the same everywhere if the material has the same thermal expansion coefficient. Only where the thermal expansion coefficient is different in the material, can structural differences cause deformations. Normally balances are made out of one material and only the strain gauges are a composite of different material layers. Therefore, by changing the temperature level, the difference in thermal expansion cause strain between the gauge and the base material. This effect is called apparent strain, since the strain gauge senses a strain although there is no strain in the base material. This effect cannot be compensated in any way by the design of the balance; it has to be taken into account by the gauge selection and the wiring design. This is discussed in Chap. 8.

However, if the tunnel flow temperature changes relatively fast, the model balance and support system are not able to stabilize at a constant temperature level and calculations show that there is a heat flux from the support over the sting and balance to the model. The wing and fuselage of the model follow the temperature of the flow rapidly due to forced thermal convection. Using a thermal heat flux calculation, this could be shown in the left diagram of Fig. 6.32, where the temperature distribution is shown 12 minutes after a sudden 40 K temperature change of the tunnel gas.

This change is the driver of heat flux between the model interface and the sting interface to the balance. This leads to temperature gradients along the axial direction. Then the model end part of the balance has a different temperature then the sting end (see right diagram in Fig. 6.32 computed for 27 minutes after a temperature change). As a consequence, the length of the model end part differs from the length of sting end part. This difference is compensated by a deformation of the axial force flexure system and axial force beam.

The diagrams in Fig. 6.33 show the principal deformation caused by a thermal gradient between model end and sting end. Without an axial force sensor in the middle, the system is symmetrical with respect to stiffness and the deformation of the left flexure system is identical to the deformation right flexure system. If an axial force sensor is integrated into the middle of the axial force system, the whole system is not symmetrical with respect to stiffness. One flexure system will act against the other flexure system and the axial force beam. One could argue, that if the axial force beam is connected to both sides to a metric beam, it will stay in the middle. This is only the case if the temperature gradients on the left and on the right side of the beam are exactly identical. This is rather unlikely the case, because the material distribution on the left and right side is not identical. All previous experiments and FE simulations have confirmed this statement.

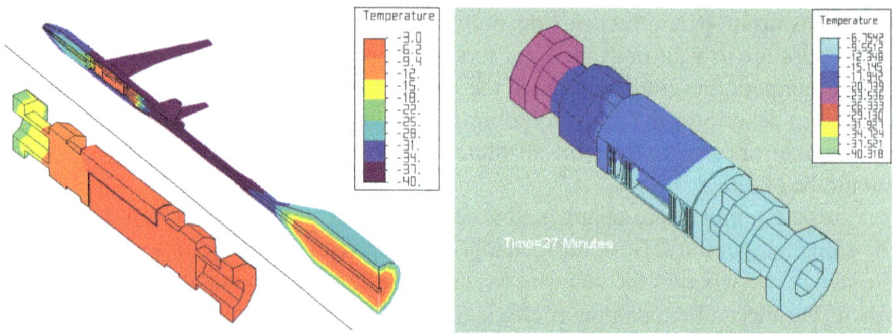

Fig. 6.32 Temperature gradients by a sudden temperature change of 40 K

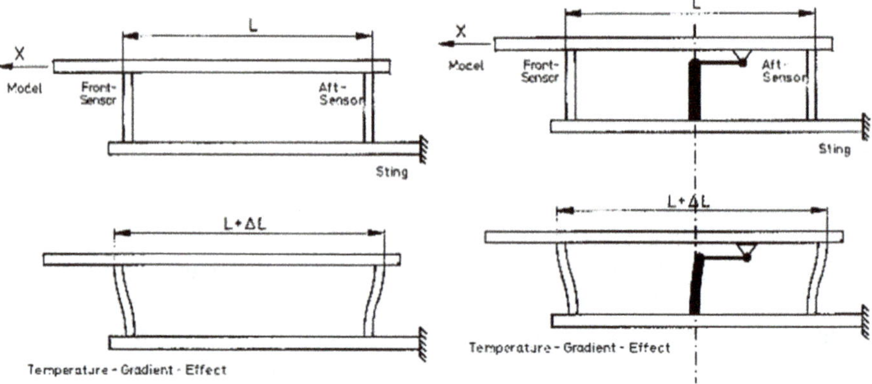

Fig. 6.33 Temperature gradient effect on the axial force system

To analyze this effect in more detail the following considerations are made. Assume that a conventional flexure system, shown exemplarily in Fig. 6.34, is separated into four parts. Parts 1 and 2 represent the model side of the balance and parts 3 and 4 represent the sting side. Moreover, it is assumed that the model side is deformed by a temperature gradient resulting in the changes of the part lengths Δl_1, Δl_2, Δl_3, Δl_4. In Fig. 6.34 no deformations are drawn in on the sting side Δl_3 and Δl_4, but they are taken into account.

For the case that no axial force is acting on the balance, the internal forces acting on the flexure system F_A, F_B and the force acting on the measuring beam F_C are in equilibrium, therefore it can be written:

$$F_A + F_B + F_C = 0 \tag{6.136}$$

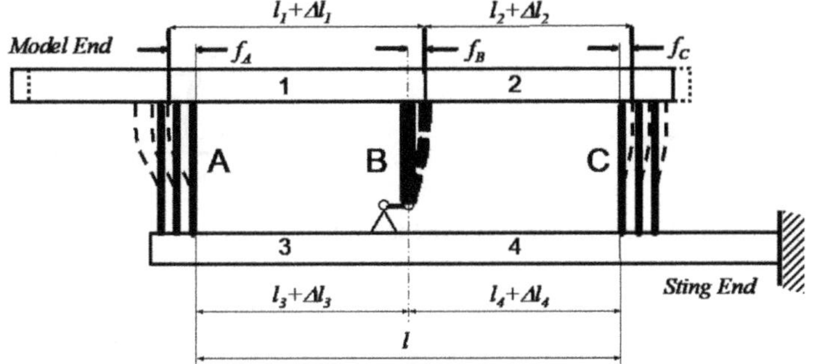

Fig. 6.34 Conventional flexure system deformed by a temperature gradient

The corresponding deformations are

$$\begin{aligned} \Delta l_1 - \Delta l_3 &= f_A - f_B \\ \Delta l_2 - \Delta l_4 &= f_B - f_C \end{aligned} \quad (6.137)$$

f_A, f_B and f_C are the flexure and measuring beam deformations according to their respective stiffness C_A, C_B and C_C.

The force equilibrium according to Eq. (6.136) can be expressed using the flexure deformation (f_i) and their stiffness (C_i) as:

$$f_A \cdot C_A + f_B \cdot C_B + f_C \cdot C_C \quad (6.138)$$

Using Eqs. (6.137) and (6.138) the force equilibrium can be expressed by the stiffness (C_i) the flexure deformations $(f_i,)$ and the temperature induced deformations (Δl_i).

$$(\Delta l_1 - \Delta l_3) \cdot C_A + f_B \cdot (C_A + C_B + C_C) + C_C \cdot (\Delta l_4 - \Delta l_2) \quad (6.139)$$

Assuming that the stiffness values of the parallelogram flexures $C_A = C_C = C_P/2$ and using $F_B = f_B \cdot C_B$, the force on the measuring beam (F_B) can be expressed as a function of the stiffness values (C_i) and the thermal expansions (Δl_i).

$$F_B = \frac{C_B \cdot (\Delta l_3 - \Delta l_1 + \Delta l_2 - \Delta l_4)}{2 \cdot (1 + C_B/C_P)} \quad (6.140)$$

The aim is to now express the force acting on the measuring beam (F_B) in terms of the axial force range (F_X) that is acting on the beam to determine the ratio between the effect of the thermal gradient and the axial force, to obtain an impression about how sensitive the axial force system reacts to thermal gradients related to the axial

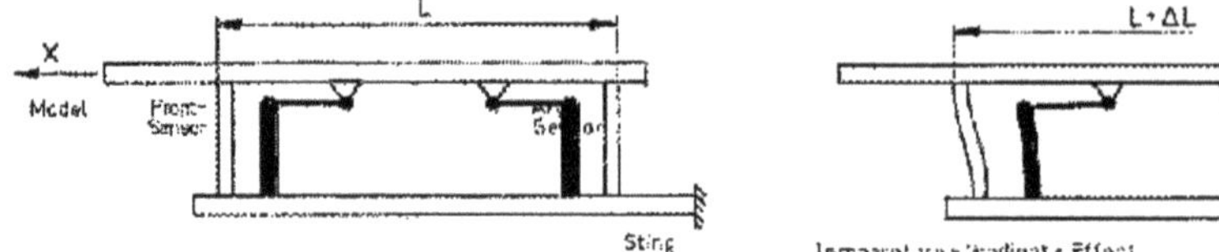

Fig. 6.35 Axial force system with double axial force element

force sensitivity. The part of the axial force F_X that deforms the measuring beam (F_{BX}) depends on ratio of the stiffness C_B and the total stiffness $C_P + C_B$.

$$F_{BX} = F_X \cdot \frac{C_B}{C_B + C_P} \tag{6.141}$$

Dividing Eq. (6.140) by Eq. (6.141), the ratio of the force by thermal gradient (F_B) and the axial force part (F_{BX}) yields the following equation:

$$\frac{F_B}{F_{BX}} = \frac{1}{2} \cdot \frac{C_P}{F_X} \cdot (\Delta l_3 - \Delta l_1 + \Delta l_2 - \Delta l_4) \tag{6.142}$$

The signals are proportional to the force acting on the measuring beam, so Eq. (6.142) provides information about the ratio of the signal by a thermal gradient to the signal by an axial force. Equation (6.142) is a remarkable result because this signal ratio (F_B/F_{BX}) depends only on the stiffness of the parallelogram system (C_P). It is not influenced by the stiffness of the measuring beam (C_B) itself. To keep C_P as small as possible is the sole measure to reduce the gradient sensitivity of a balance with a conventional axial force system. Parallelogram systems are designed to bypass all loads except F_X and so they require a certain stiffness. As a consequence, their axial stiffness remains in the order of the stiffness of the axial force system, thus, the sensitivity related to gradients is very high and cannot be reduced by a low C_P.

To overcome this problem the flexure system must be symmetrical related to stiffness. This can be achieved by integrating one axial force measuring beam into each flexure system.

Such a system was successfully tested and integrated in several sting balances for cryogenic wind tunnels. Constant stress beams as a measuring element are mandatory, because the compensation effect only works well if all four elements of an axial force section have the same sensitivity. This requirement is not only necessary for the compensation of the thermal gradient effect, but also for the compensation of the pitching moment interaction (Fig. 6.35).

Each flexure system should be designed as compact as possible because each system for itself is asymmetric related to stiffness along the x axis; hence, they are internally sensitive to temperature gradients. Full or half bridge gauge application on each beam is possible. Normally two beams at the same axial location are wired together and the signals from both stages are summed up. In case of a full bridge

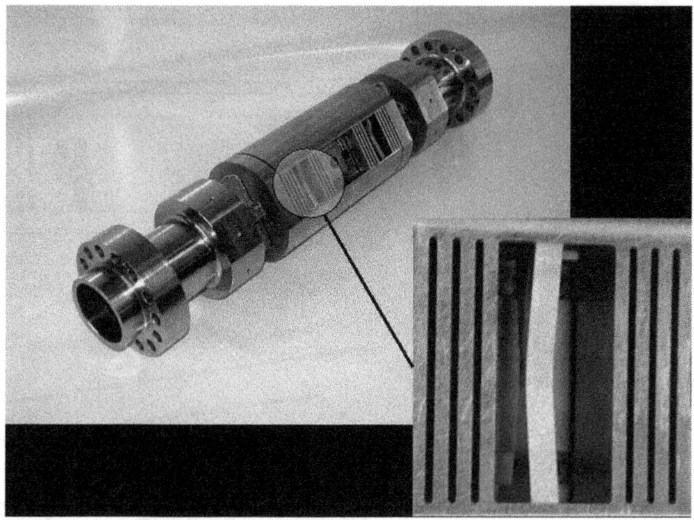

Fig. 6.36 Sting balance for ETW with four constant stress axial force beams

application at each beam, also four signals can be measured and summed up. In the latter case it is possible to take the separate signals of the beams for temperature gradient detection.

Manufacturing tolerances and strain gauge application uncertainties usually result in an uneven stress distribution of the signals, and so despite high quality manufacturing, gradient signals can still occur in axial force measurement. Nevertheless, a calibration process can compensate these influences. The first step is to calibrate the system related to force at a constant temperature. The result of this calibration is to determine a sensitivity to axial force of both axial force sections. The sensitivity to force of the model end section is S_{FM}, and S_{FS} for the sting end section (Fig. 6.36). In a second calibration, the sensitivity to the temperature gradient (S_{TM}, S_{TS}) of both sections related to the mean temperature difference (ΔT_m) between the upper part and the lower part of the axial force system must be determined. Heating the balance on one side and recording the signals and mean temperatures can achieve this.

The local shape of the temperature distribution does not influence the signal at all, because the change of the length (Δl) is an integrated value over the entire length between the measuring beams. Therefore it is sufficient to measure the mean temperatures of these lengths and calculate their difference. Several temperature sensors distributed along the distance or a long PT 100 sensor can achieve this (Fig. 6.37).

The measured signal in a test of the model side beam (U_M) and the signal of the sting side beam (U_S) each are a sum of the signals due to the gradient (U_{TM}, U_{TS}) and the signal due to force (U_{FM}, U_{FS}):

$$\text{Measured Signal} \quad \mathrm{U_M = U_{FM} + U_{TM}}; \quad \mathrm{U_S = U_{FS} + U_{TS}} \tag{6.143}$$

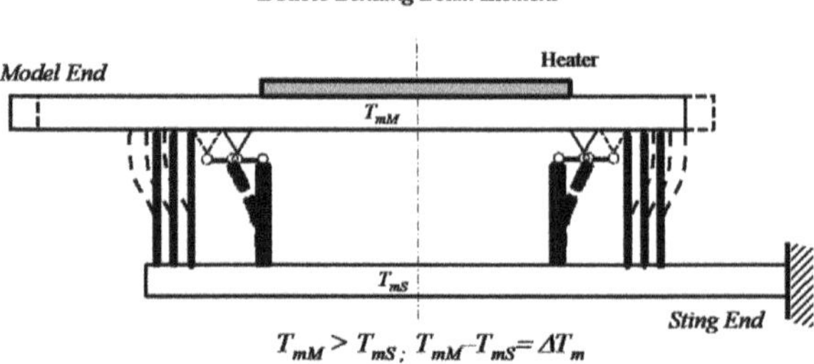

Fig. 6.37 Calibration of temperature gradient influence

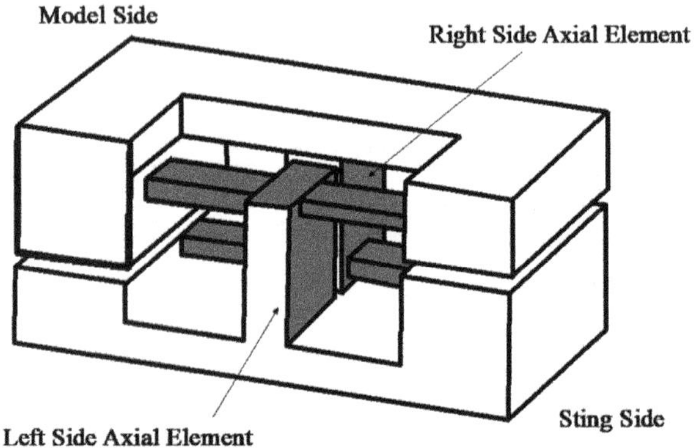

Fig. 6.38 Principal axial force element arrangement of ONERA

The results of the two calibrations are the sensitivities of the beams for force (S_{FM}, S_{FS}) and temperature gradient (S_{TM}, S_{TS}). Using the sensitivities, the signal can be described by the following equations:

$$\text{Signal by Force} \quad \mathrm{U_{FM}} = S_{FM} \cdot F; \quad U_{Fs} = S_{FS} \cdot F \tag{6.144}$$

$$\text{Signal by } \mathrm{T_m} \quad \mathrm{U_{TM}} = S_{TM} \cdot \Delta T_m; \quad U_{Fs} = S_{FS} \cdot \Delta T_m \tag{6.145}$$

The signals by the gradient expressed in Eq. (6.145) must be subtracted from the measured signals given in Eq. (6.143). By using Eq. (6.144) the axial force can be determined:

$$\begin{aligned} & U_{FM} = U_M - S_{TM} \cdot \Delta T_M \\ \text{Axial Force F} \quad & \mathrm{U_{FS}} = U_S - S_{TM} \cdot \Delta T_M \\ & F_x = F_{xM} + F_{xS} = \frac{U_{FM}}{S_{FM}} + \frac{U_{FS}}{S_{FS}} \end{aligned} \tag{6.146}$$

Another structural approach to reduce the temperature gradient sensitivity of the axial force system was proposed by ONERA in France [2], as illustrated in Fig. 6.38.

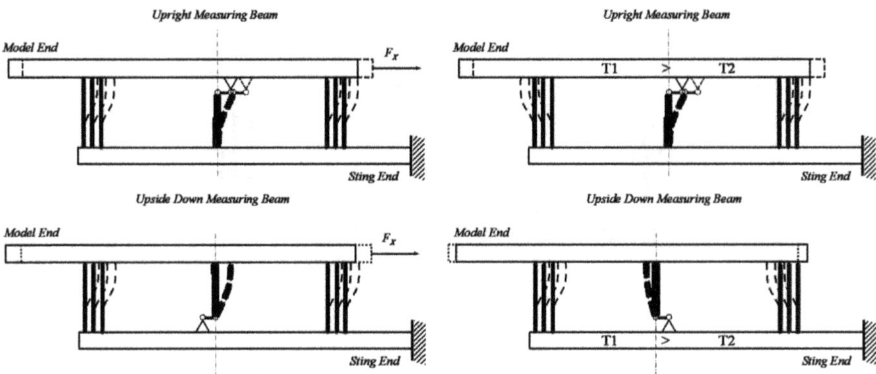

Fig. 6.39 Deformations of ONERA axial force elements design

In the schematic drawing of Fig. 6.38 it can be seen that the axial force element on the right side is built into the balance in an upright arrangement and on the left side the axial force element is arranged in the upside down position. Both are built as close as possible to the center of the balance, so that a center hole through the balance is not possible, but the interactions by pitching and yawing moment are minimized. To explain how this configuration is able to compensate for the effect of longitudinal thermal gradients the following pictures in Fig. 6.39 are used.

For certain temperature gradients the signal of the upside down beam is negative related to the positive signal under axial force, while the signal of the upright beam remains positive. By summing up both signals, the signal from the temperature gradient cancels out. The functionality of the compensation effect is limited to a special temperature distribution inside the balance; if others occur the asymmetry in the distribution of stiffness also results in temperature gradient induced axial force signals.

Beside structural measures to compensate the influence of temperature gradients, it is also possible to compensate the effect by other methods. These methods are primarily of interest for conventional balances with a measuring beam in the center of the balance. In conventional wind tunnels the temperature gradients are small and so the influence on the balance are much smaller than in a cryogenic balance.

The idea for the first method is based on Eq. (6.142). This equation gives the relation between the ratio of the force by the temperature gradient (F_B), the axial force part (F_{Bx}) and the axial force range (F_x) in dependence of the thermal expansions (Δl_i) of each arm (see Fig. 6.34). The ratio F_{Bx}/F_x is given by Eq. (6.141); unknown are the thermal expansions of the four arms. If the mean temperature change is measured in the four arms, the expansions of each arm can be determined by multiplying the difference with the thermal expansion coefficient of the material (α_M) and the arm length (l_i). The thermal expansion coefficient is constant for small temperature ranges so that Eq. (6.142) for the force by a gradient (F_B) can be expressed as followed:

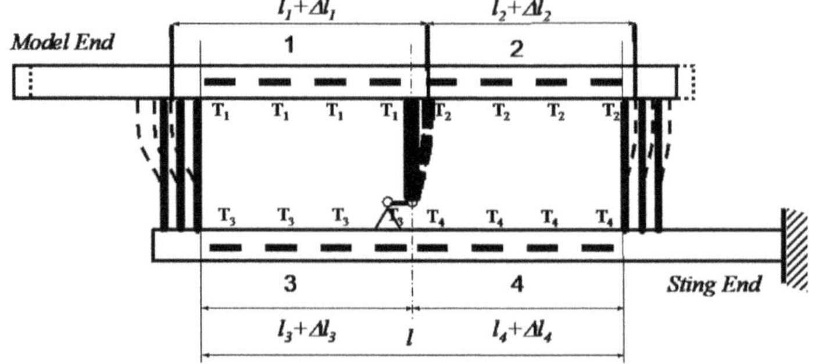

Fig. 6.40 Temperature sensor arrangement to determine the gradient influence Eq. (6.147)

Fig. 6.41 Wiring scheme for temperature sensors to compensate gradient influence

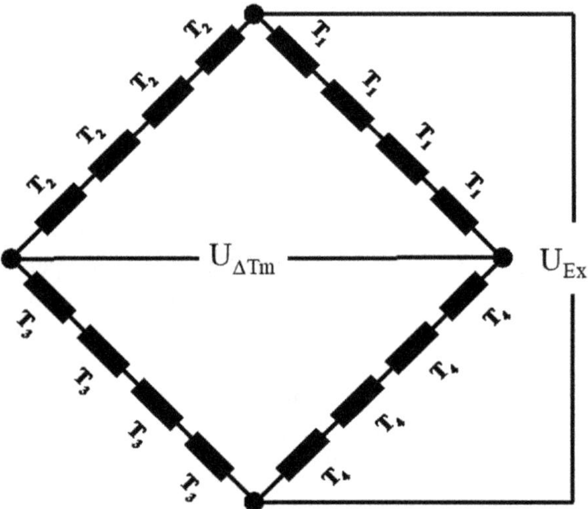

Force by Gradient F_B

$$F_B = \frac{1}{2} \cdot \frac{C_P \cdot C_B \cdot a_m \cdot l/2}{C_B + C_P} \cdot (\Delta T_{m3} - \Delta T_{m1} + \Delta T_{m2} - \Delta T_{m4})$$

$$F_B = C \cdot (\Delta T_{m3} - \Delta T_{m1} + \Delta T_{m2} - \Delta T_{m4}) \qquad (6.147)$$

The value of the mean temperature differences can be measured using a Wheatstone bridge with temperature sensors in the different arm (see Fig. 6.40).

The wiring scheme of the bridge can bee seen in Fig. 6.41.

The signal ($U_{\Delta T_m}$) is directly proportional to the sum of the mean temperatures in Eq. (6.147) and directly proportional to the signal error arising from the temperature gradient. The sensitivity to temperature gradients can now be determined by a calibration test measuring the sum of the mean temperatures and the corresponding signal of the axial force beam. Important for this kind of temperature gradient compensation is that the mean temperature of the parts between the flexures and the axial force sensor are measured. This can be done either using number of sensors like it is shown in Fig. 6.41 or by one long temperature sensor in each arm.

Because a Wheatstone bridge can yield the influence on the axial force, the idea is similar to using temperature sensitive wires directly in the Wheatstone bridge to compensate for the temperature gradient on the hardware side. To do this the following procedure must be followed. One part length must be heated to determine the gradient sensitivity. Assuming l_1 in Fig. 6.40 is heated so $\Delta T_m = \Delta T_1$. All others are zero. The measured axial force signal is then proportional to this mean temperature difference (ΔT_m).

$$\frac{\Delta U}{U} = C \cdot \Delta T_m \tag{6.148}$$

Only the change in a quarter bridge is taken into account, so the relative change of the bridge resistance is:

$$\frac{\Delta U}{U} = \frac{1}{4} \cdot \frac{\Delta R_1}{R_1} \tag{6.149}$$

Using Eqs.(6.148) and (6.149), the necessary change of the resistance of one bridge arm can be determined to be:

$$\Delta R_1 = 4 \cdot C \cdot \Delta T_m \cdot R_1 \tag{6.150}$$

The chance of the wire resistance is:

$$\Delta R = \frac{\alpha \cdot \rho \cdot l}{A} \cdot \Delta T_m \tag{6.151}$$

where α is the coefficient for the change of the specific resistance (ρ) arising from the temperature of a conductor with the length l and the cross-section A. To calculate the necessary wire length for the compensating resistor the wire must change its resistance ΔR exactly with same value as ΔR_1. So the necessary wire length can be calculated using Eqs. (6.150) and (6.151).

$$l = \frac{4 \cdot C \cdot R_1 \cdot A}{\alpha \cdot \rho} \tag{6.152}$$

One of these wire resistors must be integrated electrically into every arm of the axial force measuring bridge and each of them must be applied along the length l_i.

6.2 External Balances

Some description of force and moment separation for special external balances is given in Sect. 3.1; however, it is not feasible to give detailed design equations for an external balance because there are simply too many varieties. For this reason only some general rules will be given that apply to all types of external balances.

The first general rule is that all parts in-between model and load cells should be as stiff as possible, because any deformation will cause an error in the orientation of the model related to the wind. If the deformations are too large, a model position measurement device is necessary.

Lateral movements of the model are not a problem, because with lateral shifts of only millimeters, the flow in a wind tunnel does not change significantly. Changes in the angle of attack within a hundredth of a degree on the other hand can have a significant influence, especially on the drag measurement. As a rule of thumb the difference in the deformation from one side to the other side of the balance should not exceed 0.2 mm per meter length. The value of 0.2 mm is a deformation in the order of the total load cell deformation. That means that the load cell itself is responsible for the allowed deformation. Therefore, almost no deformation is allowed for the balance structure between model and load cell. Of course this is not possible, but as a general design rule it is necessary to fulfill this requirement as well as possible.

One very basic difference in the design of an external balance compared to an internal balance is how the turntable system is integrated into the balance system. One possibility is to turn the entire balance and the other is to integrate the turntable into the metric end of the balance.

In both designs the load cells that are used for the vertical loads, must carry the weight of the metric balance structure and model setup. If the turntable is integrated into the metric side of the balance its weight must additionally be carried (Fig. 6.42).

The range of the load cells for the vertical loads must be dimensioned with respect to loads and weight. The heavier the metric part of the balance, the less sensitive is

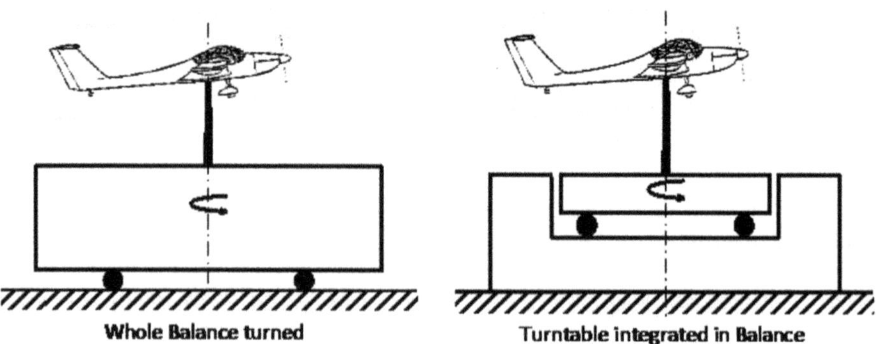

Fig. 6.42 External balance turning options

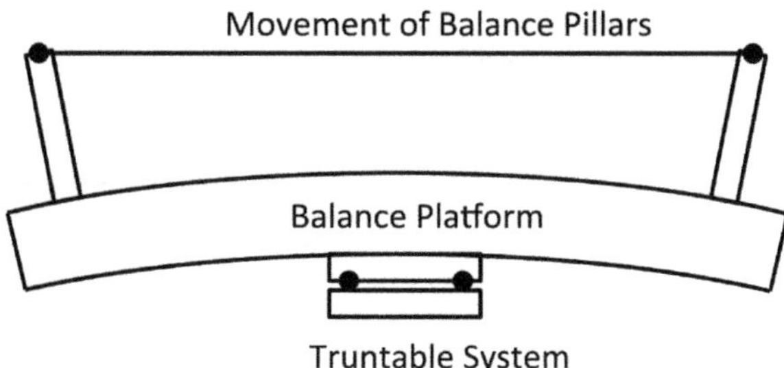

Fig. 6.43 Balance platform deformation under vertical loads

the balance for the vertical loads and the lower is the eigenfrequency. With respect to these characteristics the integration of a heavy turntable system is a disadvantage.

To minimize the weight of the metric end, the entire balance can be turned. In this case the turntable system must be designed very stiff because it has to carry all loads. A further requirement in this case is a very stiff platform over the turntable system for the load cells. If there is any deformation in the base plate this, deformation will result in signals when the balance is turned. An example of this is if the base plate bends by the vertical loads, as outlined in Fig. 6.43. the maximum elongation of the axial link between the pillars must be below 0.0002 mm. Then the axial force measurement is within a 0.1% error. The specified uncertainty of an external balance is usually much below this value and the balance plate must be even much stiffer.

This brief calculation demonstrates how stiff the base plate of an external balance has to be to guarantee high precision. This requirement for stiffness can only be achieved by a very heavy construction and so the turntable system must be very solid to carry all the weight. In any case, the zero load signals of most balances of this concept do change while the balance is turned. If the changes can be correlated to the turn angle they can be corrected.

A strict recommendation for one or the other design cannot be given because every solution has pros and cons and it is up to the tunnel managers to decide which design is best for their local space and environment.

The general remarks made here for the design of an external balance are also valid for the design of a calibration machine.

6.2.1 Load Cells and Load Cell Arrangement

For the load cells the general requirements are:

- they should fit to the required load range as good as possible
- their repeatability, long-term stability and low hysteresis should be the best that is available on the market

Therefore, the load cells are often the most expensive components of the balance, but they limit also the overall characteristics of the balance; thus, it is in general a

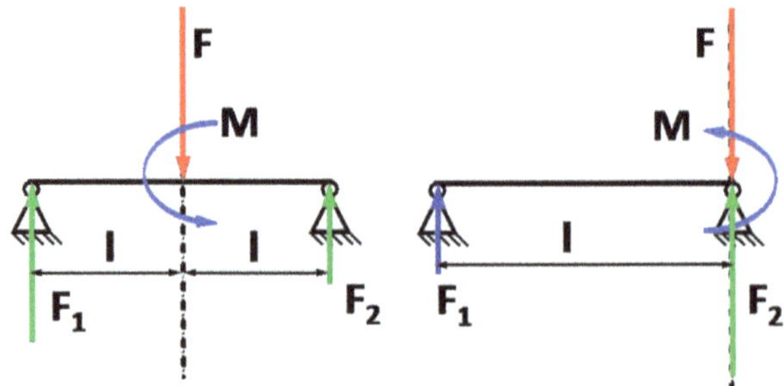

Fig. 6.44 Load cell arrangement for external balance

good investment. If several load cells with the same long-term quality are available, the stiffest one should be selected.

In one plane there is always one force and one moment acting, so this is a similar situation as for the internal balance. Two force transducers must measure a force and a moment. In contrast to the internal balance, the available space is much larger so that an adaption to the required load range can be achieved much easier by adapting the distance between the load cells. There are now different possibilities. The first is to place two load cells at equal distances to the load center (see Fig. 6.44).

In this case the force always generates reaction loads or signals corresponding to half of the load in the same direction and the moment generates two reaction forces with the identical values but in the opposite direction (see Fig. 6.44 left side). That is exactly what is described in Sect. 6.1.1. If the ratio for the signal is specified, the length l can be determined using Eq. (6.5).

The other possibility is to place one load cell exactly in the axis of the acting force. In this case this load cell has to carry the force and the reaction force of the moment load cell. But the reaction force in the moment load cell depends only on the moment. The larger available space for an external balance now allows installing a load cell with a small load range far away from the load center, where the reaction force by the moment is very small. So most of the full-scale load range of the force load cell can be used to measure the force, which at least results in a higher resolution for the force load range. This is expressed in Eq. (6.153) in the relation for U_F. For a long arm, the moment part (M/l) in the signal that is proportional to F_2 will be almost zero and so the signal is almost proportional to the force F. The relations in Eq. (6.153) can be taken to determine the load range for the load cells that must be purchased for the balance.

$$\begin{aligned} U_M &\sim F_1 = \frac{M}{l}; \quad U_F \sim F_2 = F - F_1 = F - \frac{M}{l} \\ \frac{U_M}{U_f} &\sim \frac{F_1}{F_2}; \quad \frac{U_M}{U_F} \sim \frac{\frac{M}{l}}{F - \frac{M}{l}} \end{aligned} \tag{6.153}$$

For the arrangement of the load cells for the moment measurement there are two directions possible for every moment. For example the rolling moment M_x can be

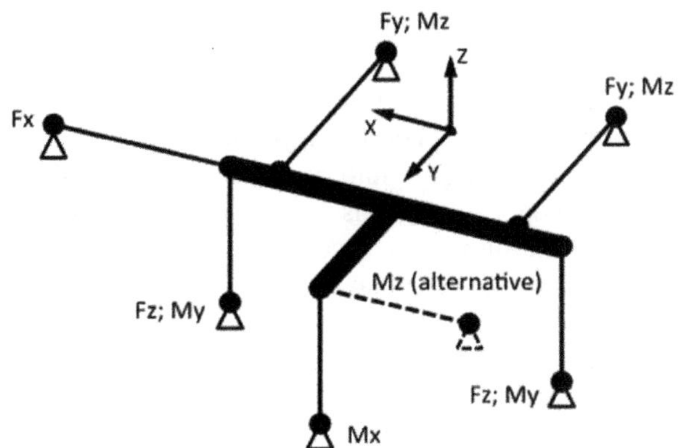

Fig. 6.45 Yawing moment options

measured in the Y-Z plane either in Z direction on a Y lever or in Y direction on a Z lever. Depending on the orientation that is chosen there will be an interaction either on the Y-force (Z lever) or on the Z force (Y lever). If one force is more sensitive than the others the interactions on this force measurement can be minimized if no moment is measured in the same direction. For example if the rolling and pitching moments (M_x, M_y) are measured in the Z direction (Y lever, X lever) and yawing moment (M_z) is measured in the Y direction (X lever) in X direction, only the axial force F_x will be measured. That means the axial force load cell has to carry only loads that are caused by the applied axial force and there is no "interaction by design" by a moment measurement (see Fig. 6.45). Instead, if the alternate position for the yawing moment (M_z) is used, there is an interaction of M_z on F_X.

There are two possibilities to realize the geometry of an external balance as precise as possible. The first way is to specify the tolerances in the parts as small as they can be manufactured and to assemble them in a precise position, but this naturally increases the costs significantly.

The second way is to allow mechanical play for the fixation of the load cells to the main frame or the ground resulting in an adjustment during installation of the balance. If all connecting rods are aligned as perfectly as possible, the load cells can be screwed tightly and dowel pins must fix this position. This is the most practical way to insure a balance with high orientation accuracy. The aim for the geometric precision should be less than 0.01 degree in angularity and less than 0.02 mm in lateral offset to the force axis if the load cells are placed in the axis, or a tolerance of less than 0.02 mm for symmetrical distance if the load cells are placed out of axis.

6.2.2 *Weighbridge*

For all horizontal loads only the acting loads must be taken into account. For the vertical loads the weight of the weighted masses must be included into the considerations for the dimensioning of the load cells. The weighted mass consists of the mass of the weighbridge, the connecting rods, flexures and the mass of the model. The weighbridge must be designed very stiff to minimize the interactions by bridge deformations. As a consequence, the mass of the weighbridge becomes very large. Taking these weights into account will obviously decrease the resolution for the loads to be measured. It is useful to limit the weight of the weighbridge to half the value of the sum of the necessary vertical forces. The mass of the weighbridge can now be distributed so that every load cell for the vertical loads is pre-loaded to half of their full-scale range. The advantage of this design philosophy is that the load cells are loaded only in one quadrant instead of going to the plus-minus full-scale value. This decreases the resolution to half of its nominal value, but it also minimizes the non-linearities by zero crossing. With the given weight for the weighbridge the stiffness of the design should now be optimized so that the overall deformations are below 0.005 mm. This is a rather tough requirement and hard to achieve, but some uncertainty considerations show that this minimizes the interactions. For example, if the load range for the axial force is 30% of the full-scale of the normal force, a deformation of 0.005 mm in axial direction by the normal force will generate a signal in the axial force load cell of 2.5% of axial force full-scale (see also the introduction to Sect. 3.1). If a construction with less weight fulfills the stiffness requirement, the remaining mass should be used to make it stiffer. The design load ranges F_i for the load cells are correspondingly the sum of the design load F_{iL} and the part of the construction weight F_{iW}.

6.2.3 Connecting Rods and Flexures

Next to the weight beam and the load cells, the connecting rods (rocking pier) are the most critical component. To enable high stiffness at low weight, of course steel pipes are one of the best choices. To transfer the loads without transferring moments to the load cell at each end, an elastic flexure must be installed (as described in Sect. 9.2.13).

To fulfill the stiffness requirements for the flexures Maraging steel should be used. A criterion for the stiffness of the rod is a safety factor against buckling of more than 10 and for this standard strength calculation relations can be used. The connection to the load cell and the weighbridge should have a centering to guarantee that they always remain in the exact position. This makes the precise alignment during assembly much easier, because the position does not change while locking the nut (Fig. 6.46).

The metric flange of an external balance should be as variable as possible to enable all kinds of model fixation. For this purpose a support on linear dovetail guides is a good choice. Higher quality guides have a spindle and a clearance free slide so that the model does not move under load.

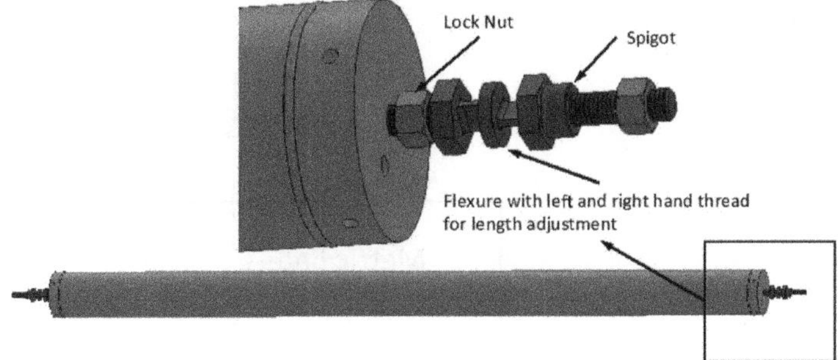

Fig. 6.46 Connecting rod with flexures

6.2.4 Temperature Problems of External Balances

External balance are normally not affected by the change of flow temperature directly because they are mounted outside the tunnel, nevertheless temperature differences between flow and outside temperature cause temperature gradients that affect the stability of the balance. Very seldom are external balances placed in a temperature controlled compartment and so the change of room temperature causes errors.

- If the temperature in the area of an external balance changes slowly and homogeneously only the temperature characteristics of the load cells have an influence on the measurement. The load cells of an external balance are compact sensors and usually of high quality so that temperature changes do not have a strong influence on the load cell itself. However external balances are normally large instruments and so they are very susceptible to temperature gradient effects. If there is airflow inside the balance compartment, the load cells and the structure can take on slightly different temperatures and this leads to deformations inside the balance that are sensed by the load cell and lead to measurement errors. Here is an example of how serious this problem can become. A change of temperature of 1 K in a structural length of 10 cm will change the length for steel by 0.001 mm. If this change causes a structural deformation, the signal of the load cell will change about 0.5% of the full-scale signal. This is far too much for a measurement with an external balance and it shows how sensitive external balances can react to temperature changes. For good measurement performance, a temperature stabilized compartment around the balance is required. Slow, homogeneous temperature changes do not have such a significant influence. To detect how sensitive the balance reacts, an overnight record of the zero signals of the balance can provide the required information, especially if the normal heating and air conditioning systems are active or sunlight influences the test section area.
- The problem is also directly related to the stiffness of the load cells. The total deformation of a load cell can vary between 0.3 and 0.05 mm for full scale. Stiffer load cells are good for a perfect position repeatability of the model, but the sensitivity to temperature gradients is up to six times higher. Therefore, the general

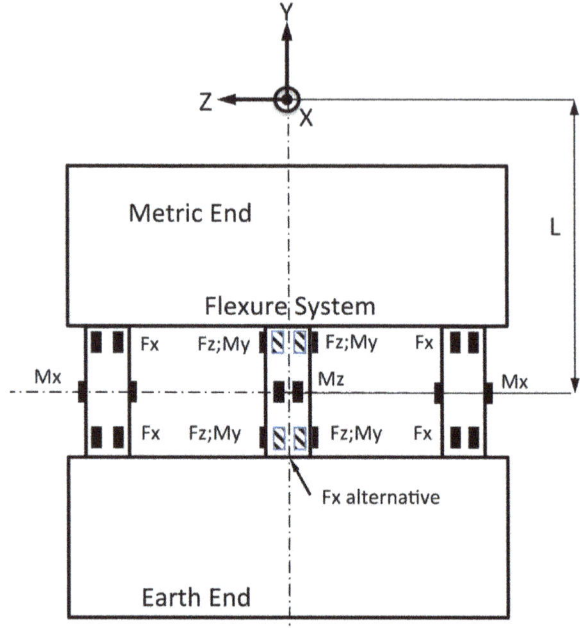

Fig. 6.47 Principle design of a semi-span balance with an example of gauge arrangement

requirement for a precise external balance is a compartment around the balance that protects it from temperature differences. This is not an easy requirement because the necessary clearance to the tunnel enables inflow and outflow from the tunnel to the balance cavity, which can be a source of temperature gradients inside the balance. As a consequence, the compartment temperature should be the same as in the tunnel flow.

6.3 Semi-span Balances

The classical sidewall balance is a balance for half-model testing and usually measures only five components. The spanwise component is usually not measured because this load is seldom of relevance, although sometimes this load is measured to detect possible interactions of the model load on the other components, in the special case that the weight of the model changes significantly during the test campaign. The principle design of a semi-span balance with the strain gauge positions for the different loads is shown in Fig. 6.47.

The massive earth end is connected via a symmetrical arrangement of flexures to the massive metric end of the balance. The major design problem of a half-model balance is the dominating rolling moment generated by the lift of the half wing. This moment must be carried by the balance flexures and it also generates high interaction

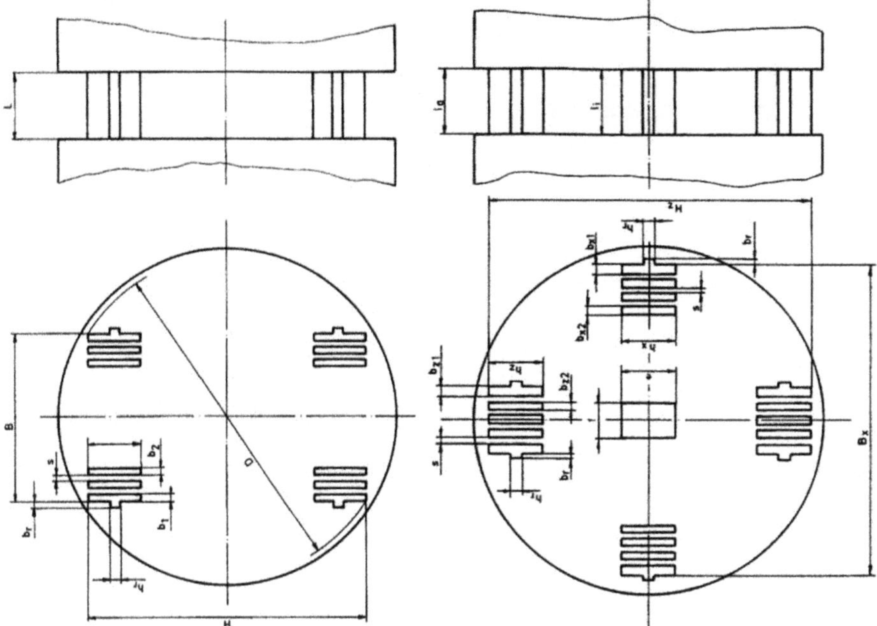

Fig. 6.48 Possible cage arrangement in semi-span balances

loads. The large interaction with the rolling moment is one of the reasons to measure this moment because its magnitude is needed to correct especially the lift of the wing. The other reason is to obtain the root bending moment of the wing. Almost all semi-span balances use a cage system to carry and to measure the loads. This cage system can be arranged in two different ways. The first is an arrangement of the flexures in the 45° position relative to both axes, as shown in Fig. 6.48 on the left side.

The second possibility is an arrangement with the flexures directly in the axes. A picture of a semi-span balance with the flexures in the axes is shown in Sect. 3.1 or in Fig. 6.51. One example of a semi-span balance with the flexures in the 45° position is a balance built by NASA Langley, as pictured in Fig. 6.49.

With the flexures in the axes, the sensitivities can be better adapted to the load ranges because different shapes can be used in the different axes. While the flexure system in the 45° position can only have different stiffness, depending on the orientation,all four single systems itself must be identical to guarantee the symmetry.

The design task is now to dimension the shape of the flexures, such that they carry all the loads and that there are areas which are particularly sensitive to the load to be measured. In principle this is exactly the same design task that has to be accomplished for a cage system of an internal balance and this is the reason why all the equations that have been given for a cage system can be used to determine the stresses in a semi-span balance (see Sect. 6.1.5). To do this, first the loads must

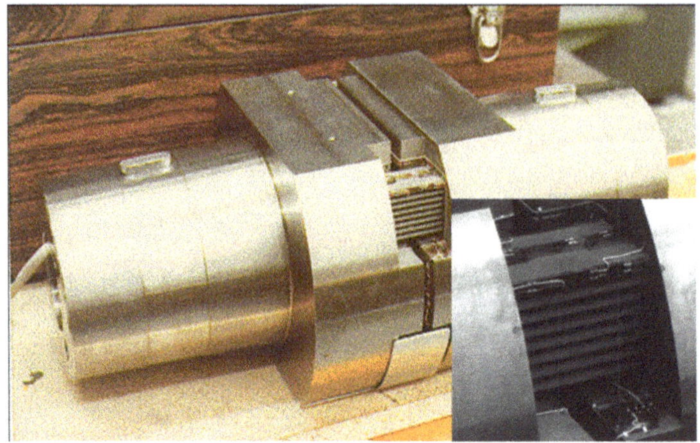

Fig. 6.49 Semi-span balance 804S of NASA Langley, detailing the flexure design

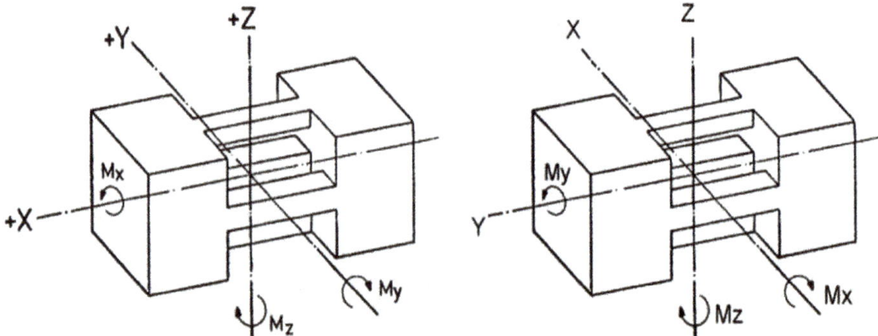

Fig. 6.50 Cage design calculation axes, internal balance left, semi-span balance right

be transferred into the axis system used for the calculation of the cage system of an internal balance. For the axis system used in Sect. 6.1.5 the loads related to the X axis of the semi-span balance become the loads related to the Y axis in the internal balance calculation. The loads related to Z axis remain unchanged, i.e., $F_{X_{semispan}} = F_{y_{internal}}$; $M_{y_{semispan}} = M_{x_{internal}}$; $M_{x_{semispan}} = M_{y_{internal}}$ (Fig. 6.50).

Normally the loads for a semi-span balance are given with a certain distance L to the geometric center of the balance. This distance L corresponds to half the distance between the bending sections $l/2$ for the internal balance calculation. Some semi-span balances use simple cages with one flexure in every quadrant. The large rolling moment M_x generated by the lift of the wing anticipates a sensitive axial force measurement because they must be measured on the same flexure (see Fig. 6.47). In this setup the rolling moment M_x is measured by tension and compression stress in the flexures and axial force F_x is measured by bending stress on the same flexure. Alternatively, the axial force F_x can also be measured by using the bending stress

Fig. 6.51 Semi-span balance and flexure systems for the Cologne cryogenic wind tunnel

of the hatched gauges on the flexure in the X axis (see Fig. 6.47), but the sensitivity due to bending of this flexure is also restricted by the flexure in the Z axis to carry the rolling moment M_x. To distribute the sensitivity better, an alternative design was developed and optimized. The balance and the isolated flexure geometry is shown in Fig. 6.51.

The flexure system in the X axis consists of two separated springs. Their shape is optimized to measure axial force F_x, lift F_z, pitching moment M_y and yawing moment M_z. The rolling moment M_x is carried by a flexure system in the Z axis that consists of two packages of leaf springs, which provides more flexibility in the X direction but enough total cross-section to enable a good sensitivity for the rolling moment by tension and compression measurement. Additionally, for thermal gradient compensation procedures, M_y and F_x are also measured by the bending stress on the leaf springs. Figure 6.53 shows the flexure systems in detail and a description which load the gauges are used for. The primary design calculation for this balance was made using the equations for the cage design and the optimization of the beams was carried out using finite element calculations.

The results of the finite element calculation for the measuring flexures in the X axis are shown in Fig. 6.53. The root section of the large flexure shows a large part of uniform stress by the normal force F_z. This was intended by using the trapeze shape to insure a large area for the strain gauge application where the stress does not change very much under F_z and to also obtain a sensitivity that is not influenced by the normal force. The same was done with the small flexure for the axial force F_x.

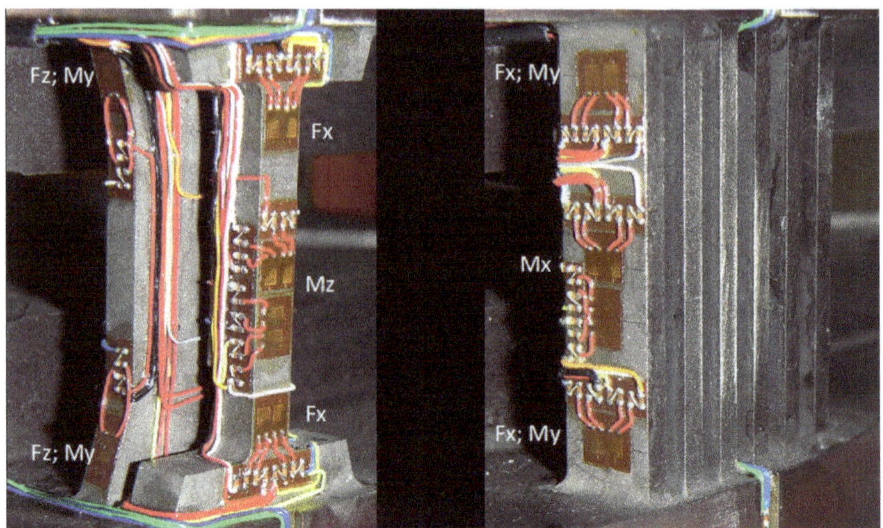

Fig. 6.52 Flexure system in X Axis Flexure system in Z Axis

The main focus of the beam design is to adjust the stress in the gauge area to the given range of the force to be measured to insure enough sensitivity. The main flexure dimensions, length, thickness and width can be adjusted to fulfill this target. The second intent is to insure a constant stress distribution in the gauge area to make the sensitivity of the flexure independent of gauge alignment tolerances. Best performance was achieved by the configuration shown in Fig. 6.51 with two separate flexures or by a rib design as shown in Fig. 6.52. The uniform stress distribution under the gauges achieved by this design is shown in Fig. 6.53 .

With this main flexure, axial force F_x, normal force F_z, pitching moment M_y and yawing moment M_z can be measured. The parallelogram flexure system is used to determine the rolling moment M_x. In some cases it is also possible to measure the yawing moment M_z on the parallelogram flexure system, depending where the higher stresses occur.

The joints of a semi-span balance are normally flanges. The requirements for the design of the flanges are the same as for that of an internal balance and flanges fulfill these requirements on a semi-span balance best. To insure the optimal repeatability for the position, two designs have been proven to be excellent. One is the serration or *Hirth* tooth system and the other are four crosswise keys (Figs. 6.54 and 6.55).

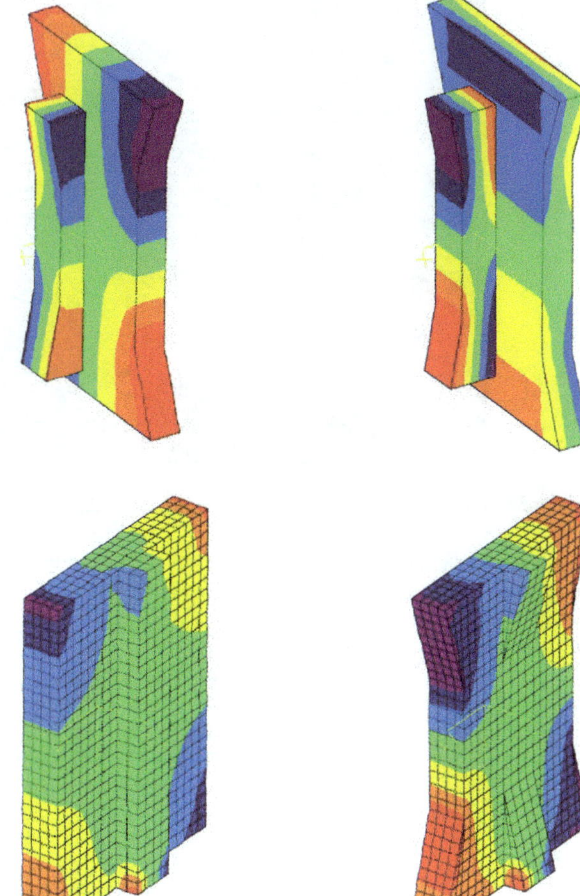

Fig. 6.53 Stress in flexure by normal force F_z stress in flexure by axial force F_x

Fig. 6.54 Rectangular flexure with rib Trapeze flexure with rib

6.3.1 Thermal Problems of Semi-span Balances

Temperature problems related to zero output and sensitivity changes are the same as for internal balances. Due to the larger size, no bridge internal wires should cross the balance because these are directly sensitive to temperature changes. Therefore, strain gauge measurements should be locally configured as local full bridges and each bridge should be measured separately. When tension or compression stress is used to measure a component, then *Poisson* bridges should be used to achieve local concentration (see Figs. 6.52 and 6.51 as well as Sect. 8.6). The signal proportional to the load must then be calculated by summing or subtracting the individual signals.

The larger size of a semi-span balance related to an internal balance makes it more sensitive to temperature gradients. A detailed study was performed in which a finite

Fig. 6.55 Flange with serration and clamp Flange with crosswise keys and screws

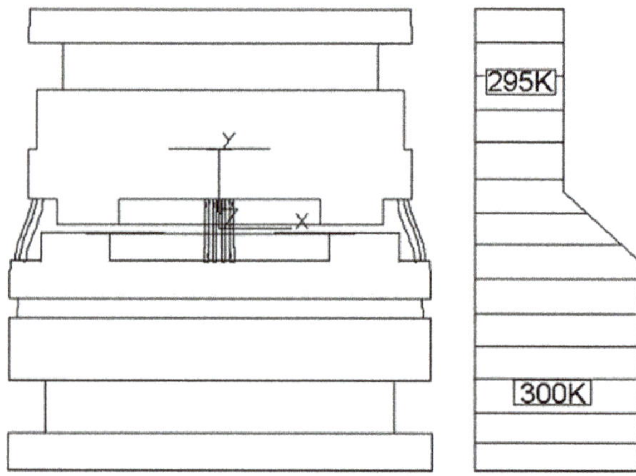

Fig. 6.56 Deformation by temperature difference between top and bottom of the balance

element method was used to determine the balance deformation when temperature changes were assumed at several positions on the balance.

The first temperature gradient analyzed was a temperature difference between both flanges of 5 K (see Fig. 6.56). This is the most common case because normally there is a temperature difference between the tunnel flow and the cavity of the balance. The outcome of this case study was: if the balance has a symmetrical stiffness distribution in the flexures, the flexures will deform with a reasonable magnitude, but in opposite directions. This opens the possibility for correction of temperature gradients along the center axis of the balance. If the deformations are measured separately at both flexures, then the sum of the signals cancels out the gradient effect. To make this work, the sensitivity of both flexures must be identical.

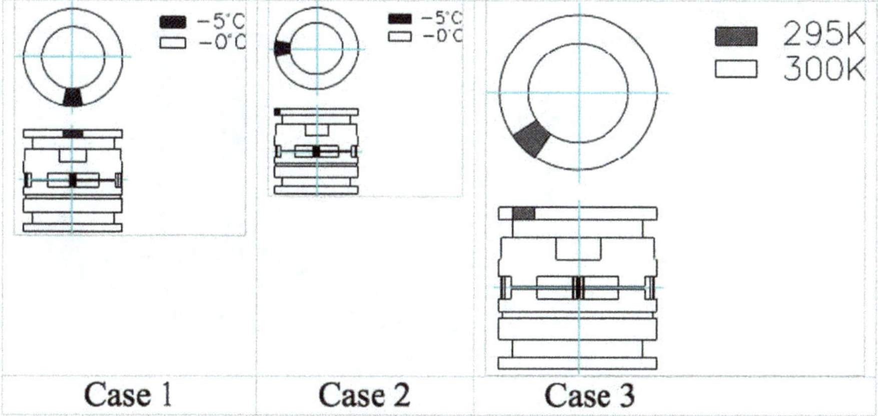

a. Temperature Difference in the Flange

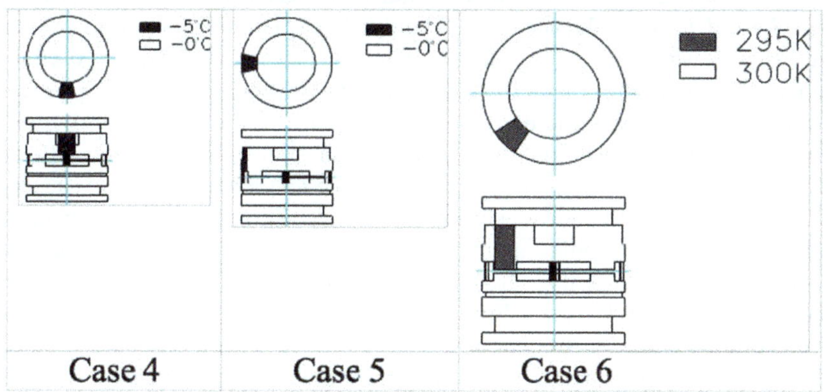

b. Temperature Difference in the Middle

Fig. 6.57 Assumptions for temperature gradients

Identical sensitivity is possible if the geometry of the corresponding flexures and the electrical characteristics of both flexures are identical. With modern CNC milling machines it is possible to build a balance with small tolerances (0.02 mm) for the flexure dimensions. However, to apply the strain gauges with the same tolerance is nearly impossible. To maintain the same sensitivity, the stress area underneath the gauge must be homogeneous and constant. To achieve this, again constant stress beams were used, onto which the strain gauges were applied (see Fig. 6.54).

At TU Darmstadt a study with finite element calculations was performed by Zhai [24] to investigate how temperature differences in the different parts of a semi-span balance can influence the signals of the balance. The assumptions for the different cases can be seen in the Fig. 6.57

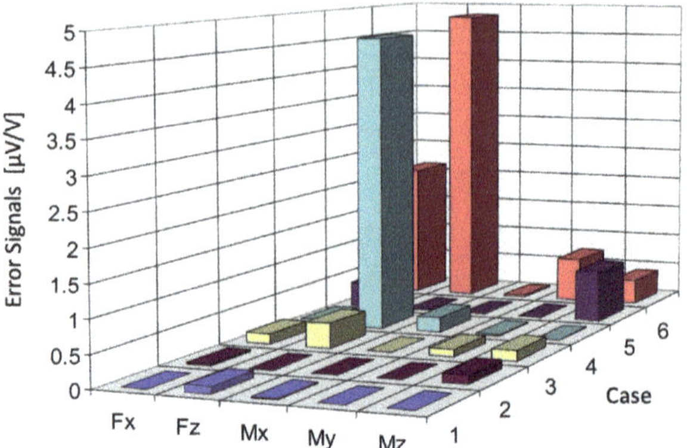

Fig. 6.58 Error signal in the components due temperature gradients (case 1 to 6)

The results of this investigation are shown in the Fig. 6.58. The results show that the semi-span balance is sensitive to radial temperature gradients. Especially the forces F_x and F_z are influenced significantly. To explain this, one has to visualize what happens when the temperature gradients of the cases 3 - 6 are exposed to the balance. One ring body of the balance expands tangentially and the flexures are deformed to compensate the deformation difference in the rings. These deformations are proportional to error signals. This error signals are also proportional to the diameter of the balance, so with respect to the temperature gradient sensitivity, the balance diameter should be as small as possible. This requirement is in contrast to the influence of the rolling moment, which increases with decreasing diameter. So always a compromise must be found between a large diameter to decrease the forces on the flexures by the rolling moment and a small diameter by decreasing the temperature gradient. Another possibility to improve the balance with respect to radial temperature gradients is to use slotted interfaces or basic shape cross instead of a ring. These two options are shown in Fig. 6.59. With both balance designs the sensitivity related to radial temperature gradients could be reduced significantly.

Another method to minimize the influence of temperature gradients is to insulate the balance as well as possible and to stabilize the temperature distribution inside the balance. However the effectiveness of this method is limited, because the contact over the model flange is one of the largest sources for the axial temperature gradients in the balance. Airflow through the clearance between tunnel wall, model and the balance cavity is a source for radial temperature gradients. This flow can be limited only to a certain extent, because wall contact or model contact will corrupt the measurement. For example at Cologne Cryogenic wind tunnel (KKK) it was intended to keep the temperature stable within 0.5 K, because this was the margin to maintain the gradient signal below the required uncertainty without compensation precautions. At least 5 K were the measured differences in the test and fortunately the balance was designed to compensate for this.

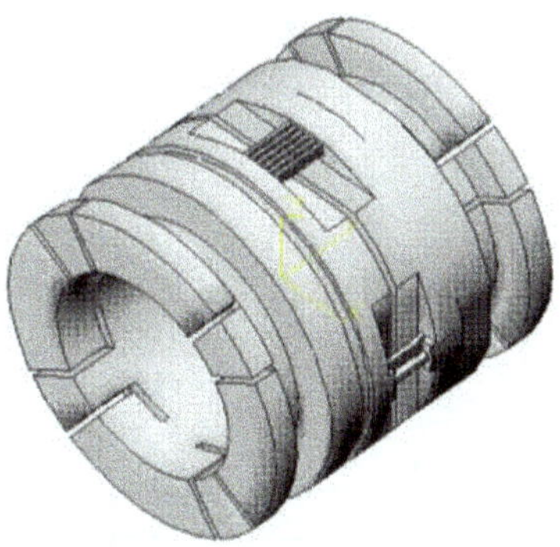

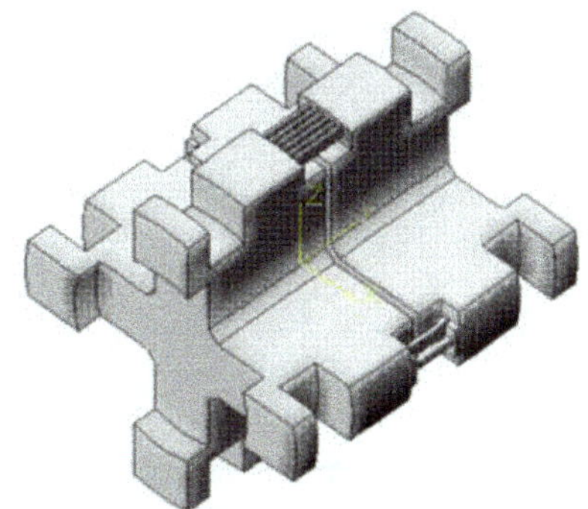

Fig. 6.59 Slotted flange Basic shape cross

Temperature stabilization by heating the cavity is a very problematic solution. The balances are sensitive to small temperature differences and in some cases the situation can become much worse due to the heating system instead of getting better.

For the half-model balances in cryogenic tunnels, stabilization of the temperature in the range of the tunnel temperature and some precise control of the heat flow from the balance to the model improved the situation. The setup for the Cologne Cryogenic wind tunnel (KKK) is shown in Fig. 6.60.

The main heat flux is from the model to the balance or vice versa. To minimize this heat flux a solid carbon fiber plate works as an insulation layer. After this plate there is a thin heating copper foil that stabilizes the radial differences and a moderate temperature rise. Additionally, an insulating spacer made out of thin walled stainless steel reduces the heat flux further. Inside the balance and the spacer there is an infill of insulating material to prevent airflow inside. A super-insulating tube covers the outside of the whole setup. The free volume inside and around the balance is minimized to keep the airflow as small as possible. The total temperature inside the model cart is a little higher than the tunnel temperature.

6.4 Bridging

A problem that often is disregarded is the need to feed wires, tubes and other systems from the metric part to the non-metric part of the balance. Modern wind tunnel models are equipped with numerous measurement systems and these systems must be connected in some way to the data acquisition system. Wireless data transfer is becoming more accessible; however, a power supply is still needed in the model and moreover, the balance wiring itself have to cross the balance. How serious the effects of wires can be is shown in Fig. 6.61, where the force generated by wires of different size and insulating material at different temperatures has been tested.

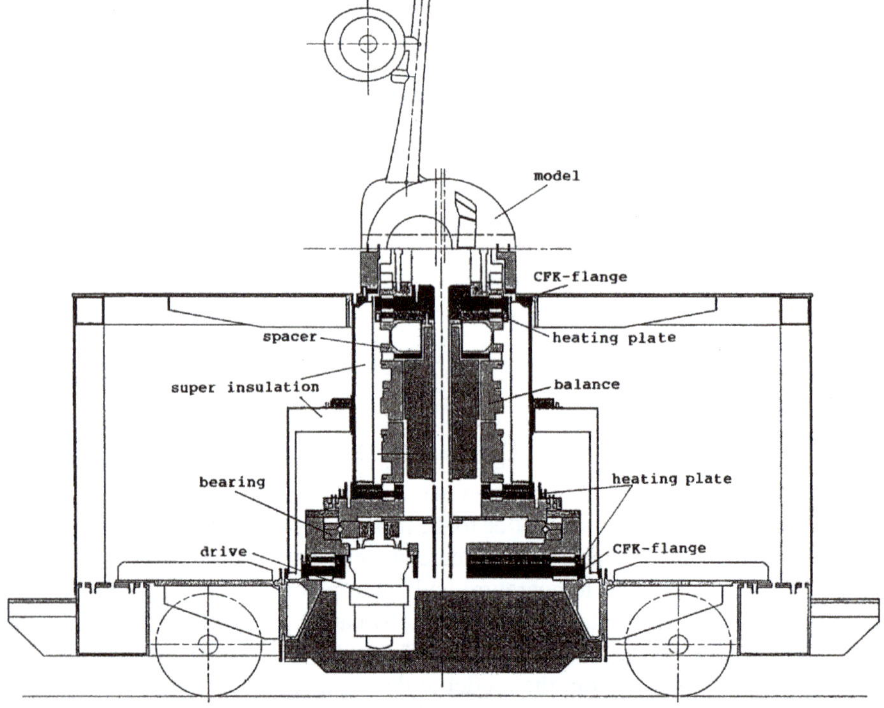

Fig. 6.60 Semi-span balance integration to model support of KKK

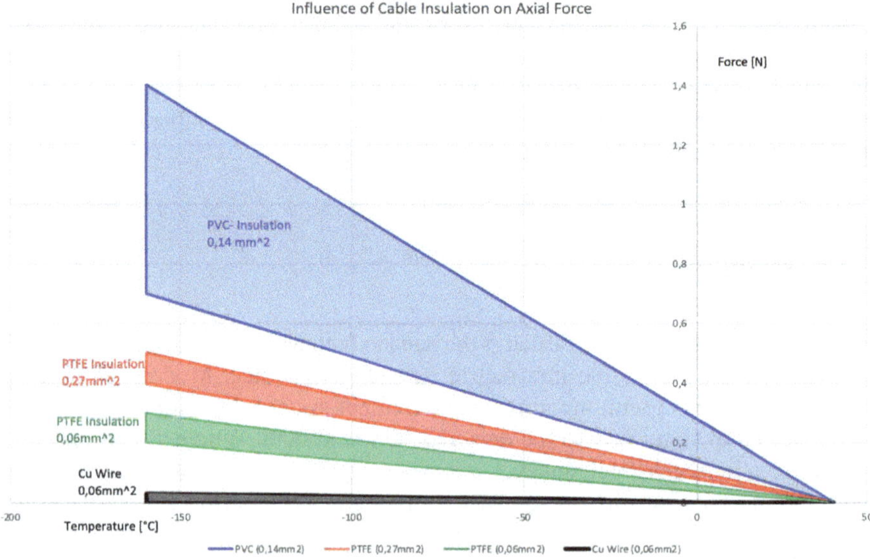

Fig. 6.61 Axial force of wires with temperature change

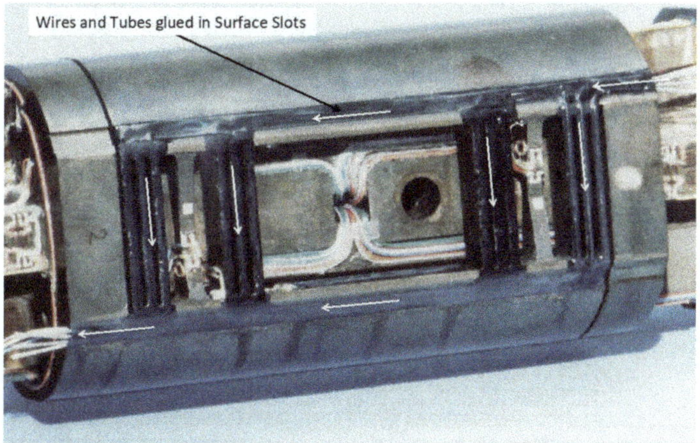

Fig. 6.62 ETW balance with additional wires and tubes glued in surface slots

This investigation was performed for a balance with an axial force range of 1000 N. Only one cable can affect a change of axial force of 1 N, which is equal to a systematic error of 0.1%. About 40 cables cross this balance only for the strain gauge and temperature measurement and so the parasitic force is far above the required uncertainty of less than 0.05%. If additional wires or tubes are needed across the balance the problem becomes even more severe. To overcome this problem by calibration, the basic requirement for all these crossings is a pure linear elastic behavior of the systems. Unfortunately, wires with plastic insulation do not fulfill this requirement and so they can be a serious source of hysteresis and creep.

6.4.1 Model Data Bridging

One way to cross the balance with tubes for the pressure scanner and wires for auxiliary equipment in the model, is to implement them in slots that are milled into the surface of the balance flexure system. In this manner, the auxiliary wires are guided and always fixed to the surface. The additional forces are therefore repeatable and are taken into account during calibration. Additional hysteresis and creep cannot be fully eliminated but reduced to a minimum (Fig. 6.62).

6.4.2 Air Supply Bridging

Tests with turbine power simulators (TPS) require a high volume compressed air supply to run the TPS. To achieve this, pipes must cross the balance (see Fig. 6.64).

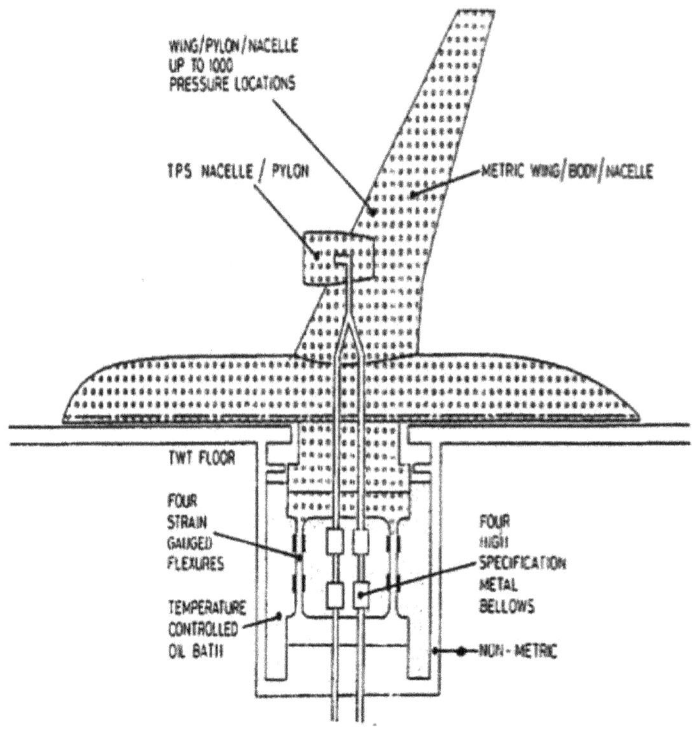

Fig. 6.63 Principle semi-span test setup with turbine power simulator at Aircraft Research Association (UK)

For semi-span tests (see Fig. 6.63), there is normally enough space inside the balance for the pipeline crossing. However, with an internal balance there is no space inside the balance so the pipes must cross the balance on the outside. To minimize the reaction force by the force shunt of the pipes, elastic compensators are built into the pipeline. The reaction forces by these compensators can be measured by an additional calibration of the balance with the airline system mounted on the balance. Additional to the elastic reaction forces, forces by the mass flow and pressure must be taken into account and calibrated.

Figure 6.63 shows the half-model test setup of Aircraft Research Association (ARA) in Bedford, United Kingdom, to measure aerodynamic forces together with the thrust by a TPS. The reaction forces are reduced by metal bellows that are elastic and exhibit highly repeatable reaction forces. Noteworthy with this setup is that the entire balance is filled with oil to stabilize it at a certain temperature.

When jet engine simulators are used in the model the air leaves through the jet engine. For tests with turboprop engines the compressed air is expanded in the engine and the expanded air must also be returned over the balance to an external outlet. For the tests with the A400 in the DNW-LLF tunnel the balance has an air line bridge and an air return bridge (Fig. 6.64 and [15]).

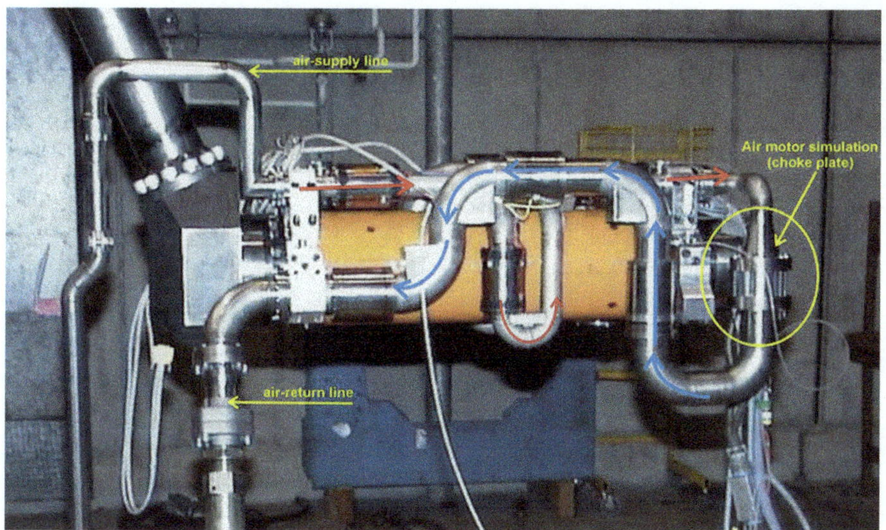

Fig. 6.64 Airline bridge on DNW-LLF balance (DNW, Airbus)

6.5 Life Time and Fatigue Calculations

Wind tunnel balances are instruments which are used over decades; hence, stress by static and dynamic load accumulates over the years and this can lead to cracks and fracture by fatigue of the material. Usually balances are designed for *infinite lifetime* if only the static and dynamic design loads occur during the test. With moderate overloading or continuous dynamic loading, partial plastic deformation can occur and this causes cracks that propagate under the dynamic loading until the balance totally fails. Cryogenic balances can also be "dynamically" stressed by temperature changes and this has to also be taken into account for lifetime calculations.

A lifetime calculation has to be performed to prevent a balance fatigue failure during a test. For this calculation the load history over the service life of a balance and the intended load history for the test must be known. Both together are used to calculate the stress history (stress values σ_a, S_a and number of cycles N) including the next test. The stress history must be calculated separately for all the critical parts of the balance because the combined stress for a dedicated balance section is different on every combined load. This stress history must be compared with the *Wöhler* curve (S-N-Curve), which gives the values for the material fatigue strength of the material. If the stress collective is below the *Wöhler* limit, the balance can be used for the test (see Fig. 6.65).

To determine the load history, a balance load monitoring system must be installed in the tunnel data acquisition system so that the loads can be recorded.

If such a system is not installed, a significant change in the zero signals can be a sign of damage and the balance must be checked. Visual inspection or other methods for

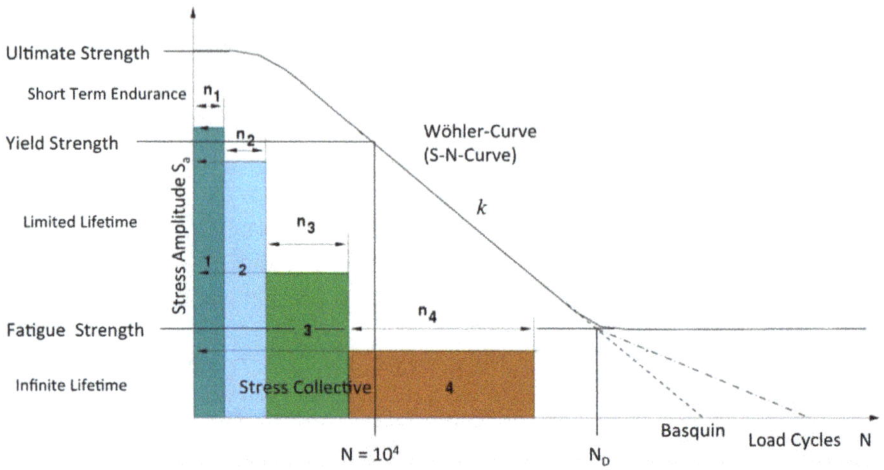

Fig. 6.65 Typical Wöhler diagram with lifetime regions

crack checking can only reveal whether if the balance is beyond repair. A calibration of the load where the changes occur, to control the stability and repeatability of the signal, shows if the balance can still be used, but a prediction of further lifetime is not possible.

To design balances for a "*limited lifetime*" the "*load future*" of the balance must be known. This means for the intended lifetime the collective of the static and dynamic load must be known. For a wind tunnel balance this is nearly impossible because normally which tests will be performed in the tunnel for a long period and what will happen during these tests is not known in advance.

Special balances like rotary balances to measure the trust of an engine have a certain load spectra in the test and so for this kind of balance a design for limited lifetime can be performed.

In general, a cost reduction is not a valid reason to design for a limited lifetime, because the material costs for a balance are low related to the other manufacturing costs. Space constraints may lead to a design with higher local stress and thus, to the risk of static and dynamic failure in such a case. The recording of the loads is absolutely necessary to predict the lifetime of the balance prior to a test and to prevent a failure.

Summarizing these introductory remarks for lifetime or fatigue calculation it is necessary to have:

1. The material fatigue characteristic, *Wöhler* curve (S-N-curve)
2. The load history and load future or load collective over intended time of use
3. The relation of load to stress in high stressed balance areas

6.5.1 Determination of the Wöhler-Curve (S-N-Curve)

The *Wöhler* curve characterizes the material fatigue properties, which means the endurable stress over the number of load cycles. The maximum endurable stress for static load only (zero load cycle) for a material is the ultimate stress (R_m). So this is the first point of the curve on the ordinate. The next value that is needed is the value of the fatigue strength (S_D) for the material and the number of load cycles (N_D) for the fatigue strength. According to the approach of *Basquin* or *Stüssi* the slope k for the section with limited lifetime of the *Wöhler* curve now can be calculated. Most commonly used is the *Basquin* approach.

$$\text{Basquin} \qquad N = N_D\left(\frac{S_a}{S_D}\right)^{-k} \quad ; \quad \sigma_a = S_a \tag{6.154}$$

This equation is only valid for a material without an existing crack. For a wind tunnel balance any further use of the balance with an existing crack is useless so the *Basquin* equation is a sufficient approximation for the *Wöhler* curve (S-N curve). The material constants S_D and N_D must be determined by a fatigue test. For infinite lifetime calculations of a balance made from Maraging steel S_D can be assumed as $R_m/2$. This is always the case for balances with a good measurement performance. Some wind tunnels specify the allowed maximum stress level according to this assumption [23]. In this case the maximum stress in the balance under combined and dynamic load is not allowed to exceed $R_m/2$.

6.5.2 Stress Collective

The applied loads over time and the intended load over time are the basic requirements to calculate the stress collective for the lifetime and fatigue calculations. The second requirement is the relation of the loads to the maximum stress in the balance. For this requirement the results of the finite element calculation can be used. Using a finite element calculation the areas where the highest stresses occur are well known and the correlation to the load is assumed linear because the balance will be used only in the linear stress range. Using these linear relations the stresses according to the load combinations can be calculated. The result is a stress versus time function with many different stress levels acting more or less often. For further use this function must be sorted into stress classes and the number of stresses in each class must be counted. To do this a so-called *Rainflow* counting algorithm is used. This algorithm is a standard now for the fatigue analysis. This algorithm is described in detail in [1].

The result of the procedure is a step function of dynamic stress levels over the number of load cycles. An example for the step function can be seen in Figs. 6.65 and 6.66.

6.5.3 Linear Damage Accumulation (Miner Rule)

The basic assumption of the Miner Rule is that the damage D of a component by dynamic loading is the sum of the fractional damages ΔD of the different load classes.

$$\text{Miner Rule} \qquad D = \sum_{i=1} \Delta D_i = \sum_{i=1} n_i / N_i \tag{6.155}$$

The fractional damage ΔD is a function of the fractional stress S_{ai} and the fractional mean stress S_{mi}. A further assumption for the application of the *Miner Rule* is that only fractional stresses S_{ai} greater than the fatigue strength S_D and smaller than the ultimate strength R_m contribute to the damage accumulation ($R_m > S_{ai} > S_D$). This assumption can be made because stresses smaller than the fatigue strength S_D can act infinitely without having an effect on the damage and stress greater than the ultimate strength R_m, which would destroy the component immediately.

Up to now only alternating stresses S_a have been accounted for in the damage accumulation. Normally the dynamic stress ($S_m \pm S_a$) contains also mean stress values S_m. The influence of the mean stress values S_m can be taken into account by transforming the mean stress S_m plus the amplitude $S_a(S_m)$ to an equivalent stress amplitude $S_a(S_m = 0)$ according to the following equations.

$$\begin{aligned}
S_a(S_m = 0) &= S_a(S_m) \cdot (1 - M) \quad for \quad S_m/S_a \leq -1 \\
S_a(S_m = 0) &= S_a(S_m) \cdot (1 + M \cdot S_m/S_a) \quad for - 1 \leq S_m/S_a \leq 1 \\
S_a(S_m = 0) &= S_a(S_m) \cdot [\frac{(1 + M)}{1 + M/3}] \cdot (1 + M/3 \cdot S_m/S_a) \\
&\quad for + 1 \leq S_m/S_a \leq 3 \\
S_a(S_m = 0) &= S_a(S_m) \cdot [\frac{(1 + M)^2}{1 + M/3}] \quad for \quad 3 \leq S_m/S_a
\end{aligned} \tag{6.156}$$

In these equations the mean stress sensitivity M depends on the material and for high strength steels it is about $M = 0.6$.

After that the *Wöhler* curve, which is valid only for alternating stress amplitude S_a, can be taken for comparison. To estimate whether a balance can still be used or not, an additional criterion for the maximum damage is needed. With respect to total failure the value for the damage accumulation must be smaller than one ($D < 1$).

The continued use of a balance according to this criterion is not sufficient because the quality of the measurement certainly becomes worse before the balance is destroyed. So these criteria can only provide the information, whether the balance with a certain load history will survive the next test with an intended load schedule. A damage criterion for the loss of measurement quality is not yet defined and depends on the user and his quality requirements.

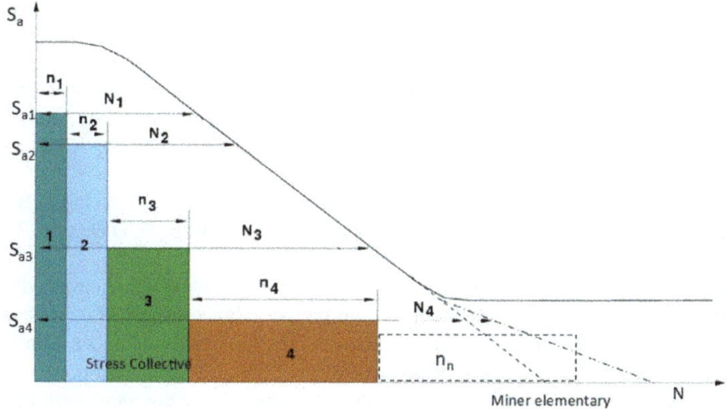

Fig. 6.66 Wöhler curve with stress collective for Miner Rule

6.5.4 *Summary*

The above description for lifetime and fatigue calculation is only a rough demonstration of how to perform such a calculation and what tools are required. A detailed fatigue life analysis has to take many more influencing factors into account. The calculation of the allowed fatigue stress values (*Wöhler* curve) must also consider fabrication methods, like EDM or surface finish influences and temperature procedures, which can reduce or increase the fatigue strength. Mostly for the fatigue strength, only un-notched specimens are tested and the influence of notches and the surface condition reduce these values.

For the stress determination it has been shown that the ratio between load and real stress at the highest stress area is not an easy task because of the complex deformation of a balance. Finite element calculations provide good estimates if the geometry and the load application are simulated correctly. Otherwise additional stress concentration by sharp edges etc. increases the real acting stress.

The damage criterion for the total fracture of a balance is $D = 1$. For other failures like the unwanted increase of uncertainty in measurement, no criterion is available at the moment.

Consequently, a lifetime prediction for the correct use of a balance is rather vague with the available information. The effort for the investment into a load monitoring and lifetime prediction system seems therefore not very attractive. Nevertheless, such a system offers some other useful information that cannot be obtained otherwise.

If loads and signal of a balance are recorded continuously, high loads and high signals provide information for the reason of changes in zero output. Changes in zero output of the balance always must be analyzed because they are an indication of plastic deformation somewhere in the balance. To localize such deformations, the overload and the corresponding signal can be used to identify the area of overload in the balance.

References

1. ASTM, E.: 1049-85. Standard practices for cycle counting in fatigue analysis (2005)
2. Bazin, M., Milliat, F., Girard, D.: ONERA balances and dynamometers. In: International Symposium on Strain Gage Balances, Hampton, VA, 22–25 Oct 1996, ONERA, TP, 1996-189 (1996)
3. Beitz, W., Küttner, K.: Dubbel, Taschenbuch für den Maschinenbau. Springer, Berlin (1983)
4. Ewald, B.: Axialkraftfehler durch Temperaturgradient. Tech. Rep. TUD Bericht, TU Darmstadt (1984)
5. Ewald, B.: Interne Windkanalwaagen. Tech. Rep. Report A38/99, TU Darmstadt (2000)
6. Grenat, G.: Real time PC monitoring of strain gauge balance in ONERA large wind tunnels using finite element design tools. 4th Symposium on strain gauge balances, San Diego (2004)
7. Haibach, E.: Betriebsfestigkeit. Springer (1989)
8. Hufnagel, K., Ewald, B.: Temperature effects and temperature compensation of internel balances for cryo wind tunnels. Tech. rep, TU Darmstadt, Darmstadt (2001)
9. Hufnagel K.; Zhai, J.E.B.: Advanced Half Model Wind Tunnel Balances. AIAA 2000-2385; AIAA Conference Denver Colorado (June 2000)
10. Hufnagel, K.: Untersuchung von Axialkraftelementen. Tech. Rep. Institutsbericht A32/87, TH Darmstadt, FG Aerodynamik und Messtechnik, Darmstadt (1987)
11. Johnson, G.: FFA Measures for Control of Wind Tunnel Balance Load and Fatigue Life. Bedford (1999)
12. Karkehabadi, D.J.: The structural integrity of balances modifications and factors affect them. 4th Symposium on strain Gauge balance, San Diego (2005)
13. Köhler, M., Jenne, S., Pötter, K., Zenner, H.: Zählverfahren und Lastannahme in der Betriebsfestigkeit. Springer-Verlag (2012)
14. N.N.: VDI Guide Line 2230, Systematic calculation of highly stressed bolted joints or joints with one cylindrical bolt. VDI (2015)
15. Philipsen, I., Hoeijmakers, A., Alons, H.: A new balance and air-return line bridges for DNW LLF models. National Aerospace Laboratory NLR (2004)
16. Pyttel Varfolomeyev, B.: Fracture Mechanics Proof of Strength for Engineering Components. VDMA, FKM-Guideline (June (2004)
17. Rhew, R.D.: A fatigue study of electrical discharge machine (EDM) strain-gage balance materials. In: International Congress on Instrumentation in Aerospace Simulation Facilities, pp. 477–487. IEEE (1989)
18. Ritzel, F.: Konzipierung eines Verfahrens zur laufenden Kontrolle der Betriebsfestigkeit von Mehrkomponenten Kraftsensoren. Diplomarbeit 363/95, TU Darmstadt, FG Aerodynamik und Messtechnik (1995)
19. Roak R.J.;Young, W.: Formulas for Stress and Strain. McGraw Hill (1985)
20. Schimansky D.; Vohy, T.: ETW Model Design Handbook. ETW Report Nb. ETW/D/95004,Revison A (August 1999)
21. Soeder, R.H., Roeder, J.W., Stark, D.E., Linne, A.A.: NASA Glenn Wind Tunnel Model Systems Criteria (2004)
22. Vos, H.: Including fatigue aspects in balance design. 6th Symposium on Strain Gauge Balance, Zwolle (2004)
23. Wigely, D.: ETW materials guide. ETW GmbH, ETW/D/95005 (1996)
24. Zhai, J.: Analyse und Optimierung der internen Windkanalwaagen mit FEM. Ph.D. thesis, TU Darmstadt (1996)

Chapter 7
Balance Material and Fabrication Methods

7.1 Balance Material

The basic requirements for the balance material are

- Yield and tensile strength should be as high as possible
- Distribution of elasticity should be as homogeneous and isotropic as possible.

These basic requirements ensure that the material characteristics are as linear and stable as possible in the range over which they are used.

The requirement for linearity is not that stringent, because non-linearity can be taken into account by calibration, but linearity of the material itself is mostly combined with its stability and hysteresis. Stability is the ultimate requirement for the repeatability of a balance and repeatability is the most prominent precondition for a good calibration.

A rule of thumb is that the ratio between the stress under the strain gauge area and the tensile strength should be 3–8. The stress in the strain gauge area should be the maximum stress in a balance or force transducer. The last requirement is particularly hard to fulfill in modern balance designs because the available space for the balance in a wind tunnel model has become increasingly smaller, resulting in a higher required material stress level in the balance.

Together with the choice of the strain measurement sensor, which determines the practical measurable strain level (ε_{max}) and this rule of thumb, the ratio of yield strength (σ_{yield}) to the modulus of elasticity (E) of the material is determined.

$$\frac{\sigma_{\text{yield}}}{E} = 3 \dots 8 \cdot \varepsilon_{max} \tag{7.1}$$

If metal foil strain gauges are used, only metals for the balance body material will fulfill this requirement. Hence, usually high strength steels or high strength aluminum, copper-beryllium or titanium are used for force transducers in this case.

The primary selection of the material is focused on the mechanical properties for the measurement. On the other hand balances must be fabricated and this is not

K. Hufnagel, *Wind Tunnel Balances*, Experimental Fluid Mechanics,
https://doi.org/10.1007/978-3-030-97766-5_7

possible without some kind of machining. The properties of a metal for good machining often contradict the ideal properties for measurement. High tensile strength is combined with a high hardness of the material and hard material cannot be easily machined. Therefore the material must be machined in softer conditions and hardened somewhere in the manufacturing process. However, hardening and annealing procedures influence the mechanical properties of the material and the tolerances of the geometric dimensions. Heat treatment processes therefore must be integrated into the fabrication sequence to optimize the final mechanical properties. Determining the optimized sequence is one of the most challenging tasks when a new material for a balance is used. This is the reason why well-known materials are generally used to build a balance. Often balance manufacturers prefer special materials they have used in the past and avoid changing the material due to incalculable risks.

7.2 Material Characteristics

The characteristics of the balance body material limit the quality of the calibration. Assuming that the strain sensor is contacted perfectly to the balance body, the sensor should copy exactly the stress situation in the bonded area, so the material itself determines the limit of repeatability. The material characteristics that limit a perfect repeatability are hysteresis and creep. While there are several methods to compensate creep by the design of the strain gauge, there is no chance to compensate hysteresis using the strain gauge.

In the following subsections some further requirements that should be taken into account for the choice of the material are discussed.

7.2.1 Tensile Strength and Yield Strength

A linear correlation between stress and strain over the entire range of encountered strain is only possible if there is no remaining deformation after relaxation. This means that many metals, especially steels, are not well suited for balance material because their exact linear behavior stops at levels of approximately 1/3–1/5 of their tensile strength. The yield strength normally used to define the end of the elastic behavior is set to a remaining strain level of 0.2%. That means for a transducer designed for tensile stress under the gauge, the remaining signal is about 4 mV/V. The sensor does not return to zero, but indicates twice the full-scale output of a properly designed sensor. The transducer has become a trailing pointer for the remaining strain under the gauge and is far away from being usable for force measurements. Therefore steels should be used with a tensile strength as high as possible. For some other reasons (e.g. small hysteresis, low creep or small loads) other materials can be more suitable for a specific application. Some typical alternatives are mentioned in

Table 7.1 Balance body material properties at ambient temperature

Material	Yield strength [N/mm^2; MPa]	Tensile strength [N/mm^2; MPa]	Young's Modulus [N/mm^2; GPa]	Usable range under gauge [N/mm^2; MPa]
Maraging steel	1430 ...2300	1500 ...2400	181 ...202	270 ...300
Copper-Beryllium	980	1310	123	185
Titanium alloy	875 ...1170	1000 ...1240	108 ...114	160 ...175
Aluminum alloy	270 ...450	300 ...515	703 ...710	105 ...110
Nickel alloy	1200	1400	199	270 ...300

Table 7.1. The final choice of most appropriate material requires a detailed study of the specific data sheets of the materials.

7.2.2 *Dynamic Stability and Fracture Toughness*

A high dynamic stability and good fracture toughness are necessary to ensure a long lifetime of the balance. Most balances are not designed with respèct to notches. Normally the stress level in a balance is very low so that the design can be optimized with respect to minimum interactions and so on. Stresses by concentrations caused by such designs are normally significantly below the value of the yield strength. If the stress level in the balance approaches 1/3 of the yield stress, notch stresses can easily become higher than the tensile strength and the risk of failure is very high.

Wind tunnel tests with a model in separated flow conditions are always combined with high or very high dynamic loads, so a good fracture toughness of the material ensures sufficient buffers for dynamic loading and a long lifetime of the balance.

7.2.3 *Young's Module*

The modulus of elasticity (E) should be stable in the specified temperature range. If this is not the case the temperature dependency should be linear. This linearity enables a compensation of the temperature dependent drift of the sensitivity caused by the change of the modulus of elasticity. To achieve a balance with a high stiffness is one of the most wanted characteristics. For a given available space the material with the highest Young's modulus can achieve this. A high modulus of elasticity also enables to build a stiff balance at a higher allowed stress under the gauge. Materials with a high modulus of elasticity should be preferred. Unfortunately the modulus of elasticity of steels decreases with an increase in tensile strength. However, if the lateral space does not limit the balance dimensions, the lowest possible Young's modulus allows designing a balance with high stiffness and sensitivity.

7.2.4 Thermal Expansion Coefficient

The thermal expansion coefficient should be isotropic and constant. This enables a good compensation of zero drift of the strain gauge bridges.

7.2.5 Hysteresis

Hysteresis of the body material cannot be compensated by the application of the strain gauge; therefore, the material with the lowest hysteresis will enable to build force transducers of the highest precision. There is very little information about the hysteresis of materials, especially about their heat treatment to achieve low hysteresis. An investigation of hysteresis properties for the balance materials was made at the TU Darmstadt. The results can be found in [3]. A general result of this investigation was that heat treatment processes that refine the grain of the material can significantly reduce hysteresis of some steels. The materials with the lowest hysteresis are titanium alloys and copper-beryllium.

7.2.6 Creep

Creep and hysteresis properties are linked together. In [3] all measures that reduce hysteresis also reduce creep of the material. Related to hysteresis there is one major difference. If the creep behavior of the material is well known, the shape of the strain gauge can compensate this. Some gauge manufacturers offer creep compensating gauges, but the disadvantage is that only a few sizes of gauges offer this possibility and the balance designer is limited to these geometries.

7.2.7 Heat Treatment

In general it can be said that a well adapted heat treatment process is necessary to obtain the best material quality after the finish of the balance manufacturing process. There is no common recommendation for all materials, but all materials require more than one heat treatment process during the fabrication. Thus, the manufacturing process must be interrupted by one or more heat treatments and becomes rather complex.

For example, annealing procedures cannot be performed with the final dimensions because the internal stresses disappear and deformations of the balance result. To return the balance into a symmetrical and well-defined shape, some oversize has to be foreseen.

Hardening processes are normally combined with a certain shrinkage of the material and close tolerances at the interfaces must be fabricated after the hardening process. Here also some oversize must be foreseen. Machining in hardened conditions is relatively expensive and some of the machining at hardened conditions reduces the strength properties of the material. This is counterproductive to what has been achieved at earlier stages and must be reduced to a minimum.

For some materials the heat treatment process is well documented. If a new material is selected for balance manufacture, an comprehensive investigation must first be conducted to obtain the best mechanical properties at the end of the manufacturing process.

7.3 Maraging Steels

For internal wind tunnel balances this is the most important group of steels used. Maraging steels offer a very high tensile strength of up to 2400 N/mm^2 with a yield strength of about 5% below tensile strength and a modulus of elasticity of about 190 GPa (Table 7.2).

The first two steels in this list cannot be used for cryogenic applications because the fracture toughness is poor at low temperature, but their yield strength is the highest of the suitable balance materials. The other two steels (1.6357 and 1.6359) are well tested and approved for ambient and cryogenic applications. The detailed material properties should be taken from standards or handbooks given in the literature list of this chapter.

Maraging steels of this quality and in the quantity needed for a balance are not easy to purchase. The quantities are small and the price is normally slightly higher then the price of titanium alloys. If a distributor is found, the small quantity of material is normally cut from a large block and than forged to the required dimensions. This process leads to highly inhomogeneous material with respect to its measurement qualities and therefore a heat treatment process is mandatory before the manufacturing process starts.

Table 7.2 List of maraging steels used by TU Darmstadt

Material Nb.:	Name	Name	Yield strength at room temp.	Temp. range
	USA	Europe	MPa	°C
1.6354; 1.6358	Maraging 300	X2NiCrMo 18 95	1900	−40 ...60
1.6356	Maraging 350	X2NiCrMo 18 12 4	2300	−40 ...60
1.6357	Maraging 200	X2NiCrMo 18 83	1430	−196 ...60
1.6359	Maraging 250	X2NiCrMo 18 85	1760	−196 ...60

7.3.1 Heat Treatment of Maraging Steels

There are differences in the temperatures and the time periods that must be set in the different phases of the heat treatment of Maraging steels. The detailed values must be taken from the individual material specifications. What all Maraging steels have in common are sequences of heat treatment sections in combination with other manufacturing states to obtain the best material properties after the final manufacturing process. The combination of these phases is shown in Fig. 7.1.

In the state of delivery, the material is normally inhomogeneous with respect to its measurement quality and an annealing process followed by a water quenching is mandatory before any other step is made. To insure a fine grain structure, several heating and water quenching cycles should be conducted. The number of cycles and the holding temperatures differ slightly with the material, but are in the range of 1 h at around 600 °C. The grain refinement process is followed by another annealing step (≈800 °C for 2 h) with air cooling to soften the material for the machining of the parts. Every machining step, i.e. milling, drilling, electro discharge machining (EDM) and welding, has to be made in the soft condition. If welding is done, there must be an additional annealing (≈800 °C for 2 h) after the welding. To keep the material free of oxide scale this process should be performed in a vacuum oven, or in a closed box filled and ventilated with an inert gas (e.g. argon). After this annealing the balance can be machined to its final dimensions, except for all surfaces that demand tight tolerances to fit into the adapters of the model or sting, since during the hardening process (≈500 °C 6 h) the material will shrink. The shrinkage ratio is about 0.07%. This is too much for fittings, but this is not of any significant influence on the functionality of the balance. So final grinding to obtain the correct tolerances

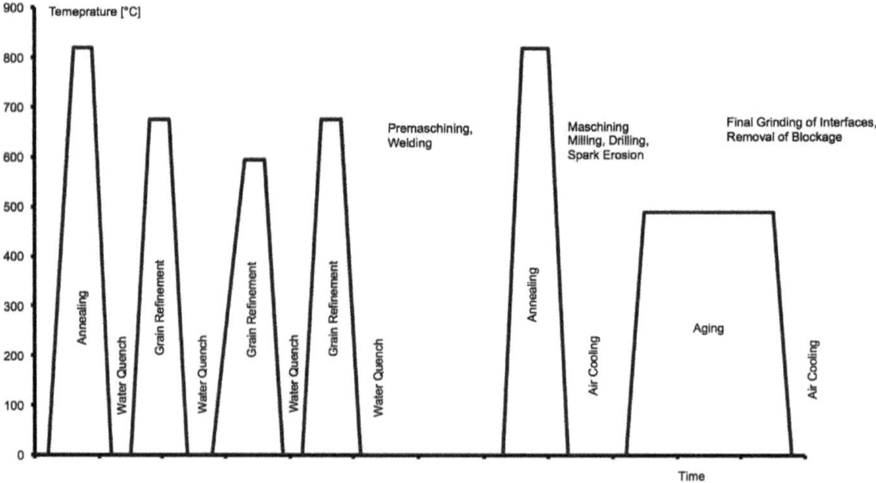

Fig. 7.1 Integration of heat treatment into manufacturing process for Maraging steel

for the fittings is the only mechanical process performed after the hardening. If it is handled this way, the balance will have maximum tensile stress at the best fracture toughness.

7.4 Stainless Steels

One of the disadvantages of high strength Maraging steels is that they corrode under the influence of water or in a high humidity environment. This was one of the difficulties that had to be overcome during the development of the cryogenic balances, where the high strength of the balance material was necessary to enable balances with the required dimensions. If the specified dimensions allow using steels with a yield strength over 1000 MPa it is appropriate to use a high strength stainless steel. Most of these steels are used for aircraft applications. A list of these steels is given in Table 7.3.

7.4.1 Heat Treatment of Stainless Steels

The rules for solution treatment, precipitation hardening and ageing differ from material to material, so these are discussed separately in the following subsections.

7.5 Copper Beryllium

The amount of beryllium is 2% in the copper-beryllium alloy (Cu Be 2) with the best mechanical properties. Copper-beryllium (1300 MPa) has much lower values for tensile strength than the Maraging steels (2000 MPa), therefore it is not a real alternative if high strength material is required. Related to mechanical properties, it is an alternative to titanium alloys. Copper-beryllium has a modulus of elasticity of

Table 7.3 High strength stainless steels

Material	German code	USA code	Yield strength [MPa]	Tensile strength [MPa]
X 3 CrNiMo 13 8 2	1.4534	PH 13-8 Mo	1150...1400	1220 ...1650
–	–	PH 14-8 Mo	1180 ...1600	1400 ...2000
X 5 CrNiCuNb 17 4 4	1.4548	17-4 PH	1000 ...1170	1070 ...1310
X 7 CrNiAl 17 7	1.4564	17-7 PH	1030 ...1310	1240 ...1720
X 7 CrNiMoAl 15 7	1.4574	PH 15-7 Mo	1170 ...1380	1310 ...1890

about 125 GPa, which is slightly more than that of titanium alloys (111 GPa). Also the tensile stress (1300 MPa) is slightly higher than that of the titanium alloys. So both materials are in the same region for these values. The main difference is in the physical properties. The heat conductivity of copper-beryllium (80 W/mK) is nearly 10 times higher than that of titanium alloys (8.17 W/mK) and even compared to Maraging steels ($\approx$20 W/mK) the heat conductivity of copper-beryllium is 4 times higher. This makes copper-beryllium very attractive for balances that are exposed to rapid changes in the operating temperature, because the balance follows the temperature changes relatively rapidly.

The mechanical machining should be performed in the solution-annealed condition, where the highest cutting ratio is possible. The cold deformation in the cutting area hardens the surface of the material and so the next cut requires a minimum cutting depth of about 0.2 mm so that the cut is made in the uninfluenced area of the material. For all other hardened conditions there are small cutting depths of about 0.02–0.05 mm.

During cutting the inner stress is released and therefore relative large deformations can occur. Normally balances require tight tolerances and therefore a symmetrical cutting is required to minimize these deformations. Symmetrical machining requires a frequent change of clamp that extends the whole machining process, but enables to stay within the given tolerances. Copper-beryllium dust and vapor is dangerous to health because it can cause cancer, so special precautions, like a fume cupboard at the EDM machine must be foreseen.

7.5.1 Heat Treatment of Copper-Beryllium

If the copper-beryllium wrought material is not delivered in the solution-annealed condition it should be brought to this condition by heating it up to 780 °C for half an hour, followed by a rapid water quenching. Then the material is in the soft condition and can be machined easily. After finishing most of the machining, the material must be hardened by heating it up to 315 °C for 3 h and cooled by air. The hardening process causes shrinkage of about 0.6%. This has to be taken into account by a corresponding over sizing to enable interfaces with close tolerances.

7.6 Titanium Alloys

Among the titanium alloys, the most suitable for balance or transducer manufacture is the alloy TiAl6V4 because it is possible to perform a precipitation hardening. So it can be machined in soft condition and hardened and aged after finishing the mechanical work. The material can also be machined at hardened conditions at lower cutting speeds. Tests with the hardened material [3] showed that TiAl6V4 has the best performance related to creep and hysteresis of all tested materials. If TiAl6V4 is used for a force transducer it should be possible to build it with very high precision.

7.6.1 Heat Treatment of TiAl6V4

The material is forged so a solution annealing is mandatory before any other machining. This can be done by heating the material to 700–800 °C, staying there for 1–8 h, followed by cooling down in air or oven cool down. The alloy is weldable (electron beam welding is preferred) and a stress relieve annealing should follow the welding: 480–600 °C for 1 to 4 h followed by cooling down in air. Hardening is done between 84 and 940 °C for 15 min to 1 h, followed by water quenching afterwards, ageing by 480–600 °C for 2–8 h followed by a cooling down in air.

7.7 Aluminum Alloys

Two aluminum alloys are usually used for force transducer applications. One is Al-Cu 4 Mg 1 (2024 T4), which is widely used in the production of aircraft and has the highest strength of all aluminum alloys. The two alloys can be used for force transducer fabrication if low loads related to the necessary dimensions are required. The modulus of elasticity is about 70,000 MPa, nearly 1/3 of the modulus of elasticity of steel so the signal on a given structure is nearly three times higher than that on steel. The disadvantage is that also the deformation is three times higher.

If a transducer design requires very thin dimensions when steel is used and the available space allows a larger design, then aluminum alloys are also an option to realize more practical dimensions of the transducer.

7.7.1 Heat Treatment of Aluminum Alloys

First Al-Cu 4 Mg 1 (2024 T4). The identification "T4" is a shortcut for the heat treatment precept and it means that the material is solution heated (495 °C ...505 °C), water quenched and naturally aged for 5–8 d.

Second Al-Zn 6 MgCu (7075 T6) is solution heated at 470 °C ...480 °C, water quenched and subsequently artificially aged at 165 °C ...185 °C for 4–6 h.

7.8 Balance Body Fabrication Methods

To build a wind tunnel balance a wide variety of metal cutting, erosive and welding methods are used. The widely used method is a combination of milling, turning and electric discharge machining. In the future, rapid prototyping of metals may be an alternative that enables the fabrication of complex structured balances.

Manufacturing internal balances from one piece of material is probably the most complicated case. For multi-component balances, the manufacture of the balance body and sensitive elements are separate. The most complicated part is then the assembly of the different parts into one balance.

7.9 One Piece Fabrication

One-piece balances offer the possibility to obtain a very homogeneous balance body. This means the structure of the material can be tuned to the best measurement qualities. The structural quality of the material sets the limits for the repeatability and stability of the measurement. In this respect material hysteresis and material creep are the characteristics of interest. The whole manufacturing must be optimized to achieve at the end of the process the best value for these two characteristics. So the whole manufacturing process comprising mechanical machining and thermal treatment of material has to be optimized. For mechanical applications, optimized processes exist and are commonly known, but for the optimization of measurement characteristics only little knowledge exists. That is the reason why every new material requires significant research to optimize the process and thus new materials are used sparingly. The integration of heat treatment into the fabrication of a balance is described in Sect. 7.3.1. Now some influences of the machining on the material characteristics are described.

In principle the machining effect on the material characteristics depends on how much and what kind of surface stresses are introduced by the machining process or if there are chemical variations introduced to the surface. While turning and milling introduce only surface stresses that can be removed by an annealing process, electro-discharge machining also carbonizes the surface, which leads to a harder and more brittle surface layer. Additionally, by the partial melting and quenching of the material, surface tension stresses are introduced. This effect is strongly correlated to the material removal rate. The higher the rate, the more and deeper the carbonizing of the surface layer and the higher the tension surface stresses. Electro or chemical polishing can remove carbonization. It is more efficient to reduce the material removal rate for the last few tens of a millimeter at the end of the EDM removal. This leads to a high surface quality with a minimum of surface stresses [2, 4]. Some attempts to enhance quality by shot peening failed because shot peening successfully reduces the surface stress but does not remove the carbonization. To enhance the material characteristics for the measurement has the positive side effect to improve also the fatigue life properties of the material.

Laser sintering of balances will be an option for the near future and certainly enables design variations that are not possible today. Some balances that can today only be built from several parts will be built as a one piece balance and so combine the homogeneous material characteristics with the advantages of a sophisticated structure.

7.10 Multi-component Balances

Some designs are not possible to be built from as single piece of material. Especially the widely used force balance needs to be built from several pieces because the sensors are hidden in the inside structure under the outer cylinder. To assemble this balance, the ready-gauged sensors must be connected in some way with the rest of the body. This can be done only by a mechanical connection like screws and clamps. Other thermal bonding like soldering, brazing or welding is not possible because the thermal stress on the gauges will be too high. If gauging can be done after the assembly process, thermal bonding can be integrated into the assembly. Soldering, brazing and welding are used.

The design aim for soldering and brazing should be a low stress in the contact area to insure no additional hysteresis by local elastic deformation in the bonding. Soldering therefore can be used for balances for very low loads. Brazing allows a higher stress level, but at least it is lower than the possible stress in the balance material.

Welding allows the same stress level as in the rest of the material. The problem of welding is the change in the material structure and the residual stresses by the thermal process. Changes in the material composition can be avoided by using a welding process that uses no wire or a filler wire of the same material. Residual stresses are correlated to the amount of heat energy that is needed for the welding. A welding process can reduce residual stresses with low heat input like electron beam welding. At TU Darmstadt numerous balances have been successfully built by electron beam welding (Fig. 7.2).

Together with a number of annealing processes this method offers the opportunity to completely remove the changes in the material when welding. So there is no additional hysteresis or creep and the maximum stress in the weld seam can be the same as in the rest of the material.

7.11 Surface Protection

Most of the high strength steels that are used for balance fabrication are not stainless steels and therefore some surface protection is necessary to prevent the balance from rust and other corrosion.

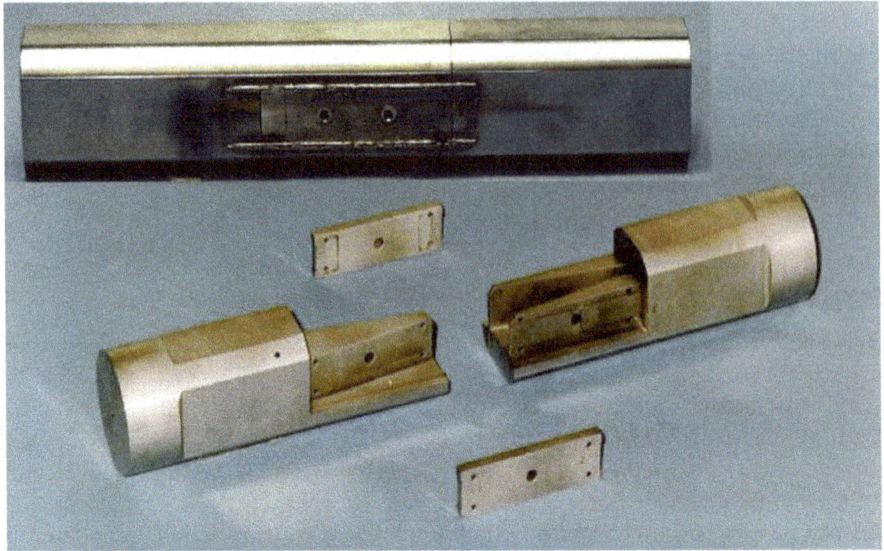

Fig. 7.2 Welded balance, parts and complete body after welding

Nickel or zinc plating is successfully used for Maraging steels. Some attention has to be given to the preparation of the balance for the chemical process. At first a coat to prevent plating must cover all areas, where strain gauges or other items are applied and the surfaces of the interfaces. The prepared balance body then must be perfectly cleaned with different solvents like MEK (Methylethylketon), isopropanol and acetone to remove all contaminants. To obtain best results the galvanizer should be advised that the electrolytic bath must be relatively new and pure. After the gauge application, the same coating as it is used for the gauges must protect the non-coated areas.

References

1. Aerospace Structural Metals Handbook. CINDAS/Purdue University West Lafayette IN (2001)
2. Jutzler, W.: Funkenerosives Senken; Verhaltenseinflüsse auf die Oberflächenbeschaffenheit und die Festigkeit des Werkstücks. Ph.D. thesis, RWTH-Aachen (1982)
3. König, H.G.: Eigenschaften metallischer Meßkörper für die Windkanalmeßtechnik. TU Darmstadt Ph.D. Thesis (1993)
4. Kumar, S., Singh, R., Singh, T., Sethi, B.: Surface modification by electrical discharge machining: a review. J. Mater. Process. Technol. **209**(8), 3675–3687 (2009)
5. Rush, H.F.: Grain-Refining Heat Treatments to iimprove Cryogenic Toughness of High-Strength Steels. NASA TM 85816 (1984)
6. Wagner, J.A.: Mechanical behavior of 18 Ni 200 grade maraging steel at cryogenic temperatures. J. Aircr. **23**(10), 744–749 (1986)

Chapter 8
Strain Measurement

8.1 Strain Gauge

The basic technique to measure force in any kind of wind tunnel balance is the measurement of the strain on an elastic spring that is deformed by the aerodynamic loads acting on the wind tunnel model.

In this chapter the fundamentals of strain measurement with strain sensors are described. For wind tunnel balances two major types of strain sensors are used. The most widely used is the wire strain gauge sensor and the second most widely used is the semiconductor strain gauge. Although in principle optical strain gauges offer even higher resolution, to date balances constructed using such sensors have not operated with the same precision as conventional balances.

8.1.1 Wire Strain Gauge Fundamentals

The wire strain gauge is based on an electro-mechanical effect *W. Thomson (Lord Kelvin)* found in 1856. *Thomson* measured that the electrical resistance of a metal wire can be correlated to the strain in the wire while stressed.

This effect was used by *E. Simmons (Caltech)* in 1936 and *A. C. Ruge (MIT)* in 1938 to invent the wire strain gauge. Simmons was the first to build a force transducer based on the wire strain gauge technique and *A. C. Ruge (MIT)* used his wire strain gauges to perform experimental stress analysis. The strain gauge of *A. C. Ruge* was very successful because it was inexpensive and easy to handle. The industry demand was immediately huge and in 1952 a technique was patented to produce the foil strain gauge. No longer was a wire glued to a carrier foil, now a thin metal foil was glued on the carrier and the contour of the wire was etc.hed out of the metal foil by a photochemical process. This technique is still in use to produce foil strain gauge sensors, a very precise sensor with high resolution at a low price.

K. Hufnagel, *Wind Tunnel Balances*, Experimental Fluid Mechanics,
https://doi.org/10.1007/978-3-030-97766-5_8

The physical principle of a wire strain gauge is the change of the electrical resistance by the strain applied to the gauge. The electrical resistance of a wire can be written as:

$$R = \frac{\rho \cdot l}{A} \tag{8.1}$$

with R = resistance of wire; l = length of the gauge wire A = cross section of wire; ρ = specific electric resistance.

The specific electric resistance is given as:

$$\rho = \frac{2 \cdot m \cdot v_0 \cdot A \cdot l}{N_0 \cdot e^2 \cdot \lambda} \tag{8.2}$$

where m = mass of an electron; v_0 = velocity of the electron; N_0 = number of free electrons; e = charge of an electron; λ = free wavelength of the electrons. With this equation for ρ the resistance of a wire can be written as:

$$R = \frac{2 \cdot m \cdot v_o \cdot l^2}{N_o \cdot e^2 \cdot \lambda} \tag{8.3}$$

The relative change of the resistance of the wire is given by the partial differential equation:

$$\frac{dR}{R} = \underbrace{\frac{2dl}{l}}_{\text{Influence of size}} + \underbrace{\frac{dv_o}{v_0} + \frac{dm}{m} - \frac{dN_0}{N_0} - \frac{d\lambda}{\lambda} - \frac{2de}{e}}_{\text{Piezoresistive effect}} \tag{8.4}$$

The relative change of the length is defined as the strain: $\frac{dl}{l} = \varepsilon$ and Eq. (8.4) can be written as:

$$\frac{dR}{R} = 2 \cdot \varepsilon + \frac{dv_0}{v_0} + \frac{dm}{m} - \frac{dN_0}{N_0} - \frac{d\lambda}{\lambda} - \frac{2de}{e} \tag{8.5}$$

The relative changes of the mass (*dm/m*) and the charge of the electrons (*de/e*) are zero and the sensitivity (k) of a strain gauge is defined as the relative change of the resistance divided by the strain. Then the sensitivity of a wire is given by the following equation:

$$k = \frac{dR}{R \cdot \varepsilon} = 2 + \frac{1}{\varepsilon}\left(\frac{dv_0}{v_0} - \frac{dN_0}{N_0} - \frac{d\lambda}{\lambda}\right) \tag{8.6}$$

From this equation it is seen that the sensitivity of a wire is 2, plus a term which itself is dependent on the strain. Actual strain gauges exhibit a constant gauge factor around 2. Thus, for an ideal grid material the second term of Eq. (8.6) must be zero and then the gauge factor will be nearly 2.

$$\frac{dR}{R} = k \cdot \varepsilon \tag{8.7}$$

8.1.2 Semiconductor Strain Gauge Fundamentals

While in a metal foil strain gauge the "Piezoresistive Effect" contributes only about 5% [2.1 instead of 2 referring to Eq. (8.6)] to the gauge factor, in a semiconductor the "*Piezoresistive Effect*" dominates the change of the electrical resistance with the strain level (Eq. 8.4). This effect is so strong that the sensitivity to strain characterized by the gauge factor k is up to one hundred times higher than the gauge factor of a metal foil strain gauge. Nowadays, commercial semiconductor strain gauges have a gauge factor of about $k = \pm 100$ to ± 150. Even though Eq. (8.6) is derived for a metal wire, it can be seen in this equation that the *Piezoresistive* part of the gauge factor depends on the strain itself and as a consequence, the sensitivity is not constant with strain [7]. While the *Piezoresistive Effect* dominates the gauge factor of a semiconductor gauge, the speed of a free electron v_0 and the number of free electrons N_0 and the free mean path λ_0 now characterize the specific resistance of the semiconductor instead of the geometry change. These characteristics are highly isotropic and their contribution to the gauge factor depends on the orientation of the material to the strain. As a consequence, the gauge factor of a typical N-type semiconductor strain gauge is highly non-linear (see Fig. 8.1).

The other feature shown in Fig. 8.1 is that a positive change of strain results in a negative change of resistance. That is why this kind of gauge is called N-type (N = negative). These gauges are made out of silicon doped with phosphorus. The P-type gauges are made out of silicon doped with boron and this material has a positive change of resistance under a positive change of strain (see Fig. 8.2).

The actual characteristic of a semiconductor gauge is measured by a test and the relative change of resistance can be described by the following equation:

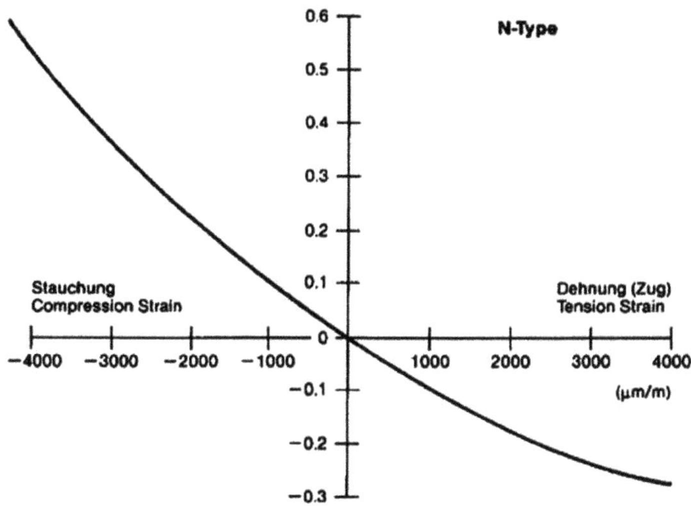

Fig. 8.1 Relative resistance change of N-type-gauges versus strain

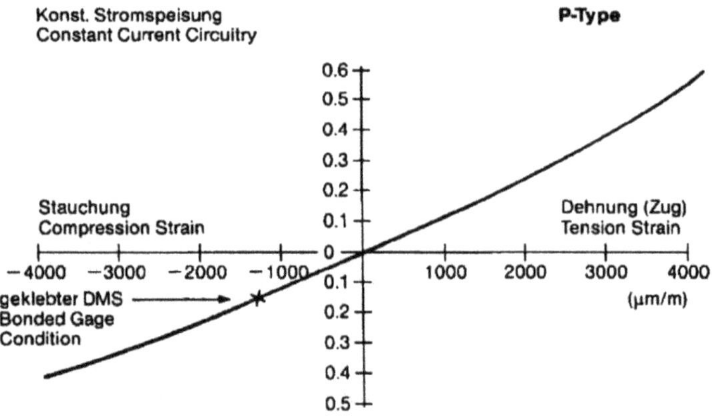

Fig. 8.2 Relative resistance change of P-type gauge versus strain

$$\frac{dR}{R} = k_1 \cdot \varepsilon_1 + k_2 \cdot \varepsilon^2 + \ldots. \tag{8.8}$$

The two gauge factors are constants and they are obtained by approximation of the measured curve. The non-linear factors represent the effect of a non-linear change of the specific electrical resistance $\Delta\rho/\rho$. Normally a second-order fit is enough to describe the non-linear behavior of the gauges.

Another peculiarity of the semiconductor strain gauges is the extreme dependence of the sensitivity to temperature, so that even for small changes in temperature the gauge factors k_1 and k_2 will change. The change of sensitivity with temperature has to be taken into account in Eq. (8.8) and it can be written now in the following way:

$$\frac{dR}{R} = k_1 \cdot (\frac{T_0}{T}) \cdot \varepsilon_1 + k_2 \cdot (\frac{T_0}{T})^2 \cdot \varepsilon^2 \ldots. \tag{8.9}$$

T_0 is the reference temperature where the gauge factors k_1 and k_2 are measured and T is the temperature where the gauge operates. The strong influence on the sensitivity of a semiconductor gauge and thus, the poor stability of the gauges in an unstable temperature environment like a wind tunnel are the major disadvantage of semiconductor gauges. That is why they are seldom used for wind tunnel balances.

For short term measurements, where the change of temperature is not significant, the high gauge factors offer the possibility to build a much stiffer balance. This fact qualifies the semiconductor gauges for balances in shock tunnels and other short-term applications with extremely high sample rates. The semiconductor strain gauge manufacturers offer gauges that compensate the disadvantage of non-linearity and sensitivity drift in a reasonable way, nevertheless still far removed from the stability and the adaption capabilities of a metal foil gauge. For applications with temperature changes, metal foil gauges are preferred.

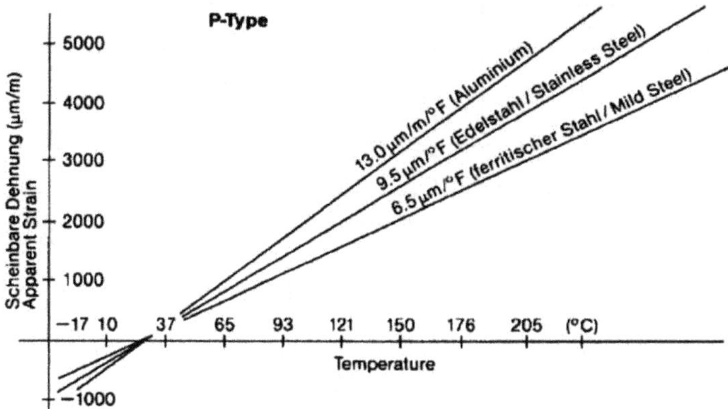

Fig. 8.3 Apparent strain of P-Type semiconductor gauges on different materials (Vishey Micro Measurements)

In Fig. 8.3 the apparent strain of a typical P-type semiconductor gauge on several materials is shown. The output related to temperature is up to three times higher than the output related to normal mechanical strain, thus, the effects of temperature changes on a semiconductor gauge circuit have to be taken into account very carefully to insure reliable measurement results [3].

Instead of using P-type gauges it is also possible to use "Self-Compensating N-Type Gauges". The thermal behavior of N-type gauge can be adapted to the thermal behavior of the material in some ways and this results in a much lower output of apparent strain, as it is shown in Fig. 8.4. In this case a N-Type gauge is bonded onto an aluminum alloy AL 2024 [4].

On the other hand, semiconductor gauges have nearly no creep or hysteresis. In a constant temperature environment this makes them usable for extreme long-term applications or long-term fatigue lifetime applications. If semiconductor strain gauges are being considered for use, a comprehensive description for their application and temperature compensation is given by *James Dorsey* in his *Semiconductor Strain Gage Handbook* [3].

Due to the limited use of semiconductor gauges for wind tunnel balances, no further descriptions of compensation methods and application problems are given here.

8.1.3 Fiber Optic Strain Gauge Fundamentals

In the field of wind tunnel force measurement there are only rare applications of balances using fiber optic sensors. The most experienced group in this respect is found in South Africa at CSIR (Council for Scientific and Industrial Research Pretoria). They have built wind tunnel balances using fiber Bragg grating gauges [10, 11].

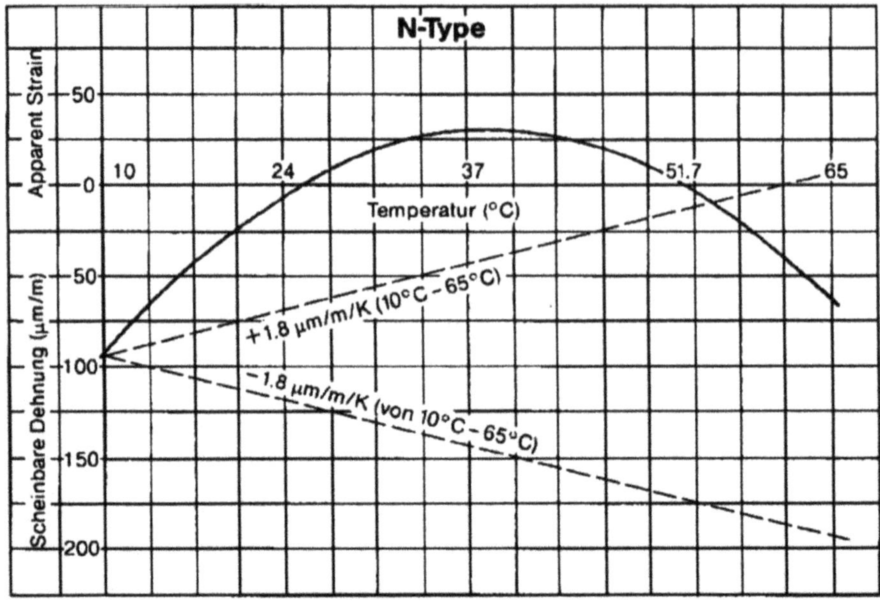

Fig. 8.4 Apparent strain of a compensated N-Type semiconductor gauge on Al 2024 (Vishey Micro Measurements)

Others have conducted tests using bending beams for comparison with wire strain gauges. Although the characteristics of the fiber optic gauges promise high resolution to build much stiffer balances with the same sensitivity, they have not yet established themselves in the field of wind tunnel balances. The reason is the fact that there is no other spring material available that can really utilize the full capabilities of the gauge. In other words, even the perfect strain gauge would only describe the strain situation with all the uncertainties given by the spring material they are gauged on. A perfect force measurement can only be achieved with a gauge that is perfectly adapted to the spring material. For the wind tunnel balances there is no alternative to high strength steels at the moment and their properties do not match ideally to the characteristics of the fiber gauges.

The fundamental mechanism used is the transport of light through a standard optical fiber, that consists of a light guiding core fiber with a diameter around 4–9 μm and a so-called cladding around the core of about 125 μm. Around this cladding there can be an additional cover of acrylate or polyamide. The core is of fused silica that is doped with germanium or boron. The cladding is pure fused silica. Caused by the different refractive indexes of doped and pure silica, there is almost total internal reflection of the light between core and cladding and so the light is guided through the fiber with rather low damping.

To measure strain by the use of optical fibers two principal physical phenomena are used—interferometry and scattering, whereby inteferometry is the more practical

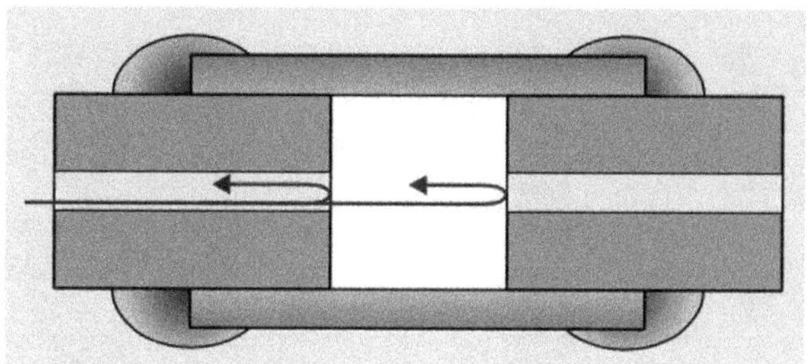

Fig. 8.5 Principle of Fabry-Pérot strain gauge

in use. There are two types of strain interferometers, one is the *Fabry-Pérot* sensor and the other is the Fiber Bragg Grating sensor. Both sensors use the interference of a light beam with a reflected light beam to detect the change in strain.

In the *Fabry-Pérot* sensor the fiber is interrupted by a cavity. Part of the light is reflected at the end of the fiber and another part of the light is reflected at the end of the cavity. This is illustrated in Fig. 8.5. The two parts of the light beam interfere and the interference pattern is proportional to the length of the cavity. If the length of the cavity changes under the influence of strain, the interference pattern changes. The resulting phase change is a measure for the strain [12].

The most commonly used type of optical strain gauge is the *Fiber Bragg Grating* sensor (FBGS). FBGSs are commercially available as "ready to bond" sensors and so the first hurdle to use them is eliminated. This makes the application as easy (or complicated) as the application of wire strain gauges. Equipment to analyze the signal at an affordable price is becoming available, but is still more expensive per channel than that for the metal wire gauge.

The physical working principle of a *Fiber Bragg Grating* strain gauge is the change in wavelength of the UV-light that is reflected by the Bragg grating due to the change of grid period by the strain.

A Bragg grating defines periodic changes in the refractive index in the core fiber. This pattern is introduced by strong UV-laser lighting either after or during the spinning of the fiber. When it is done on a ready fabricated fiber, the cladding has to be removed before and closed in after the modification. The advantage of this process is a relative precise grid, because a mask can be used to generate the grid. The process is not time dependent and so the application of energy can be rather strong. With this method, changes of the refractive index of 0.01–0.1% are possible, establishing reflection factors of up to 90%. Such grids are called strong reflecting grids.

To burn the grid during spinning a pulsed laser must be used. The pulse for one refractive index abnormality has to be very short and so the energy impact is much smaller. The result is a reflection factor of about 10–20% and therefore the grids are called weak reflecting grids. The difference in the signal of both types can be seen in the wavelength distribution.

A Bragg grating has normally a length between 6 and 9 mm with a few thousands of grid lines (Fig. 8.6).

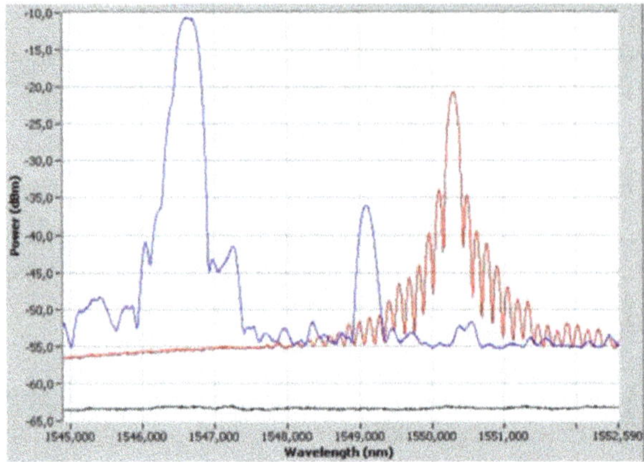

Fig. 8.6 Wavelength distribution of fiber Bragg strain gauges, blue = strong reflecting, red = weak reflecting (HBM)

The sensitivity of a FBGS, depends on the strain caused by a change in elongation due to temperature and mechanical strain. Analog to the metal wire gauge it can be written as:

$$\frac{\Delta\lambda}{\lambda_0} = k \cdot \varepsilon + \alpha_\delta \cdot \Delta T \tag{8.10}$$

In this equation the parameters are defined in Table 8.1.

Table 8.1 Parameters used in defining sensitivity of FBGS sensors in Eq. (8.10)

Parameter	Description	Dimension	
$\Delta\lambda$	Change of wavelength	nm	
λ_0	Reference wavelength	nm	
$k = 1 - p$	k-factor; $k = 0.78$		(8.11)
p	Photo elastic coefficient; $p = 0.22$	–	
$\epsilon = \epsilon_m + \epsilon_T$	Strain	–	
ϵ_m	Mechanical strain	$\mu m/m$	
$\epsilon_T = \alpha_T \cdot \Delta T$	Thermal elongation		(8.12)
α_T	Thermal elongation coefficient	1/K	
$\alpha_\delta = \frac{\delta n/n}{\delta T}$	Change of fraction index by temperature δT		(8.13)
n	Fraction index of the fiber	–	

Using Eqs. (8.11)–(8.13) and inserting these into Eq. (8.10) yields

$$\frac{\Delta\lambda}{\lambda_0} = k \cdot (\varepsilon_m + \alpha_T \cdot \Delta T) + \alpha_\delta \cdot \Delta T \tag{8.14}$$

Equation (8.14) describes the behavior of a FBGS by a change of the mechanical strain ε_m, the change of strain by temperature ($\alpha_T \cdot \Delta T$), and the change of fraction index by temperature ($\alpha_\delta \cdot \Delta T$). The FBGS is therefore sensitive to both temperature change and mechanical strain. These effects must be separated to obtain a measurement for the stress in the material. This is very similar to the situation with metal strain gauges, which also react to both temperature and strain. In contrast to metal strain gauges, self-compensating FBGSs for temperature are not yet available and the compensation must be achieved by fixing the FBGS to a piece of material or at a place on the same material that has the same temperature but no mechanical stress. The advantage is that this can be done within the same fiber. Doing this, the temperature induced signal can be subtracted and Eq. (8.14) can be written as:

$$\begin{aligned} \frac{\Delta\lambda_1}{\lambda_{10}} - \frac{\Delta\lambda_2}{\lambda_{20}} = k \cdot \varepsilon_{m1} + k \cdot \alpha_{T1} \cdot \Delta T + \alpha_{\delta 1} \cdot \Delta T \\ - k \cdot \varepsilon_{m2} - k \cdot \alpha_{T2} \cdot \Delta T - \alpha_{\delta 2} \cdot \Delta T \end{aligned} \tag{8.15}$$

If the temperature at both FBGSs is the same, then $\alpha_{T1} \cdot \Delta T_1 = \alpha_{T2} \cdot \Delta T_2$ and $\alpha_{\delta 1} \cdot \Delta T_1 = \alpha_{\delta 1} \cdot \Delta T_2$ and the mechanical strain at location $2 : \varepsilon_{m2} = 0$; the difference of the signals is only sensitive to the mechanical strain at location 1 and Eq. (8.15) can be written as:

$$\epsilon_{m1} = (\frac{\Delta\lambda_1}{\lambda_{10}} - \frac{\Delta\lambda_2}{\lambda_{20}}) \cdot \frac{1}{k} \tag{8.16}$$

For applications at very high temperatures, additionally a change in the reference wavelength of the light (λ_0) has to be taken into account. The application of FBGSs for wind tunnel balances is still in its infancy. The decision whether to use FBGSs can be evaluated by considering the following advantages and disadvantages.

The main advantages of FBGSs are:

- very high strain, larger than 10,000 μ m/m can be measured
- insensitive to electric or magnetic fields
- many FBGSs in the same fiber (usually about 15 per fiber, more are possible)
- excellent long term stability
- use in cryogenic environment possible
- use at very high temperatures (up to 700 °C)
- good corrosion stability
- long transmission line possible.

The disadvantages of FBGSs are:

- the diameter is small but the required length for installation is rather long
- a strong sensitivity to temperature, no self-compensating FBGSs for temperature
- transverse sensitivity high, they are very sensitive to pressure changes
- reinforcement effect caused by a high *Young's modulus* (only in case of stress determination)
- radius of curvature is limited to values greater than 15 mm.
- higher cost per measurement equipment, potentially smaller per gauge when many FBGSs per fiber are used, but higher compared to wire gauges.

8.2 Strain Gauge Selection

The choice of the strain gauge type (wire, semiconductor, fiber optic) is determined by the operation conditions. For highly dynamic load conditions semiconductor gauges exhibit advantages, or for applications in composites fiber gauges could be the best choice. For quasi-static, high precision measurements of loads in a wind tunnel, test metal wire gauges are still the best choice. Therefore in this section only the selection of metal wire strain gauges is addressed. The strain gauge selection process is very important in the design phase of a force transducer or a balance and many problems can be avoided by a careful selection of strain gauges.

The selection of the optimal strain gauge for a certain transducer is an iterative process, comparing the requirements and the characteristics of the offered strain gauges, generally ending in a compromise.

As outlined in Sect. 8.6.2, strain gauge balances should always gauge with a chrome-nickel alloy (*Karma, Modco*) to obtain a good compensation for the sensitivity shift. The gauges should be ordered from the same lot, which means they were manufactured at the same process on one sheet. Additionally, they should be pre-selected to a difference of less than 0.05% in nominal resistance.

The type and size of the stress field that should be measured by the gauge are the first determining factors of size and type for the gauge. To minimize the influence of stress concentrations, low gradient strain fields are required. Usually the areas that fulfill this requirement are relative small, and so small gauges have to be used. Sometimes on small balances even the smallest gauge does not fit into the calculated area without cutting the carrier foil.

On the other hand the excitation voltage should be as high as possible to insure a high output. However, high excitation voltage will heat up the gauge and this will produce errors. The strain gauge manufacturers provide diagrams, from which the maximum excitation voltage for the combination of the selected strain gauge area and the spring material can be read. In general, transducer materials with high heat conductivity and gauge patterns with a larger grid area allow higher excitation voltages than materials with low heat conductivity and small grid patterns.

To choose gauges with a high nominal resistance is always a good choice. The self-heating effect is relatively low, because at a constant excitation voltage the current through the bridge is lower and therefore the heating power of the gauge is lower. Another advantage is the lower influence of the bridge wire thermal effects on the signal, because the ratio of bridge resistance to the bridge wire resistance is high. Finally, the high resistance gauges are small and so they fit into strain fields with small strain gradients. A disadvantage of high resistance gauges is that the wires that are needed to compensate zero, zero drift and sensitivity shift must also have relatively high resistances. This usually requires very thin wires, which are not easy to handle.

As gauge patterns, double gauges should be preferred, because the effort for the application is reduced and the two gauges are perfectly parallel to each other.

As backing foil, fiber reinforced epoxy laminate is the best selection for a wide temperature range and a high fatigue life.

The solder dots on the gauge should be either pre-soldered or copper plated, because the chrome-nickel grid alloy needs an aggressive soldering flux to be soldered. This makes the attachment of the wires complicated and increases the risk of damage during wiring of the bridge. Welded lead wires out of copper beryllium have a higher fatigue life than copper lead wires.

8.3 Strain Gauge Application

To bond the strain gauges onto the balance would appear to be a relative simple process, but to obtain a high quality transducer, much experience is necessary. The strain gauge manufacturers therefore offer courses where all the techniques can be acquired and they give precise instructions about how to use their adhesives and accessories. For the application of a balance or force transducer the use of the adhesives supplied by the manufacturers is highly recommended, even if they are much more expensive than comparable products on the market. Experience shows that the expiry data should also be adhered to. Solvents and cleaning fluids have to be very pure and not contaminated with other ingredients. Following all the manuals for a careful strain gauge installation is the best premise for reliable and stable measurements.

A study using experimental and numerical methods showed that most of the nonlinear interactions of an internal wind tunnel balance are caused by alignment errors of the strain gauges. As a consequence, perfect strain gauge positioning should be a target condition.

To minimize the errors by the wire quality and guarantee the best lifetime, the wires should be as thick as possible and stranded copper wire should be used instead of solid copper wire. As insulation *Teflon*® or *Kapton*® is preferred.

8.3.1 Bonding

The bonding technique is determined by the operating conditions. For normal conditions, like temperatures between −10 °C and +80 °C and normal ambient humidity hot bonding is required for a long-term use balance. This produces the best results and best stability. Hot bonding requires more work and so it is more expensive.

On balances used for only a few tests, the bonding can be performed with cyanoacrylate bonds (superglue). These bondings can be performed very quickly and the repeatability for short term is sufficient.

For cryogenic applications some hot bonds have been tested and deliver reliable results over a long period. In any case, the use of certified glues from the strain gauge manufacturers is highly recommended. The cost of re-bonding necessary due to bad glue or glue that has passed the expiry date is much more expensive than the glue sold by the manufacturers.

For hot bonding some requirements have to be fulfilled very carefully. The first is to realize the required bonding pressure as precisely as possible. For this, special clamping devices must be designed individually for every application. Springs can be used to generate the bonding pressure. The pressure has to be adjusted before the bonding by a calibration test. To perform this test the clamping device must be counterbalanced to correct the force (bonding pressure) using weights and then set to the correct clamping dimension. This position must be marked and reset during the bonding.

The aim of all this work must be to achieve a glue film under the gauge that has a constant thickness and a consistent distribution. The best values are tested by the gauge manufacturers and best results are obtained when their recommendations are fulfilled. The portion of glue that is put under the gauge should be near the amount that is needed for the film (recommended thickness x gauge area) so that the surplus that is squeezed out is a minimum. When too much glue is put under the gauge, the gauge tends to drift out of position, even if it is fixed with *Kapton*® tape.

After curing the surplus of glue around the gauge should be scraped away carefully. This is a very tricky step and must be done with great care as to not damage the gauge. All bonding must be aged several times according to the requirements of the manufacturers at elevated temperatures, that are usually slightly higher than the curing temperature (Fig. 8.7).

8.4 Wheatstone Bridge Wiring

The relative changes of the electric resistance of a strain gauge are very small. For example a signal of 1 mV/V using a strain gauge with a resistance of 350 Ω will be 0.0875 Ω and this change must be measured with an accuracy of better than 8.75 Ω. For direct measurement this will require a resistance measurement instrument with a resolution of more than 40×10^6 parts. Even nowadays such an instrument does not

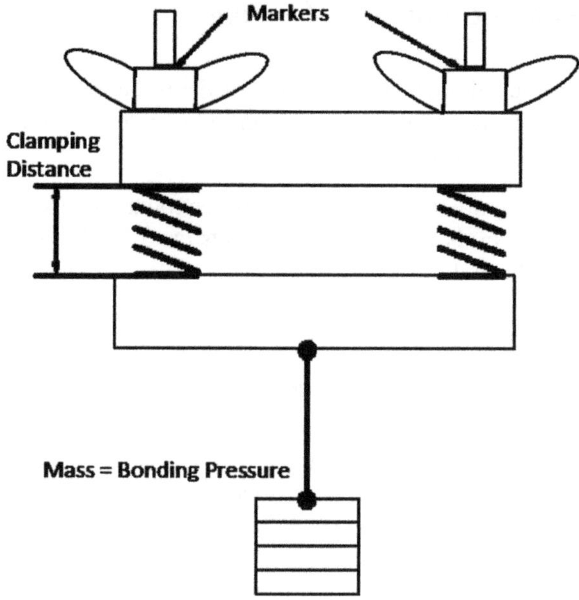

Fig. 8.7 Example scheme of bonding pressure calibration

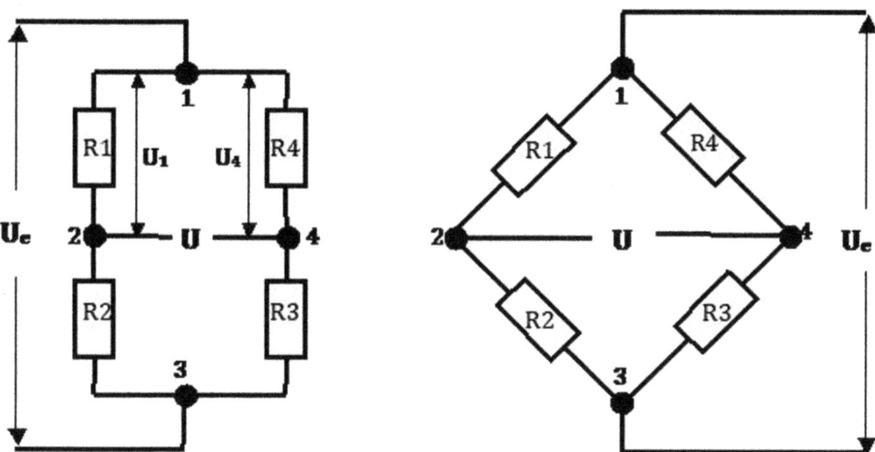

Fig. 8.8 Wheatstone bridge circuit

exist. In 1843 *Charles Wheatstone* described a method to measure directly the small relative changes of an electrical resistance, the so-called Wheatstone Bridge. Four resistors of the same nominal resistance value are arranged as it is shown in Fig. 8.1. Two of the resistors are in series and these two series resistor pairs are wired parallel to one another.

8.4.1 *Notation of Gauging and Wiring*

There is no common rule for the numbering of the resistors in Fig. 8.8, but all the equations that follow are based on the numbering given in Fig. 8.8 and they remain valid as long as the following rules are fulfilled:

1. The order of the numbers is either clockwise or counter clockwise, not mixed.
2. The excitation voltage is applied between points *1* and *3* and signal (U) is measured between points *2* and *4*.

8.4.2 *Relation Between Signal and Resistance*

When an excitation voltage is applied at the points 1 and 3, the resistors R_1, R_2 and the resistors R_3, R_4 each form a voltage divider. The voltages U_1 and U_4 can be described by the following equation:

$$U_1 = \frac{R_1}{R_1 + R_2} \cdot U_e; \qquad U_4 = \frac{R_4}{R_3 + R_4} \cdot U_e \tag{8.17}$$

The difference between these two partial voltages is the voltage $\boldsymbol{U}$, which is given by

$$U = U_1 - U_4 = (\frac{R_1}{R_1 + R_2} - \frac{R_4}{R_3 + R_4}) \cdot U_e \tag{8.18}$$

The relative change of the voltage U to the excitation voltage U_e is:

$$\frac{U}{U_e} = (\frac{R_1}{R_1 + R_2} - \frac{R_4}{R_3 + R_4}) \tag{8.19}$$

$$\frac{U}{U_e} = \frac{R_1 \cdot R_3 - R_2 \cdot R_4}{(R_1 + R_2) \cdot (R_3 + R_4)} \tag{8.20}$$

It can be seen that the ratio of these two voltages U/U_e are in some way proportional to the ratios of the resistances, but $U = 0$ if:

1. Every resistor has the same value $R_1 = R_2 = R_3 = R_4$.
2. The ratio $R_1/R_2 = R_4/R_3$

In this case the bridge is called a balanced bridge.

In the case of strain gauge measurements the resistors R_1 to R_4 are the strain gauges, including the resistance changes of the gauges by the strain. Now the resistors R_1 to R_4 in Eq. (8.20) are replaced by the initial resistor plus a resistance change ΔR_1 to ΔR_4 and Eq. (8.20) can be written as:

$$\frac{U}{U_e} = (\frac{R_1 + \Delta R_1}{R_1 + \Delta R_1 + R_2 + \Delta R_2} - \frac{R_4 + \Delta R_4}{R_3 + \Delta R_3 + R_4 + \Delta R_4}) \tag{8.21}$$

The relationships between the ratio of the voltages U/U_e and the ratio of the resistances $\Delta R_i/R_i$ are non-linear. These non-linearities can be around 0.1% if only one resistance changes. This is too much for a precision force transducer and therefore these non-linearities cannot be neglected if only one strain gauge is used in a force transducer. This is the reason why most strain gauge based transducers use four active gauges.

In case of a "full bridge", as it is called when four gauges are active, the nominal values of the resistors R_1 to R_4 are the same and the changes of the resistances are very small, so that higher order terms can be neglected and Eq. (8.21) can be simplified to the linear equation:

$$\frac{U}{U_e} = \frac{1}{4} \cdot (\frac{\Delta R_1}{R_1} - \frac{\Delta R_2}{R_2} + \frac{\Delta R_3}{R_3} - \frac{\Delta R_4}{R_4}) \tag{8.22}$$

The relation between relative change of the resistance and the strain was found as

$$\frac{\Delta R}{R} = k \cdot \varepsilon \tag{8.7}$$

Now the linear relation between the relative change of the bridge output and the strain becomes

$$\frac{U}{U_e} = \frac{k}{4} \cdot (\varepsilon_1 - \varepsilon_2 + \varepsilon_3 - \varepsilon_4) \tag{8.23}$$

This is the basic equation for the strain gauge measurement.

In this equation it can be seen that the relative resistance change or the strain in the bridge arms ε_2 and ε_4 are subtracted from the relative changes in the bridge arms ε_1 and ε_3.

Tension is now defined to be a positive change of strain, e.g. a positive relative change of the resistance. In a full bridge with gauges all having the same tension ($\Delta R_1 = \Delta R_2 = \Delta R_3 = \Delta R_4$), no output will be measured at this bridge. Thus, for a force transducer a strain situation comprising two tension and two compression gauges is required. This favors bending or torsion type transducers for a measurement using strain gauges.

8.4.3 Electrical Influence of Bridge Wires

The influence of the bridge wiring can be explained by adding a wire resistance R_{wi} for every wire to every arm of the bridge (Fig. 8.9).

Assuming that the resistance of the gauges does not change ($\Delta R_i = 0$), Eq. (8.22) can be written as

$$\frac{U}{U_e} = \frac{1}{4} \cdot (\frac{R_{w1}}{R_1} - \frac{R_{w2}}{R_2} + \frac{R_{w3}}{R_3} - \frac{R_{w4}}{R_4}) \tag{8.24}$$

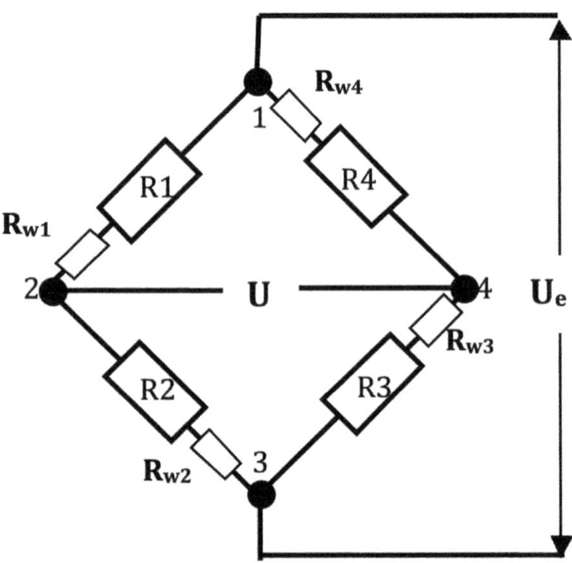

Fig. 8.9 Wheatstone bridge with circuit wire resistors

From this equation it can be immediately seen that the wires have no influence if their length is absolutely identical, then $R_{w1} = R_{w2} = R_{w3} = R_{w4}$. In practice this is difficult to achieve. Another possibility is to have two pairs of wires with equal length. For example, if $R_{w1} = R_{w2}$ and $R_{w3} = R_{w4}$. In this case the bridge is balanced by the influence of the wiring and the bridge zero output (no resistance changes in the gauges) is close to zero.

To minimize the resistance change by the wiring itself, either wires with equal length or two pairs of equal length inside the bridge circuit must be used. The position of the lead wire connections can be seen in the following circuit diagram (Fig. 8.10).

8.4.4 Mechanical Influence of Bridge Wiring

The connection of lead wires directly to the solder dot of the strain gauge can produce significant force of the wire onto the strain gauge grid, while the structure of the balance is deformed. This force can be repeatable if the wires are glued to the balance surface. Due to the large size and stiffness of the bridge wires related to the strain gauge structure, such an application normally causes hysteresis and creep. The usual measure to avoid these effects is an additional solder dot beside the active strain gauge. A thin, bended copper wire to the solder dot of the strain gauge connects to one side of the solder dot, and to the other side of the solder dot the bridge wires and the connecting wires can be soldered. So the forces induced by the wires are carried by the additional solder dot and they are decoupled from the strain gauge.

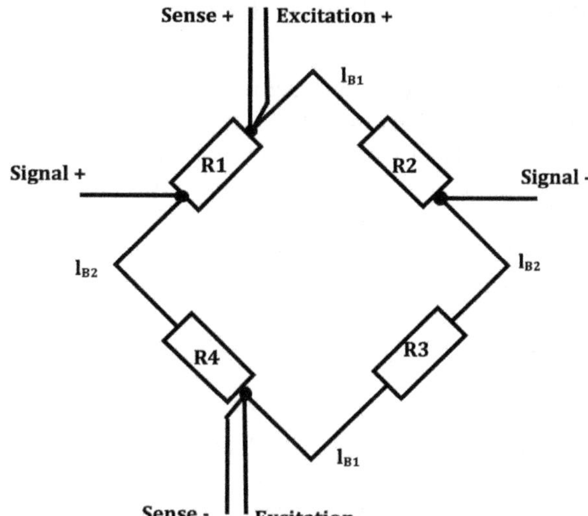

Fig. 8.10 Lead wire connection points

So far this is a standard procedure recommended by the strain gauge manufacturers and described in several handbooks. Sometimes, in very small balances, this recommendation cannot be fulfilled because there is not enough space for additional solder dots. In this case one should be aware that the application then exhibits more hysteresis and creep and the uncertainty of the measurement is higher.

To optimize the decoupling of the lead and bridge wiring one could think that very thin bended copper wires would be the best solution. Very thin wires indeed will reduce the mechanical coupling from the bridge wires to the gauge, but their resistance related to the total bridge resistance is higher than that of thicker wires. So their change in resistance by thermal effects has a larger effect on the total thermal stability of the bridge. The magnitude of this effect depends on the temperature range of the application and determines which wire size is the optimum.

For a large temperature range, thicker wires are thermally more stable and should be preferred. For small temperature variations, a thinner wire gauge can be better. No general rule can be given here, because the nominal resistance of the strain gauge itself finally determines the influence. To minimize the influence of the internal bridge wiring on the signals, high resistance gauges are preferred. The disadvantage of high resistance gauges is the compensation of zero and zero drift, because higher values for the resistance of the compensation wires are required. This results in rather long and very thin wires that are difficult to handle.

8.5 Compensation of Bridge Zero Output

An unloaded force transducer in the normal mounting position and without any load applied should have no output. Usually this cannot be achieved, even if the gauge and the wire resistances are the same, because some output is generated by the weight of the metric end of the transducer itself, which causes deformation in the transducer. However, the primary reason for the bridge zero output is the differences in the nominal resistance the strain gauges exhibit. Considering the data sheet of a normal foil strain gauge, the bandwidth of the initial resistance is up to 0.4% of the nominal value. For example, in the worst case the zero output of a bridge with a nominal resistance of 350 Ω is 4 mV/V, which greatly exceeds the full scale signal of the transducer (usually 2 mV/V). If a strain gauge amplifier is able to work in the range up to 6 mV/V or more, the offset can be set to zero and there will be no problem for the measurement. With high precision amplifiers this is not the case and the zero output must be compensated to values lower than 5% of the nominal output. To do this the transducer or the balance must be set to its normal working position. In this position the output of the bridge must be measured and a compensation resistance must be calculated according to

$$\Delta R_{\text{Compensation}} = \frac{U_0}{U_e} \cdot R_{\text{nominal}} \tag{8.25}$$

R_{nominal} [Ω] is the nominal resistance of the bridge. U_0/U_e [V/V] is the zero output of the bridge. If Constantan wire for zero compensation is used, the length of the Constantan wire must be calculated as follows

$$L_{\text{Const.}} = \frac{U}{U_0} \cdot \frac{R_{\text{nominal}}}{r_{\text{wire}}} \tag{8.26}$$

r_{Wire} [Ω/m] is the resistance per length of the Constantan wire. Alternatively, commercially available adjustable resistors can be inserted into the bridge. In which bridge arm the compensation resistor must be installed depends on the sign of the zero output. If the zero output is negative the compensation resistor must be added into the arm which generates positive signals in the bridge and vice versa if the zero output is positive (Fig. 8.11).

8.6 Compensation of Thermal Effects

8.6.1 Compensation of Zero Drift for Metal Foil Strain Gauges

Zero drift is defined as the change of the bridge zero output at different stable temperature conditions. To have stable uniform temperature conditions is very important,

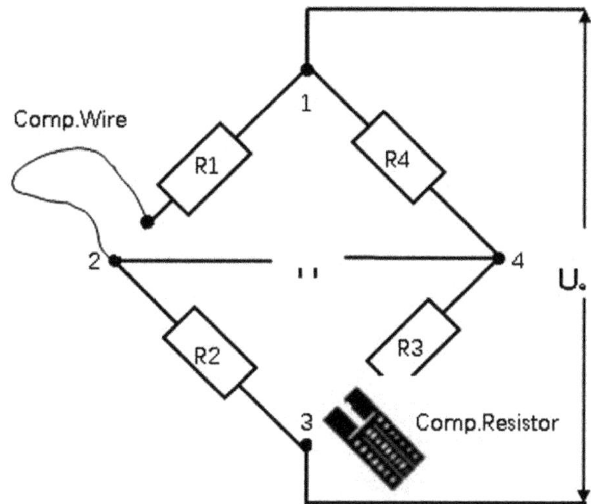

Fig. 8.11 Zero output compensation with constantan resistors

because temperature gradients inside the transducer can cause much higher bridge output than the zero drift itself. Under stable conditions the zero drift is caused by the change of the resistance of the grid and the different heat expansion coefficients of the foil and the basic material to which they are bonded. The physical reason for this temperature behavior of the strain gauge is a mix of a resistance change by temperature and a real strain, which is applied to the gauge though no load is acting on the transducer. For a single gauge this output is therefore called "apparent strain".

Apparent strain curves are given by the strain gauge manufacturer on the data sheet, as it can be seen in Fig. 8.12. For a single gauge, this apparent strain is usually much higher than the zero drift of a full bridge, because on a stable uniform temperature level all four strain gauges have nearly the same apparent strain with the same sign and so it nearly cancels out. This is another very important reason to use only full bridges for a transducer or a wind tunnel balance. In Fig. 8.13 the zero drift of the bridges of a balance are shown where "matched gauges" are used for each bridge. That means for every bridge four gauges were selected out of a large number, having nearly the same apparent strain curve, and so the zero drift becomes minimized to $0.15\,\mu V/(VK)$. In this case selection was necessary because the balance is operated over a large temperature range, but even for operating temperatures around ambient the selection process minimizes the zero drift to very low values.

This kind of self-compensation by a full bridge arrangement only works if all gauges are on the same temperature level. If a set of two gauges is placed in locally separated areas, like in a bridge to measure the bending stress of a beam (two gauges are placed on the upper side and two are on the lower side of the beam), or on a direct read balance where two gauges of a bridge are placed on the forward and two on the aft bending section, the temperature in these two areas can differ significantly and so even with well matched gauges there is no self-compensation of the zero drift. To insure self-compensation in each area, 4 gauges in the so-called *Poisson* arrangement

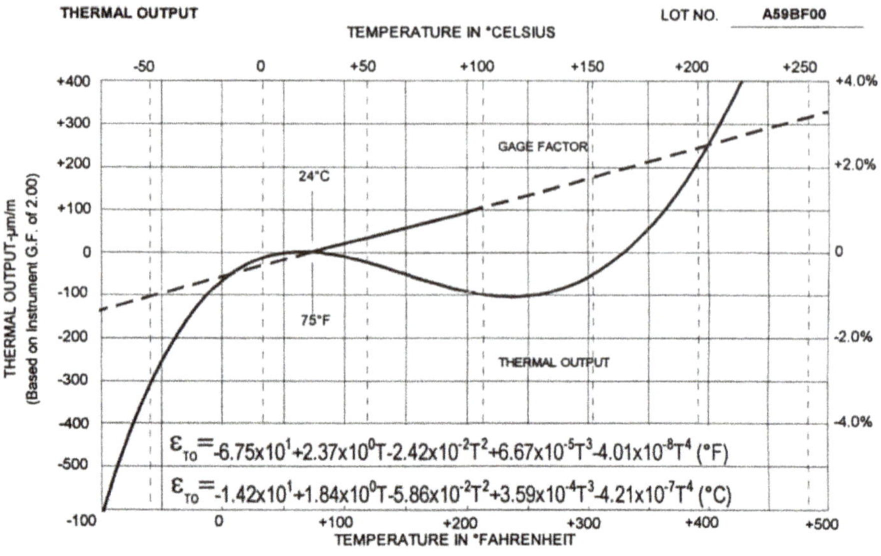

Fig. 8.12 Typical apparent strain curve of a strain gauge (Vishey)

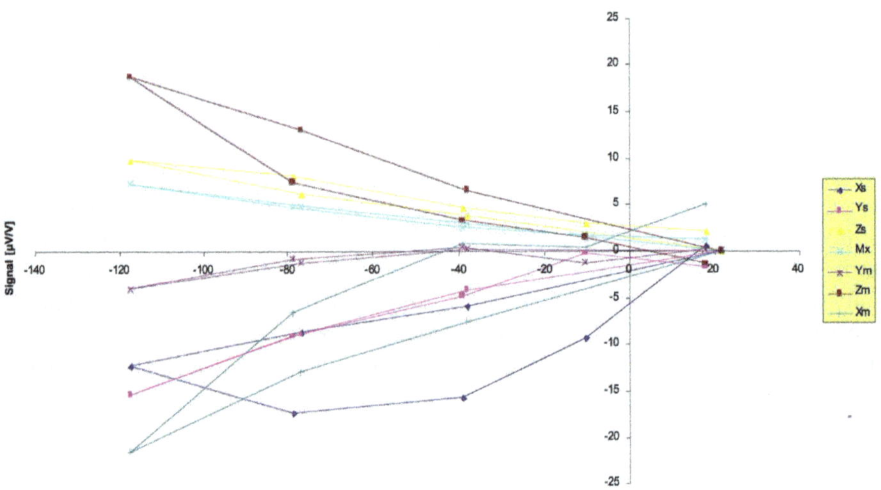

Fig. 8.13 Balance zero drift with matched gauges

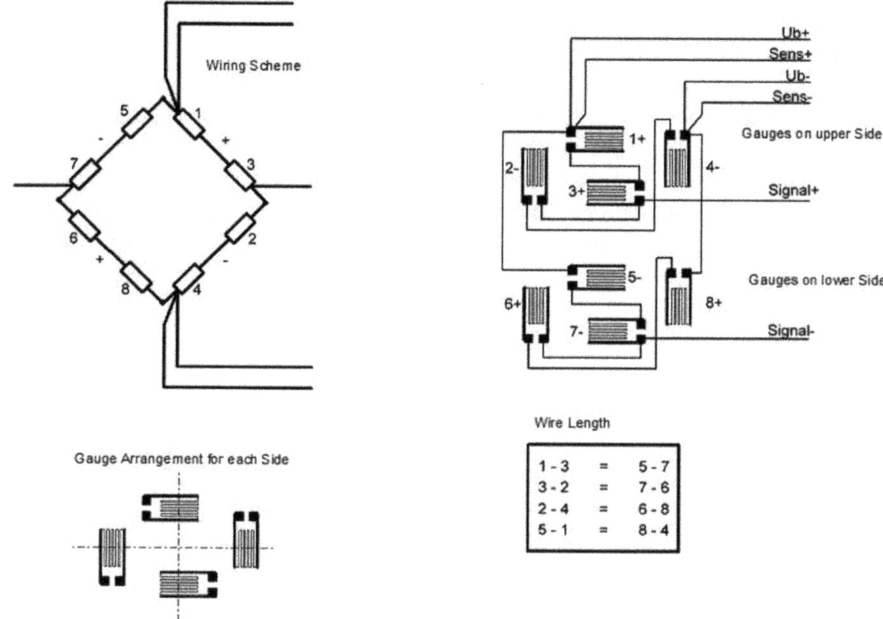

Fig. 8.14 Poisson gauge bridge compensation for different temperatures in two areas

must be installed and the 8 strain gauges of the two areas must be wired to one full bridge. The circuits for this method are shown in Fig. 8.14. In each measurement area four gauges must be installed. Two of them are placed in the sensitive direction and two have to be glued perpendicular. The wiring has to be done according to the scheme with pairs of wires having equal length, as shown in the figure.

Assuming that the gauges in the upper plane have the temperature T_1 and the gauges in the lower plane have the temperature T_2, the resistance change of the bridge can be written as follows:

$$\frac{\Delta U}{U} = \frac{1}{4}\left(\frac{R_{1T1} + R_{3T1}}{2R} - \frac{R_{2T1} + R_{4T1}}{2R} + \frac{R_{5T2} + R_{7T2}}{2R} - \frac{R_{6T2} + R_{8T2}}{2R}\right) \quad (8.27)$$

where $R_{1T1}, \ldots, R_{8T2}$ are the resistance changes of the gauges due to the temperatures T_1 and T_2. In Eq. (8.27) it can be seen that the resistance changes due to temperature cancels out if the individual gauges ($R_{1T1} = R_{2T1} = R_{3T1} = R_{4T1}$; $R_{5T2} = R_{6T2} = R_{7T2} = R_{8T2}$; $T_1 \neq T_2$) have the same apparent strain. To achieve this compensation, two sets of four matched gauges are needed so that the effort is double as high as for a single full bridge. The disadvantage of such compensation is that the bridge output is non-linear and smaller than that of a single full bridge for bending.

Using matched gauges in a Wheatstone bridge is the most complicated way to minimize zero drift. If the operating temperature is in the range of -10 to 60 °C, a compensation with pieces of temperature sensitive wires or prefabricated temperature sensitive resistors is much easier. Generally copper wires or copper resistors are used.

The first step in the compensation is to perform a temperature run with the unloaded balance in normal setup position. Each temperature level has to be very stable and the temperature distribution must be very uniform all over the balance. This cannot always be easily accomplished, but every temperature gradient inside the balance produces a signal caused by internal mechanical stresses and these signals do not belong to the zero drift of the bridge. If the model changes, gradient effects are not repeatable and so a correction which includes such gradient effects will fail when the setup is changed.

After determining the zero drift for each bridge there are two methods to continue:

The first possibility is the determination of a compensation resistor using the following equations:

$$(\frac{\Delta R}{\Delta T})_{comp} = \frac{U}{U_e} \cdot \frac{1}{\Delta T_{21}} \cdot R_{nominal} \tag{8.28}$$

$$R_{comp} = (\frac{\Delta R}{\Delta T})_{Comp} \cdot \alpha_{Comp} \tag{8.29}$$

where ΔT_{21} [K] is the difference in temperature between the lowest and the highest measured temperature, $\Delta U/U$ [mV/V] is the change of the bridge signal due to the temperature change, $R_{nominal}[\Omega]$ is the nominal bridge resistance; $\alpha_{Comp}[1/K]$ is the coefficient of the resistance change of the material which is used for compensation and $R_{nominal}$ [Ω] is the required resistance for the compensation. The compensation resistor must be included in the Wheatstone bridge wiring, as it is shown in Fig. 8.4, and the resistor must be cut according to the instructions of the manufacturer.

The second possibility is to install a temperature sensitive wire instead of a prefabricated resistor, then in a third step the length of the compensation wire must be calculated with the equation:

$$L_{Comp} = (\frac{\Delta R}{\Delta T})_{Comp} \cdot \alpha_{Comp} \cdot r_{wire} \tag{8.30}$$

where L_{comp} [m] is the length and r_{wire} [Ω/m] is the resistance per length of the compensation wire. For this purpose mostly copper, nickel or Balco® are used. To insure good results, the material data should be used for the given material and not from handbook data. Some length for soldering must be added to the calculated length, such that the active length of the wire is outside the solder dot area. Especially for short wires this is very important to achieve a good result at the first step. The compensation wire must be installed into the bridge circuit as it is shown in Fig. 8.4. The compensation resistor must be added into the bridge arm which has a lower resistance when the temperature changes.

8.6.2 Compensation of Sensitivity Shift of Metal Foil Strain Gauges

Sensitivity shift is defined as the change of sensitivity of a transducer with the change of temperature.

There are three different sources of a change of sensitivity of a strain gauge based force transducer. The first is the change of the geometry of the transducer body, caused by thermal expansion or contraction. The next is the change of the stiffness of the transducer material, since *Young's modulus* is a function of the temperature. Finally, the gauge factor can also change with temperature.

As an example, the sensitivity shift is calculated for a cantilever beam with the length (l) and a rectangular cross-section of ($b \times h$), which is loaded by a force (F) as shown in Fig. 8.15.
The sensitivity of this transducer is defined as the ratio of the signal ($\Delta U/U$) to the applied force ($\boldsymbol{F}$):

$$\frac{\Delta U/U}{F} = \frac{k \cdot \varepsilon}{F} \tag{8.31}$$

The strain ε under the strain gauge is given by the equation $\varepsilon = \frac{\sigma}{E}$, where σ is the mechanical stress under the strain gauge and E is *Young's modulus* of the cantilever beam material. The stress σ can calculated by using the following equations: $\sigma = \frac{M_b}{W_b}$; $M_b = F \cdot l_{\mathrm{DMS}}$; $W_b = \frac{b \cdot h^2}{6}$s, where l_{DMS} is the distance from the tip of the beam to the strain gauge location.

Using these relations, Eq. (8.31) becomes

$$\frac{\Delta U/U}{F} = \frac{k}{E} \cdot \frac{6 \cdot l_{\mathrm{DMS}}}{b \cdot h^2} \tag{8.32}$$

On the right-hand side of Eq. (8.32) all parameters are a function of the temperature and it can be seen that the sensitivity shift depends on the gauge factor, Young's modulus and the geometry of the transducer. The change of the dimensions of the cantilever beam can be written as $\Delta l_i = l_i \cdot \alpha_T \cdot \Delta T$ so that the dimensions after the change of temperature are $l_i(T) = l_i + \Delta l_i$. Applying this to all geometry parameters in Eq. (8.32), the equation shows the temperature dependence of the sensitivity.

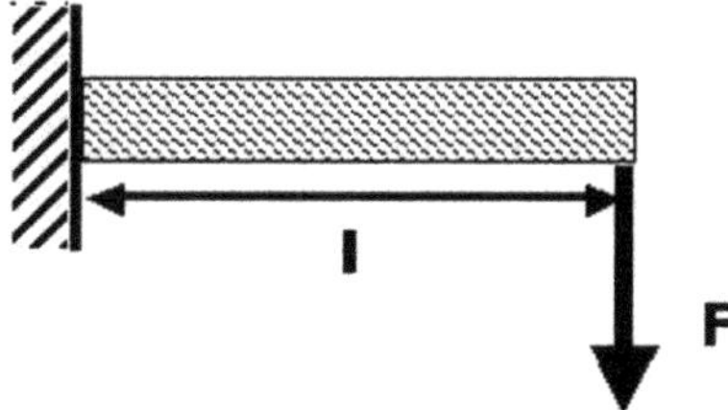

Fig. 8.15 Cantilever beam as force transducer

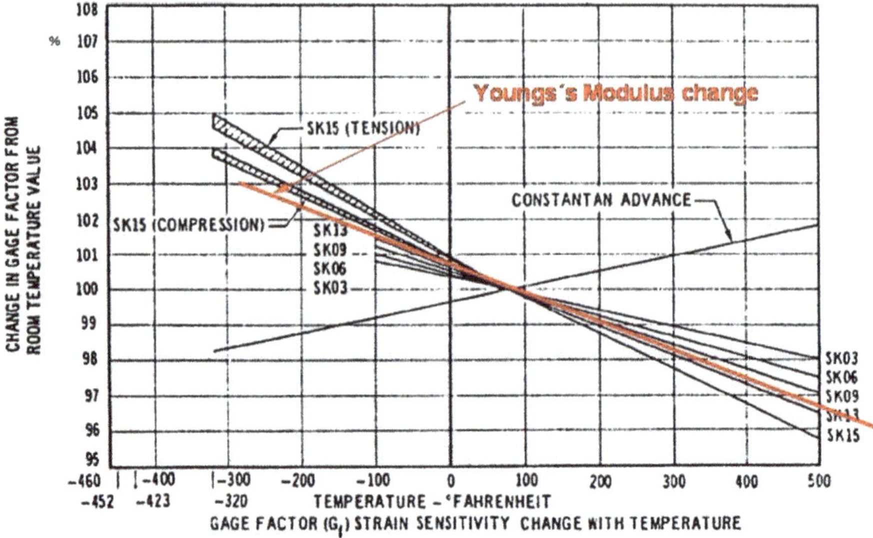

Fig. 8.16 Gauge factor shift of Karma and constantan (Vishey)

$$\frac{\Delta U/U}{F} = \frac{k(T)}{E(T)} \cdot \frac{1}{(1+\alpha_T \cdot \Delta T)^2} \cdot \frac{6 \cdot l_{\mathrm{DMS}}}{b \cdot h^2} \tag{8.33}$$

For the correction of the sensitivity shift there are two major results given by Eq. (8.33):

- First, compensation is only possible if the gauge factor changes in the same way as *Young's modulus* of the cantilever material. This requirement can only be fulfilled by strain gauges using Karma® or equivalent alloys, because *Young's modulus* decreases when the temperature rises. Constantan behaves in the opposite manner, so that it is impossible to compensate the sensitivity shift of a strain gauge transducer with Constantan gauges (Fig. 8.16).
- Second, a perfect compensation can only be performed if the geometry influence is also compensated by the gauge factor variation. As a consequence, for transducers with different geometries, different gauge factor drifts are necessary. The main difference appears if a bending stress type or a shear stress type transducer is used. The influence of the geometry on a shear stress type transducer is larger than the on a bending stress type transducer.

Conveniently, strain gauge manufacturers offer modulus compensating strain gauges for the main transducer materials to compensate the main sensitivity shift caused by the *Young's modulus* change. The disadvantage of a modulus compensation gauge is that no self-compensation option for temperature can be chosen at the same time.

Another possibility is to change the excitation voltage in the same way the sensitivity shift changes. In Eq. (8.31) it can be seen that the relative change of the signal

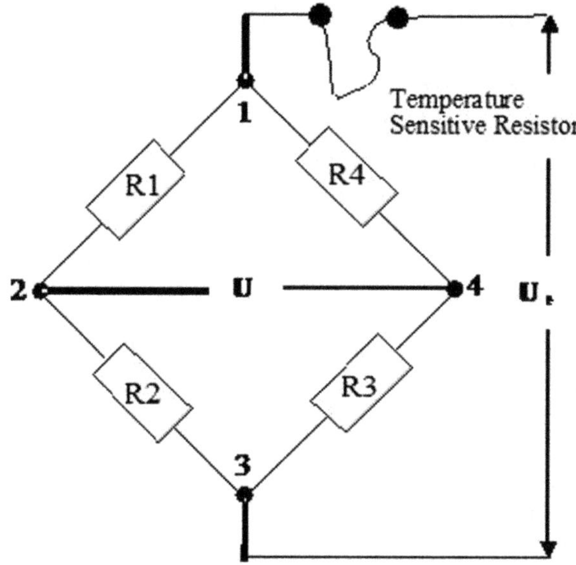

Fig. 8.17 Wheatstone bridge with sensitivity shift compensation

Table 8.2 Quantities of equation (8.34)

S_0	[mV/V]	Signal measured with force ***F*** acting at the reference temperature T_0 [K] and excitation voltage U_e [V].
S_T	[mV/V]	Signal measured with the same force ***F*** acting at the temperature ***T*** [K] and excitation voltage U_e [V].
R	[Ω]	Nominal resistance of the bridge
A	[m^2]	Cross section of the wire
α_R	[1/K]	Thermal expansion coefficient of the wire material
ΔT	[K]	Temperature difference between the two tests
ρ	[Ωm]	Specific electrical resistance of the wire material

remains constant. To achieve this, a temperature sensitive resistor is integrated into the excitation voltage line of the bridge nearby the strain gauges, as is shown in Fig. 8.17.

To determine the length of the compensation wire (l_{Wire}) the following equation can be used:

$$l_{\text{Wire}} = (\frac{S_0}{S_T} - 1) \cdot \frac{R \cdot A}{\alpha_R \cdot \Delta T \cdot \rho} \tag{8.34}$$

in this equation the quantities listed in Table 8.2 are used.

8.6.3 Computational Correction Methods of Temperature Effects

To correct the thermal effects numerically using a calibration test to determine the correction function is not as easy as it seems at first glance. The first reason that makes it rather difficult is the fact that thermal effects can take a long time to establish themselves inside the balance and the model. The second reason is that the surface-to-mass ratio of the model determines the time constant for the temperature change of the balance in the model. Fighter aircraft models will cause a different time/temperature function than transporter aircraft models. So a correction function is model dependent. A third influence that cannot normally be taken into account in a generic calibration test is the thermal radiation between model and balance surface. One measure to reduce this problem is the so-called radiation sleeve around the balance. It requires some additional space in the fuselage of the model and where it is possible such sleeves should be installed.

Common to all these problems is that it is rather complicated to perform a basic calibration test that exactly represents the wind tunnel situation. It is difficult to define by the knowledge of the temperatures from a few sensors the complete temperature distribution in a balance. Nevertheless, this information is required to separate thermal effects, like zero drift and sensitivity drift, from temperature gradient effects.

In case of a slow homogeneous changing temperature of a full bridge, the remaining zero drift of the bridge can be correlated very precisely with the temperature measured by a sensor nearby the bridge. A calibration test that gives the bridge output related to several constant temperature levels will yield the correct correction function. Such situations can occur in low speed wind tunnel tests, especially in tunnels that operate at almost constant temperature levels. In these cases, normally the hardware compensation of the thermal behavior is sufficient.

In wind tunnels with large temperature changes, the balance temperature cannot follow the tunnel temperature simultaneously and so the thermal effect in the balance is also time dependent and the problem of separating the thermal effects from a temperature gradient effect must be solved.

Zero drift and sensitivity drift can be corrected if the temperature of all four gauges is known. In cases where all four gauges are placed almost in the same area, it can be assumed that all gauges have the same temperature and one sensor directly beside the bridge can be used for the correction function. For a bending full bridge, where two gauges are on one side of the bending section and the other two are on the opposite side, this assumption can lead to a false correction, because the temperature can be different on both sides of the bending beam. In this case a calibration test is much more complicated because the temperature behavior of the gauges on each side must be known separately. This test can only be performed without the bridge wiring and so its influence cannot be taken into account and has to be corrected by a separate function. This is one reason why a numerical correction of the thermal behavior of a direct read balance is nearly impossible.

One possibility for a bending bridge arrangement to minimize the problem is to use full bridges on each side of the bending section in a Poisson arrangement (see Fig. 8.8).

8.7 Direct Read Balance Wiring

The definition of a direct read balance has been given in Sect. 3.2.1.3. The basic characteristic of such a balance is that the output of the bridges is directly proportional to the force or the moment they measure. The wiring is identical either for a force or a moment balance. Both circuits have in common that two gauges in two different sensing sections (bending or tension) are connected in such a way that one bridge yields a signal proportional to the moment and the other a signal proportional to the force. To obtain this only the connection sequence is different. In the following, the wiring is shown for a moment type balance with two bending sections.

In Fig. 8.18 strain gauges with positive signal (tension) are numbered with odd numbers and gauges with negative output (compression) are marked with even numbers. This rule is not common, but recommended to achieve a cyclic or anti-cyclic numbering of the gauges in a Wheatstone bridge. In the wiring scheme it can be seen that there are always two short wires that connect a gauge from top to bottom and two long wires that connect the gauges from left to right. If the pairs of wires have identical length and they are connected to the equipment as shown in Sect. 8.4.3, the influence of the wire length can be compensated. On balances with different temperatures in the bending section the short wires have different resistances caused by the different temperatures, and this effect cannot be compensated. This is a disadvantage and therefore a direct read wiring is not recommended for balances with large temperature gradients along the balance axis.

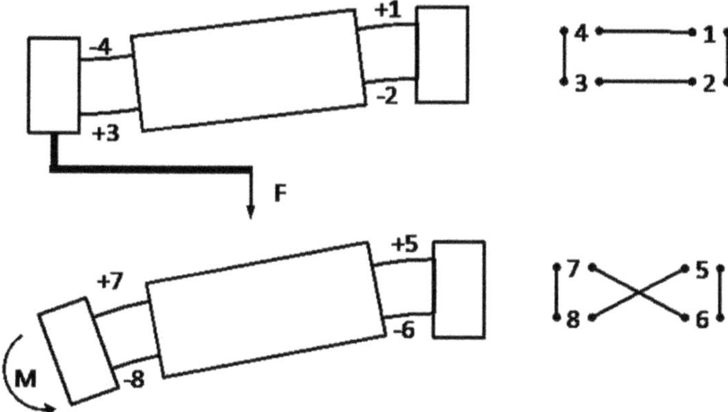

Fig. 8.18 Wiring scheme for force and moment of a direct read balance

8.8 Moment Balance Wiring

Alternative to the direct read wiring, a moment balance can also be wired so that the four gauges on the model side bending section are connected to a full *Wheatstone* bridge and the same is done on the sting side section.

Depending on the sign convention, either the sum or the difference deliver a force proportional signal or a moment proportional signal. A general rule which signal (sum or difference) is proportional to the force cannot be given, because the definition of a positive signal varies throughout the different manufacturers. In the following, the wiring is described for a moment balance and a positive signal is generated by a positive moment, so the sum of the signals is proportional to the moment and the difference is proportional to the force. This definition has the advantage that a positive moment can be easily applied to the metric end of the balance and so the control of correct signals can be easily monitored. To apply a pure force to the balance is much more complicated, because the force has to be applied exactly between the two bending sections. The gauges are numbered so that odd numbered gauges are positive (tension) for a moment and the even numbers are negative (compression) for a moment.

In the lower part of Fig. 8.19 a moment is acting on the balance. Both bridges deliver a positive signal according to the above-mentioned sign convention. The sum of the two signals is therefore proportional to the moment. In the upper part of Fig. 8.19 a pure force is acting on the balance and the bridge on the model end of the balance delivers a negative signal. So the difference of the two signals is proportional to the force.

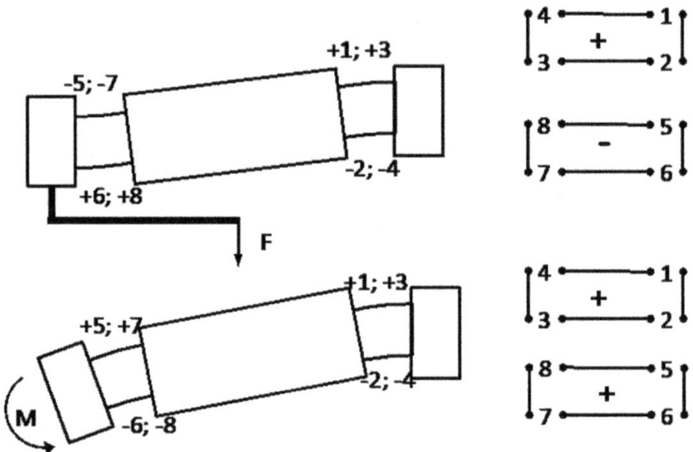

Fig. 8.19 Wiring scheme for force and moment on a moment balance

8.9 Insulation and Moisture Proofing

In a normal wind tunnel, moisture may not be a severe problem but at any time a change of moisture in the ambient air can affect the quality and stability of the strain gauge bridges. The basic requirement is that no moisture should corrode the metal grid. This would change the resistance of the grid dramatically or destroy it. If the moisture is contaminated with dust that usually accumulates on a balance over a long time, this dust together with the moisture can produce an electrolyte so that electro-mechanical corrosion can occur. To protect the gauges some sealing should cover them. Some gauges are encapsulated and thus have good basic protection. All others need something else. To cover them with glue is one of the easiest possibilities and sufficient for ambient applications with normal humidity (Fig. 8.20).

Nitrile rubber coating in also a good choice for normal ambient conditions and has the advantage that in case of repair it can be easily removed. Most of the commercially available sealing materials work well at tested environmental conditions:

Fig. 8.20 Poisson bridge with a nitrile rubber coating

Fig. 8.21 Cryogenic balance coated by CVD with silicon carbide

even a protection for operation under water is possible. A bigger problem is to seal a strain gauge application absolutely against vapor over a wide temperature range, like in cryogenic operation. Due to the temperature cycles from ambient to cryogenic temperature and back, the balance is exposed to moisture when reaching the dew point of the surrounding air. With conventional sealing a change in the zero output was always observed that can be removed by a drying process (Fig. 8.21).
The change of zero output by moisture can also be prevented by thin coatings obtained by chemical vapor deposition (CVD).

8.10 Connectors

Some tunnels do not use connectors on the balance and use a long cable instead. To install a balance in the tunnel a very long cable has to be pulled through the sting and all other installations between the model and the data acquisition system. Then they must be soldered to a connector. This extensive work has the advantage that no signal errors by contact resistances can occur that may has been caused by the bad connector quality, as was often the case in the past. However damage to the cables must be avoided at all costs. The problem of contact resistance is most severe with measurement systems that use four wire DC techniques to measure the bridge output. Modern systems with six wire AC techniques do not have this problem. On the other hand there are now high quality connectors available and even in a cryogenic environment with high temperature changes, these connectors work very well and reliably. To reduce the danger of damaging the cable during installation or dismounting, the use of connectors rather than long cables is recommended. In the past ITT Cannon connectors with up to 85 pins and a diameter of about 20 mm have been very successfully used.

8.11 Signal Conditioning Units

There are numerous systems for strain gauge signal measurement on the market and no attempt will be made here to describe all the variations. The intention of this section is to provide some recommendations about how to find the best system for a particular strain gauge application. The cost for a system strongly depends on the required accuracy of the system, therefore before purchasing a system a careful uncertainty analysis for the planned tests has to be made. It is not necessary to purchase a system that is better than the specified accuracy. The strain gauge sensor itself normally offers very high resolution and even the most inexpensive force sensor will exhibit excellent repeatability. So the data acquisition system sets the uncertainty limits for the measurements. The specified system should have a 5–10 times higher resolution then the smallest step to be detected. A good thermal stability is one of the major items that ensure low uncertainty.

It is instructive to consider the best possible solution which can be achieved with a wire gauge measurement. Obviously the first limiting factor is the strain gauge itself. The thermal noise level of the wire resistance sets the absolute minimum for the resolution. Thermal noise in an electrical resistance is created by the random movement of electrons in the metal. The thermal noise for a 350 Ω gauge at a bandwidth 1 Hz can be calculated with the following formula:

$$V_{rms} = \sqrt{4 \cdot k \cdot T \cdot R \cdot B} \tag{8.35}$$

V_{rms} = Root Means Square Noise Voltage [V] k = Boltzman constant (1.380662×10^{23}) [J/K]

T = absolute Temperature [K]

R = Resistance [Ω]

B = Bandwidth [Hz].

The result for the root mean square noise voltage is about 3 nV and for peak-to-peak it is 6 nV. A typical strain gauge transducer has a sensitivity of 2 mV/V and with an excitation voltage of 5 V the maximum signal output is 10 mV. The resolution of the strain gauge is the smallest signal difference that can be detected and this is the 6 nV. The ratio of the full scale signal of 10 mV to the 6 nV is about 1.6 million and this is the maximum resolution of a gauge. Every signal that is greater can be detected within the thermal noise of a gauge. For practical measurements some statistics have to be taken into account and so for a confidence level of $\sigma = 2$ (95%) the maximum resolution is about 800,000 parts. This simple calculation shows that the resolution of a strain gauge is not really the limiting factor for the measurement. The problem is to build measuring equipment that enables stable measurements at such high levels of resolution. This equipment exists (see [8]) but it is very expensive and obviously such a high resolution for the wind tunnel force measurement is not really necessary. A system resolution of 50,000 parts or $0.04\,\mu V/V$ or 20 bit is good enough for a precise measurement. Such systems are available at a reasonable cost.

Something that is not taken into account in this calculation is that a strain gauge is not only sensitive to strain. All the other influences like temperature effects, are sensed with the same resolution and are superimposed on the strain signal. The good news is that all the other effects are systematic errors that can be detected by the gauge and thus, can be compensated or eliminated by calibration. Lessons learned from this part is that the strain gauge offers much more resolution than required and the data acquisition system or cost is the limiting factor.

Related to measuring equipment another issue that is often discussed is, which kind of system, AC or DC, is better? Numerous pros and cons have been discussed over the years and advocates of each system defend their positions with good arguments. Nevertheless, experience shows that systems with the highest resolution and precision are AC systems. Unfortunately, they are also the most expensive. At the same cost level either one of the systems can have advantages related to the other, depending on the test requirements.

The first criterion to consider about is the limiting upper frequency. While DC Systems have a relatively high limiting frequency in the range of 10 kHz, the limiting upper frequency of AC systems depends on the frequency of the AC carrier voltage. For low carrier frequencies around 200 Hz the limiting frequency is in the range of 10 Hz. This means that unsteady data with higher frequencies cannot be correctly measured. As a rule of thumb, for unsteady aerodynamics AC systems must have a carrier frequency about ten times higher than the highest frequency to be observed. AC systems are typically available up to a carrier frequency of 5 kHz. For highly dynamic force measurements, DC systems can be better. However in this case stiffer piezoelectric sensors instead of wire strain gauge based sensors should be preferred anyway.

A totally different characteristic of the systems can be observed related to the stability of the zero output versus temperature. For a temperature difference of 10 K the change in the zero output for the AC system with a 225 Hz carrier frequency is lower than 2 mV/V (full scale = 2 mV/V). For the corresponding DC system the change in zero output is about $10\,\mu V/V$. So for steady precision absolute measurements, this is an advantage for the AC system. The same situation can be observed related to the stability in sensitivity. Here also the systems with a low carrier frequency have a far better stability than the DC Systems. With respect to the error in linearity, both systems have almost the same values.

The thermal noise of the strain gauge itself is very small and noise in the signal is the limiting factor for the resolution. Beside the thermal noise there are other sources for noise in electronic equipment, like shot noise. All the sources of noise accumulate to an input noise of the amplifying equipment. The values for the input noise are the highest for the DC equipment. For AC equipment the input noise level becomes lower by increasing carrier frequency. This is the reason why the maximum resolution for strain gauge measurement is obtained by carrier frequency systems with a carrier frequency around 5 kHz.

Summing up all these effects, the best static measurement with the highest resolution can be obtained by AC systems, while for dynamic measurements DC systems allow higher measurement frequencies.

Another limiting factor for the measurement can be influences coming from environmental error sources like:

- Thermo voltage
- Electro-magnetic crosstalk
- Electro-chemical effects
- Supply voltage peaks.

Thermo Voltage

From the strain gauge itself to the last connector on the system there are numerous solder joints or plug connections where different metals have contact. Each connecting pair forms a thermocouple and if these thermocouples are on different temperatures they generate a thermo-voltage. This voltage is a DC voltage that sums up with DC signal voltage and leads to measurement errors that can be in the range of a few percent of the measured signal.

The signal of an AC system is not affected by thermo-voltage. This is another reason why AC systems should be used for balances working over a large temperature range.

Electro-magnetic Crosstalk

Power lines parallel to the measurement wires can induce voltage into the sensing lines and distort the signal, especially near electric drives and electromagnetic relays. This can be avoided by using shielded cables and if applicable, installing wires as far away as possible from such sources.

Strong magnetic fields can also introduce voltage into the sensing lines of a strain gauge circuit. That can be avoided by using twisted pair cables with about 15 twists per meter.

Electro-chemical Effect

This effect only occurs when the strain gauge circuit is contaminated by an electrolyte. This may happen if the gauge application is exposed to salt water, other liquids or dust in combination with moisture. If this happens every combination of two metals in the circuit, like copper with tin or tin with Constantan (Karma, etc.) form together with the electrolyte an electro-chemical element that generates voltage up to several hundred millivolts, making a correct measurement impossible. On the other hand, if such a micro electro-chemical element is supplied, over longer exposure times with a constant DC voltage, electro-chemical corrosion can destroy the gauge grid.

Supply Voltage Peaks

Inside the wind tunnel numerous high voltage systems operate and switched or controlled electrical systems can cause high peaks in the supply voltage of the strain gauge measurement equipment. These peaks can reach several kilovolts in the relative short time of a few nanoseconds and they can affect the entire measurement circuit. To protect the system against such disturbances is very complicated because their characteristics are rarely known in advance. So the best way to eliminate such disturbances is to suppress them by $R - C$ elements at the respective unit.

References

1. Aerospace Structural Metals Handbook. CINDAS/Purdue University West Lafayette IN (2001)
2. Baldwin, C.S.: Optical fiber strain gages. Springer Handbook of Experimental Solid Mechanics, pp. 347–370 (2008)
3. Dorsey, J.: Semiconductor Strain Gage Handbook. Baldwin-Lima-Hamilton Electronics Division, p. 17 (1963)
4. Dorsey, J.: N-type self-compensating strain gages. Exp. Mech. **5**(9), 27A-38A (1965)
5. Hannah, R.L., Reed, S.E.: Strain Gage Users' Handbook. Springer Science & Business Media (1992)
6. Hoffmann, K.: An Introduction to Measurements Using Strain Gages. Tech. rep., Hottinger Baldwin Messtechnik Darmstadt (1989)
7. Kanda, Y.: Piezoresistance effect of silicone. Sens. Act. A **28**, 83–91 (1991)
8. Kreuzer, M.: High-Precision Measuring Technique for Strain Gauge Transducers. Internal publication of Hottinger Baldwin Messtechnik, GmbH, Darmstadt, Germany (1999)
9. Murray, W.M., Miller, W.R.: The Bonded Electrical Resistance Strain Gage: An Introduction. Oxford University Press (1992)
10. Pieterse, F.: Conceptual design of a six-component internal balance using optical fibre sensors. In: 51st AIAA Aerospace Sciences Meeting including the New Horizons Forum and Aerospace Exposition, p. 547 (2013)
11. Pieterse, F.F.: The Application of Optical Fibre Bragg Grating Sensors to an Internal Wind Tunnel Balance. Ph.D. thesis, University of Johannesburg (2011)
12. Sharpe, W.N.: Springer Handbook of Experimental Solid Mechanics. Springer Science & Business Media (2008)
13. Wagner, J.A.: Mechanical behavior of 18 Ni 200 grade maraging steel at cryogenic temperatures. J. Airc. **23**(10), 744–749 (1986)

Chapter 9
Calibration

The aim of the calibration is to find the correct relation between the balance signals and the true loads. However, this seemingly simple statement leads to numerous difficulties. First, not one quantity, but six or more quantities that interact with each other (interference, crosstalk) must be calibrated. Second, the required precision is very high, and third, each of the quantities is influenced by pressure, temperature and temperature gradients, especially in the case of balances for cryogenic wind tunnels or balances with air-line bridges.

Additional to these problems, there is no common regulated standard describing how to perform such a calibration and how to determine the uncertainty of the measurement. Thus, many methods exist, all attempting to minimize the effort for calibration and to obtain the best results.

Some basic principles for the calibration can be extracted from the "Guide to the expression of uncertainty in measurement" (GUM; [12]). In the USA a "Recommended Practice" by the AIAA Internal Balance Technology Group has been worked out that summarizes one specific numerical method, i.e. the so-called "Iterative Method", that is widely utilized in the USA for the analysis of strain-gauge balance calibration data (see [1]). This guide does not cover all possible solutions and is certainly standard, but not still a good guide to work with.

An important basic principle is the traceability of the calibration loads to a national standard. This seems to be a natural requirement, but this requirement cannot be directly fulfilled for the calibration machines that use force generators. The easiest and most frequently used calibration loads that fulfill this requirement with high accuracy are weights, but they are also not free of problems [11].

Beside the correct magnitude of the calibration load, complete knowledge of the direction and position of these loads relative to the given calibration axis system is another basic requirement that must be fulfilled.

Weights act precisely vertical in the geodetic axis system and so either the calibration axis system has to be aligned to the geodetic axis system or the misalignment of the axis system relative to the geodetic axis system has to be measured very precisely

K. Hufnagel, *Wind Tunnel Balances*, Experimental Fluid Mechanics,
https://doi.org/10.1007/978-3-030-97766-5_9

to transform the acting load into the calibration axis system. Both realignment and direction measurement are a source of errors and this has to be taken into account in the uncertainty analysis. Additionally, for the calibration of moments the correct lever arm length has to be known.

The manual calibration by the use of weights is a laborious and time consuming task and there is a natural motivation to automatize this process. A number of systems have been developed and they differ from each other in many ways. There are systems where weights are automatically applied to the balance and so the traceability is given by the weights themselves. Systems where the weights are replaced by force generators have the problem that the traceability to a national standard must be realized in another step of the measurement chain. At first glance this increases the uncertainty, but this is not inevitable, because the opportunity of performing multiple calibration runs automatically enables the use of statistical estimators arising from many repetitions. Another approach to reduce measurements uncertainty is to perform the calibration only within the range of expected test loads; hence, the uncertainty of the balance for any particular test within that range remains lower than the uncertainty derived from a calibration over the entire load range of the balance. The data collected during the calibration procedure must then be analyzed using appropriate software to yield a quantitative relation between the applied true loads and the measured signals. Many different approaches for this step have been developed over the years, according to the particular needs of individual wind tunnels. Therefore, there is no singular approach that can be considered appropriate for all situations. However, the overriding aim for the data analysis should be to minimize the uncertainty of the load evaluation for a performed test.

Summarizing this introduction for the calibration of a wind tunnel balance, the following needs have been identified:

1. Calibration hardware that enables the application of all loads, load combinations and parameters that generate basic sensitivities or systematic errors (bias) in the balance signal.
2. Calibration software that is able to find a quantitative relation between signals and calibration loads and which minimizes the uncertainty for the measurement of the loads in the wind tunnel.

9.1 Calibration Fundamentals

9.1.1 Calibration Theory and Problems

A wind tunnel balance measures up to six load components and the possible load combinations can result in interactions on the signal of any one component. These interactions (or interferences) have to be taken into account by the data processing. The fundamental relation between the loads and the signals of the balance must take the form of:

$$\vec{F} = \underline{E} \times \vec{S} \tag{9.1}$$

where $\boldsymbol{F}$ is the unknown load vector and $\boldsymbol{S}$ is the signal vector. The task of the calibration data reduction is to determine the elements of the matrix E. Matrix E is the evaluation matrix. This matrix will be used in the wind tunnel software to determine the aerodynamic loads from the measured data.

During calibration the load vector $\boldsymbol{F}$ is known and the signals are measured, so in this situation the dependent variable is the signal vector and the corresponding equation is:

$$\vec{S} = \underline{K} \times \vec{F} \tag{9.2}$$

Here the matrix K is called the calibration matrix. The usual approach to determine the evaluation matrix E from the calibration matrix is to invert the calibration matrix K.

$$\underline{E} = \underline{K}^{-1} \tag{9.3}$$

As long as the calibration matrix is a 6×6 square matrix it is called a linear matrix and the inversion of this linear matrix is not a problem. The matrix is called linear because there is only a linear relation between the interactions and the loads. If the requirements on the accuracy of a multi-component force transducer are not very high and the mechanical decoupling of the load measurement sections is good enough, such a linear matrix is suitable to describe the interactions between the components. Such a situation arises typically in some external balances.

In reality the characteristics of the interactions are more complicated and non-linear. Non-linearity may lead to a few percentage difference to the linear characteristic, but the requirements for an internal wind tunnel balance is typically an overall accuracy of less than 0.1%, so nonlinear effects on direct sensitivity and interactions must be taken into account. The result of this are matrices of the order 6×21–6×33. If asymmetrical behavior of the balance for positive and negative loading is significant, the mathematical model can be extended to a 6×84 or a 6×96 matrix using the loads, the absolute values of the loads and combinations of both as described in [1]. It is important to point out that the use of a 6×84 or a 6×96 matrix does not imply that all 84 or all 96 terms must be used. Instead, it is critical to only use the subset of the 84 or 96 term model that (1) can be supported by "physical reasoning" and (2) does not cause unwanted massive non-linear dependencies in the regression models. The resulting complex rectangular calibration matrix cannot be inverted directly to obtain the evaluation matrix. For such a calculation a numerical algorithm for an approximation or an iterative solution must be used (see, e.g. the description of the "Iterative Method" in [1]).

In addition to the large number of unknown interactions, there are unknown tare loads. In other words: the outputs of a balance under no load are not known because most of the calibration principles are not able to apply real zero load. Even the pure balance with nothing on it produces some signal caused by the weight of the balance and the moment generated by this weight. These signals are part of the tare load. Two principle methods are used to take tare loads into account, as caused by the balance itself and by the calibration sleeve. One is to find an approximation by putting the basic configuration in different positions and using a linear interpolation. The other possibility is to take the tare loads as additional unknown loads and calculate them from the data by using a tare load iteration process (see [1] for more details).

9.1.2 Mathematical Models

In Sect. 9.1.1 the equations and the general process of data reduction is described. The problem is how to determine the elements of the evaluation matrix E that are needed to calculate the loads F using the balance signals S. In any case, a linear or non-linear set of equations must be solved, and there is no direct solution. To calculate the matrix elements from calibration data, only approximation methods can be used. There are many possibilities for solving this task and almost every balance calibration lab has its own solver. Given that there is no universal solution, only some principle approaches will be presented here.

Direct approximation of evaluation matrix
The first approach is to approximate the elements of the evaluation matrix E directly. This means that Eq. (9.1) is solved using the signals S and the loads F of the calibration data sets as known variables and the elements of the evaluation matrix are the unknown variables. This approach is often criticized as mathematically incorrect, because the signal S, the dependent variable of the calibration, now becomes the independent variable and vice versa. To explain that this approach is nevertheless possible, the following consideration is made.

Figure 9.1(top) shows the principal problem of the direct approximation. Assuming the function of the calibration curve is of a pure third-order characteristic, the approximation with a third-order polynomial will deliver the exact third-order function. If a third-order fit is applied to the inverted data (the dependent variable is taken as independent variable and vice versa), differences between the fitting values and the measured values occur [Fig. 9.1 (bottom)]. In contrast to the above shown data, the characteristic of a real calibration curve is not pure third order, but rather linear with slight third-order non-linearities. What has been done with the third-order curve in Fig. 9.1 now is done with real data obtained by calibration in Fig. 9.2.

In the diagram for the approximation and the inverse approximation no differences can be seen. Only the error diagrams of both approximations show that the error band for both is of the same order. The same can be seen if error matrices for direct and indirect evaluation are investigated. For the direct approximation a Gauss algorithm is used. Before this algorithm can be used the unknown tare load and the unknown tare moment as defined in Eq. (9.7) must be known. This can be done using an iterative nonlinear Newton approximation for Eq. (9.7). The advantage of this procedure is that no approximation method is needed to invert the calibration matrix K into the evaluation matrix E.

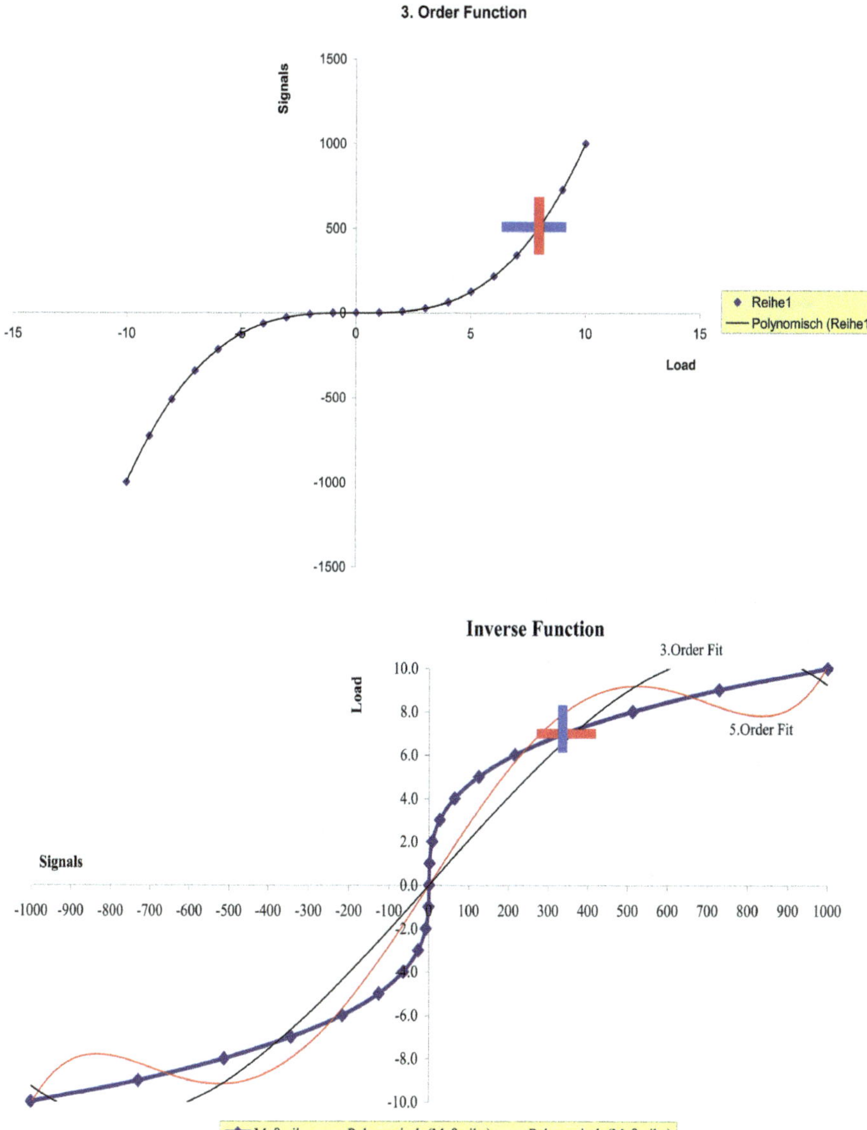

Fig. 9.1 Approximation of third-order function (top) and inverse third-order function

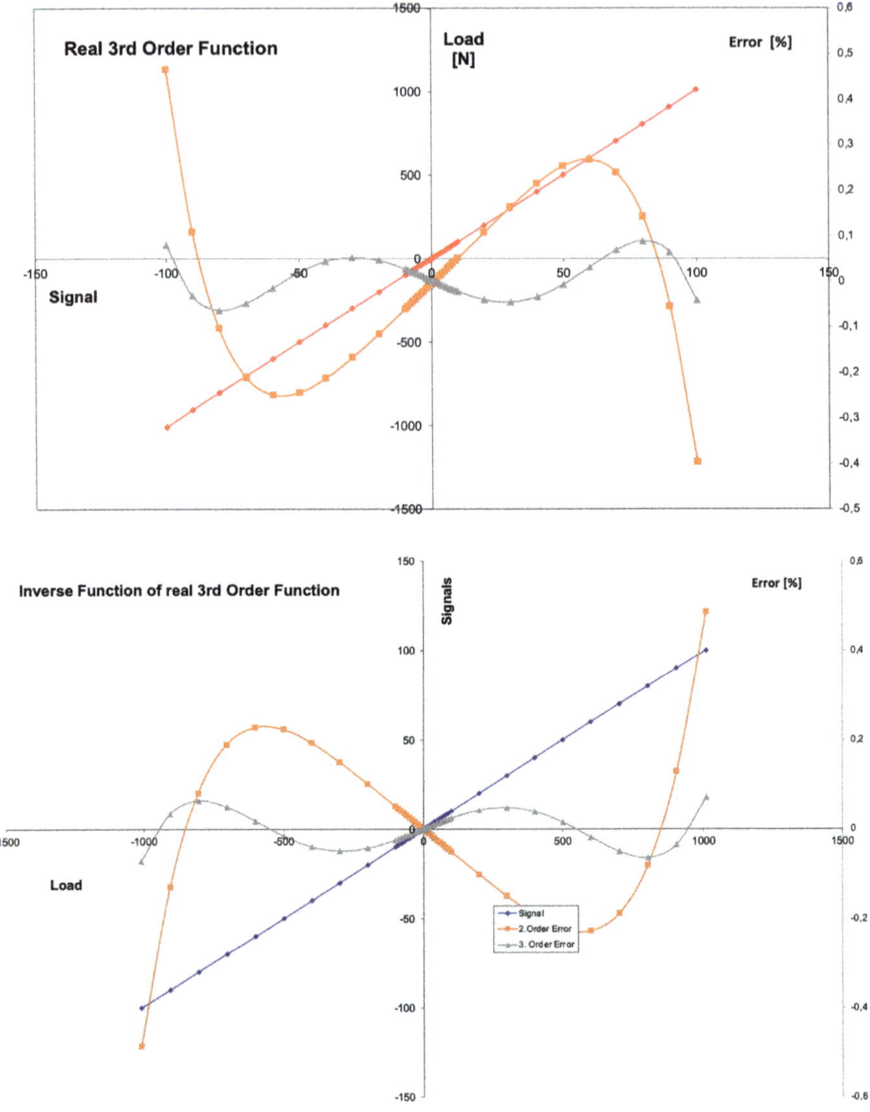

Fig. 9.2 Third-order approximation of calibration data and approximation error

Neural Networks
Another method for the direct approximation of the evaluation matrix E is the use of neural networks. The task of an artificial neural network is to determine the relation between a known input and an output. After identifying this relation, the network is able to calculate the unknown output using given inputs. That is exactly what is required. In preliminary tests of this approach, a commercial program (Membrain) was used to test the neural network for this application. Although the results were reasonable, the uncertainties of this approach were larger than that those achieved using a conventional code; however, this was only a preliminary test and presumably with more effort, comparable results can be achieved.

Indirect approximation of evaluation matrix or loads
Indirect approximation means that in a first step the calibration matrix K is calculated using the data of the calibration. In this case the signals S of the balance measured during the calibration remain the dependent variable and the calibration loads F are the independent variables. To obtain the elements of the calibration matrix K, an approximation algorithm still has to be used. After the elements of the calibration matrix K are known there are two possibilities. The first possibility is to invert the calibration matrix K into a evaluation matrix E. This cannot be done easily because the calibration matrix K usually is a rectangular matrix that cannot be inverted directly and another approximation algorithm must be used to do this.

The second possibility is to use an iterative algorithm that uses the calibration matrix K and the values of the inverted linear evaluation matrix E_{lin} as starting values. With these starting values F_{lin} and the calibration matrix K signals S_{cal} are calculated and these calculated signals S_{cal} are compared with the measured signals S. The difference $(S - S_{cal})$ between the measured signals and the calculated signals is the convergence criterion of the iteration. Once the difference is below the termination criterion, the assumed value of the loads are identical to the measured values $(F_{\text{cal}} = F_{\text{ass}})$. In the past the computationally time required by this approach exceeded limits to provide 'online' values for the loads during a wind tunnel measurement; however, nowadays the computing power is sufficient for this method to be applicable. The main literature that describes the different methods using Newton iteration in detail are: NASA-Iteration [18], VFW-Approximation [7], NSWC-Iteration [10], NLR-Approximation [24].

9.1.3 Description of Interactions

The signals of a wind tunnel balance are influenced by numerous systematic errors and one of the major systematic errors is the interaction of all loads on one single strain gauge bridge. This means that if a certain load combination is applied to the balance, each bridge will react more or less to every load. This is an undesirable situation because the original intention of the design is that any particular bridge should react only to one load. The characterization of the interaction is therefore one of the major tasks of the calibration step. To do this the balance must be loaded with load combinations, mainly with those that occur during the tests in the wind tunnel and with those that have the largest influence on a bridge. The question is, which load combinations should be applied during calibration? There are some strategies to answer this question.

The first is to take the finite element stress calculation and sum up the local strains inline with the direction of the desired load sensitive direction. These are the strains the gauge recognizes for the different loadings. This analysis will provide information about the geometrical interaction that is built into the balance by design. Zhai [26] has done this successfully. The major deficiency of this method is that the interactions of the strain gauge misalignment and manufacturer tolerances are not taken into account.

The second strategy is to take the signal of the single loading to obtain information about the sensitivity on interactions of the non-loaded bridges. In this case the interaction includes the effects by design and manufacturing. The major problem is to perform pure single loadings for all components.

The third strategy is a method called *Design of Experiments (DOE)* or *Modern Design of Experiments (MDOE)*. These methods use a statistical approach to minimize the number of tests when more than one quantity affects a process outcome. All these methods have in common that from measurement to measurement more than one quantity is changed. By analyzing the sensitivities or their variances, the variation of the multiple load change is determined. For the calibration of internal balances the methods are described by Bergmann in [3].

In any case, combinations that never occur in a wind tunnel test should not be used as calibration loads. These combinations will increase the uncertainty for the experiments. Single loads also never occur in the experiments, but they define the sensitivity for the load best, so if possible, single load cases should be performed.

For a calibration of all load ranges of a balance, all full scale loads and necessary full scale load combinations have to be calibrated. With calibration machines the balance can be calibrated with the expected loads for a specific test. This will decrease the uncertainty and this is one of the major advantages of calibration machines in a wind tunnel like *ETW*. However, it is incorrect to conclude that the calibration of a limited load range decreases the uncertainty to the same order of magnitude. For example, if only 10% of full scale is calibrated, the uncertainty is not decreased by a factor of 10, because the sensitivity of the sensor itself does not change and only 10% of the load range is used. The improvement will be achieved by a better description of the interactions and the non-linear effects, which are certainly much lower, while sensor and equipment related uncertainties remain the same.

Mathematically the interactions are the elements of the matrices with two or more different indices. The matrix Eq. (9.1) can be written as a sum of terms:

$$F_i = \sum_{j=1}^{6} A_{ij} S_j + \sum_{j=1}^{6} \sum_{k=j}^{6} B_{ijk} S_j S_k + \sum_{i=1}^{6} C_{ij} S_j{}^3 \tag{9.4}$$

This equation takes all elements of the 6×33 matrix into account. A_{ij} represents 36 linear, B_{ijk}: 126 s-order and C_{ij}: 36 third-order coefficients. The six elements A_{ij} with $i = j$ are the direct sensitivities for the six loads.

During calibration the loads and the combined loads are known and the signals are a combination of the signals due to these loads. For this case the signals are a function of the loads and the corresponding equation can be written as follows:

$$S_j = \sum_{j=1}^{6} K_{ij} F_j + \sum_{j=1}^{6} \sum_{k=j}^{6} L_{ijk} F_j F_k + \sum_{j=1}^{6} M_{ij} F_j{}^3 \tag{9.5}$$

Linear Interaction In Eq. (9.4) the elements A_{ij} for $i \neq j$ are the coefficients of the linear interaction. A linear interaction is the influence of a load component that is not intended to be measured with the transducer. For example, the side force generates a signal in a transducer which is used to measure tension and compression (Fig. 9.3).

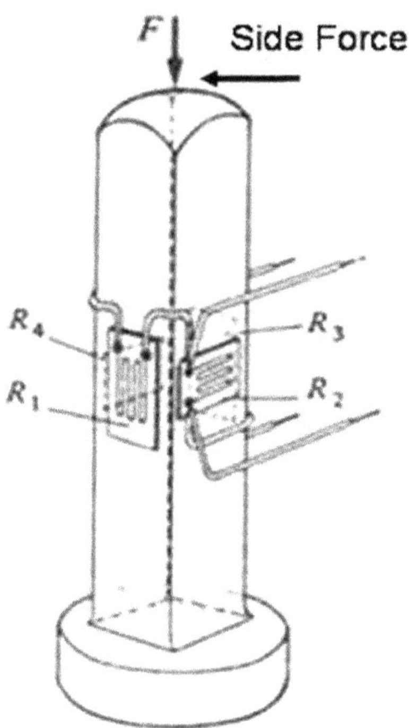

Fig. 9.3 Side force interaction

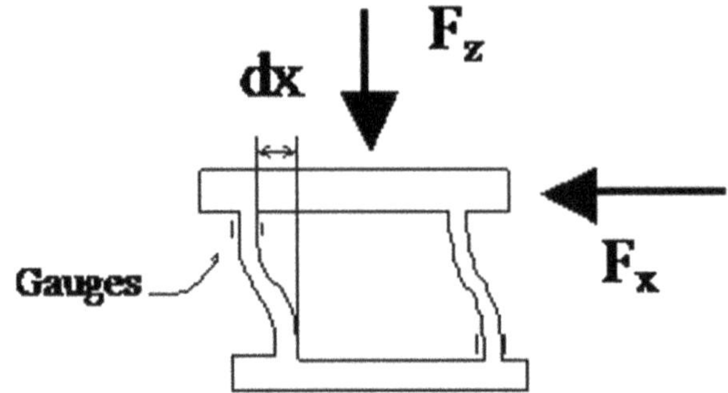

Fig. 9.4 Product interaction of F_x and F_z

The bending stress caused by the side force will also affect the strain gauges because they have a certain area and they are also strained by the bending. Another reason for the sensitivity to the side force is misalignment of the strain gauges themselves. If they are not perfectly aligned in the vertical and horizontal direction they become increasingly sensitive to strain in other directions.

Second-Order or Poduct Interactions

In Eq. (9.4) the elements B_{ijk} are the coefficients of the non-linear interaction. In case of $j \neq k$ the interactions are called product interactions. In case of $j = k$ they are called second-order interactions and describe the second-order non-linearity of the sensitivity.

Product interactions are the sensitivity of a transducer related to the product of loads not to be measured. For example, when an axial force is acting on the parallelogram section of a balance a deformation d_x occurs (see Fig. 9.4). If an additional normal force is loaded, an additional deformation d_x is measured by the strain gauges on the flexure. The additional signal is proportional to the product of F_z and F_x.

Third-Order Interactions

In Eq. (9.4) the elements C_{ij} are the coefficients for the third-order non-linearity. Third-order non-linearity is taken into account only for the third-order term of the load itself, not for any load combinations.

The main reasons for taking third-order interactions into account is the non-linearity around zero, when the load changes from positive to negative and vice versa.

The function shown in Fig. 9.5 is a typical calibration curve of a single load cell. Though the plotted data appears to exhibit a linear behavior, the error diagram indicates that for a second-order fit the errors become up to five times higher than for a third-order fit. The error is expressed as, the difference between the measured data and the value of fit function used.

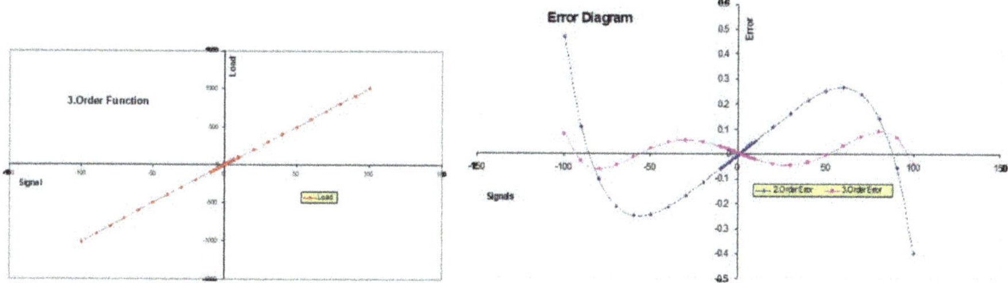

Fig. 9.5 Calibration curve and error diagram

9.1.4 Tare Load Handling

Tare loads are the unknown loads that are generated by the weight of (1) the metric part of the balance,(2) the calibration body, and (3) all other calibration fixtures that are needed so that the calibration loads can be applied. Caused by these loads, the no-load output of the balance cannot be determined as long as gravity is present. For a strongly linear behavior of a balance this is not a real problem and the no-load outputs are the mean value of the signal in normal and upside down positions. This procedure can also be performed for the side force direction using the 90° and 270° roll angle positions. Unfortunately, most internal balances exhibit a nonlinear characteristic and for these non-linear characteristics this method can only be used a a first approximation for the no-load output signal.

In Fig. 9.6 this non-linearity is has been magnified to illustrate the effect. Assuming the tare load is 100 N and the non-linearity is symmetrical (Fig. 9.6 left side), the mean value of the normal and upside down positions deliver the exact zero output, but all

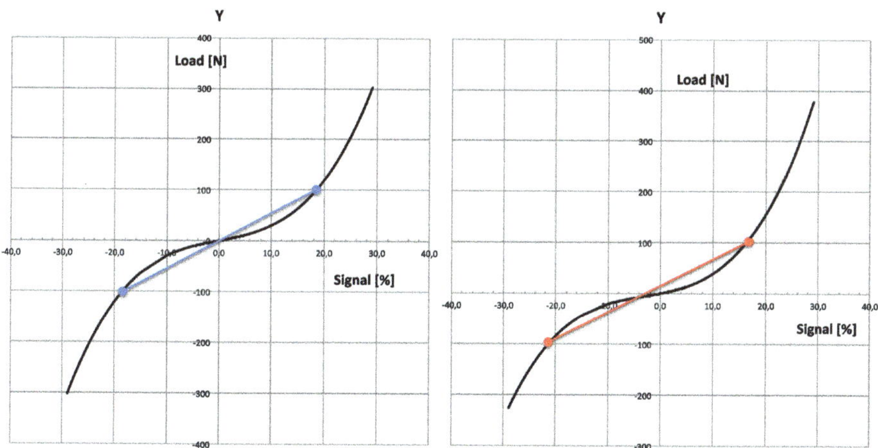

Fig. 9.6 Errors due to non-linearity and tare loads

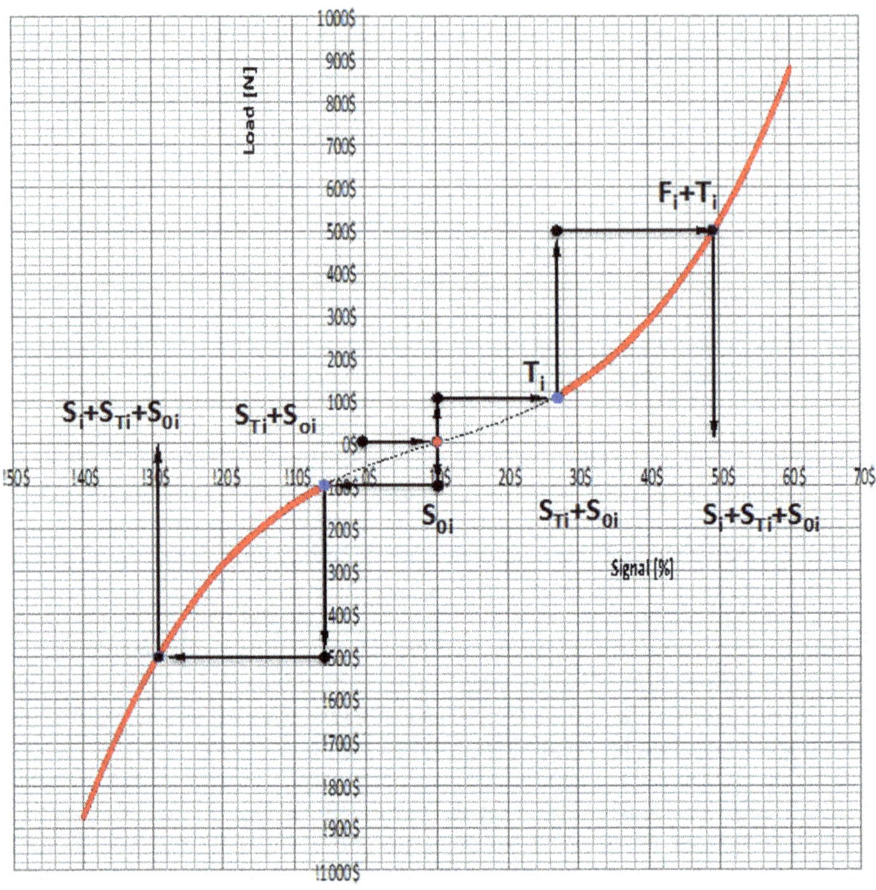

Fig. 9.7 Third-order characteristic with zero load shift

values between the tare load and the zero have errors due to non-linearity. If the non-linearity is not symmetrical (Fig. 9.6 right side) the rotation procedure does not even deliver a correct zero output. This is even more significant if the sensitivities for positive loads and negative loads are different, and this can be seen in Fig. 9.8.

Moreover, in general the no-load output will not be zero. This fact will generate a third-order curve like that shown in Fig. 9.7. This curve is shifted by zero load signals. For better understanding, the nonlinear characteristic is magnified in this figure. For real calibration curves the nonlinear effects can be seen only in the error diagram. The red parts of the curve represent the part that can be measured by applying loads on the balance. The dotted line represents the part that cannot be measured and are caused by tare loads.

T_i are the unknown tare loads and S_{0i} are the unknown zero load outputs. The unknown tare loads (sleeve weight and sleeve moment) and the unknown zero load

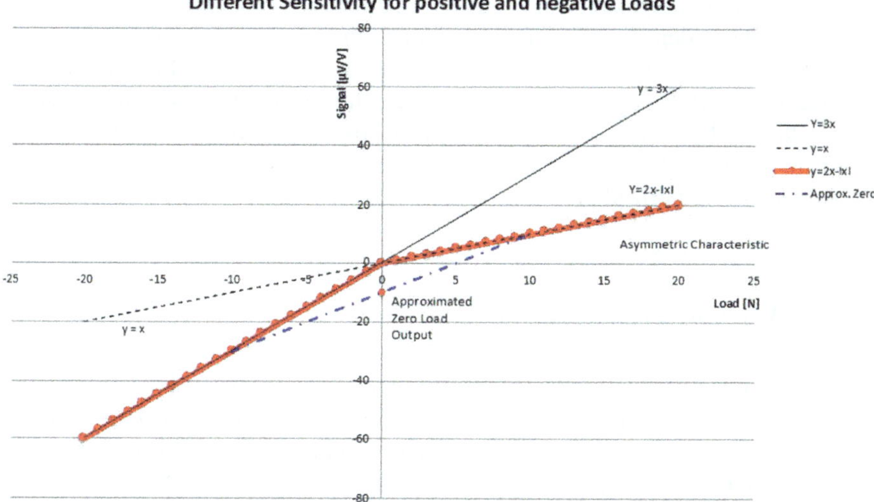

Fig. 9.8 Implementation of absolute value terms

output must be taken as additional unknowns into the matrix evaluation process. Equation (9.4) must now read:

$$F_i + \beta T_I = \sum_{j=1}^{6} A_{ij}(S_j + S_{Tj} + S_{oj}) + \sum_{j=1}^{6}\sum_{k=j}^{6} B_{ijk}(S_j + S_{Tj} + S_{oj}) \cdot (S_k + S_{Tk} + S_{ok}) + \sum_{i=1}^{6} C_{ij}(S_j + S_{Tj} + S_{oj})^3 \tag{9.6}$$

This equation cannot be solved directly and so the unknown tare loads must be derived from the equation where only the tare loads are unknown and the signals are a combination of the signals due to the applied loads and the tare loads. This equation can be written as:

$$S_i = -(S_{i0} + S_{Tj}) + \sum_{j=1}^{6} K_{ij}(F_j + \beta T_j) + \sum_{j=1}^{6}\sum_{k=j}^{6} B_{ijk}(F_j + \beta T_j) \cdot (F_k + \beta T_k) + \sum_{j=1}^{6} C_{ij}(F_j + \beta T_j)^3 \tag{9.7}$$

In this equation $(S_{i0} + S_{Ti})$ is the measured output of the balance at the tare position and T_i is the tare load, i.e. either the tare force or the tare moment. For a manual calibration, the tare load for every orientation is the same because the balance is rotated inside the calibration sleeve and depending on the orientation, the tare load acts on different components and in different directions (positive or negative). This orientation dependence is taken into account by the orientation factor β.

Equation (9.7) can be solved iteratively and the outcome of this process is the calibration matrix K and the tare loads. These tare loads now have to be used to correct the calibration loads and then these loads have to be used to calculate the evaluation matrix. Using this procedure, the non-linearity around the zero output of a balance is characterized best.

9.1.5 Asymmetric Sensitivity

In the AIAA R-091-2017 recommended practice: *Calibration and use of Internal Strain Gage Balances with Application to Wind Tunnel Testing* [1] an asymmetric sensitivity for positive and negative loads is reported for multi-piece balances. This asymmetric behavior must be taken into account by the mathematical model. The basic idea is to implement absolute value terms into the mathematical model. Figure 9.8 shows the basic mathematical approach for the implementation of absolute value terms.

As an example, the sensitivity for the positive loads is $S_{\text{positiv}} = 1\,\mu\text{V}/(\text{VN})$ and the sensitivity for the negative loads is $S_{\text{negativ}} = 3\,\mu\text{V}/(\text{VN})$. So the equation that describes the negative part of the characteristic is $y = 3x$ and the equation that describes the positive part of the characteristic is $y = x$. To combine both into one equation, the absolute value term $|x|$ is subtracted so that the difference of the positive and negative sensitivity ($S_{\text{negative}} - S_{\text{positive}}$) is the sensitivity of the real numbers. The basic approach for the description of non-linear interactions used in Eq. (9.5) can now be extended according the above described approach by the absolute value terms and the sensitivities of the absolute value terms are additional unknowns. Equation (9.5) can then be written as:

$$
\begin{aligned}
S_i &= \sum_{j=1}^{6} K_{1ij} F_j + \sum_{j=1}^{6} K_{2ij} |F_j| \\
&+ \sum_{j=1}^{6}\sum_{k=j}^{6} L_{1ijk} F_j F_k + \sum_{j=1}^{6}\sum_{k=j}^{6} L_{2ijk} |F_j F_k| + \sum_{j=1}^{6}\sum_{k=j}^{6} L_{3ijk} |F_j| F_k \\
&+ \sum_{j=1}^{6}\sum_{k=j}^{6} L_{4ijk} F_j |F_k| + \sum_{j=1}^{6} M_{1ij} {F_j}^3 + \sum_{j=1}^{6} M_{2ij} |F_j|^3
\end{aligned}
\tag{9.8}
$$

If all absolute value terms must be taken into account, the matrix K expands to a 6×96 matrix.

Asymmetric sensitivity is a special problem of the Able type balances (see Sect. 3.2.1.1) that are built from different pieces. Concentric tubes are connected with force transducers by screws. The transducers are sensitive to tension and compression.

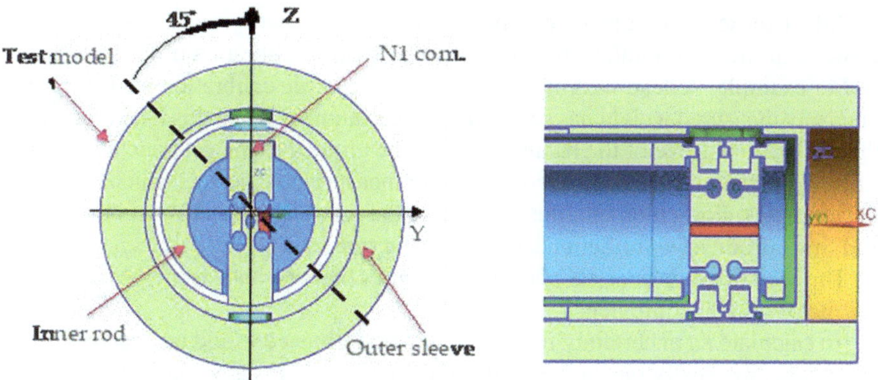

Fig. 9.9 Transducer section of finite-element model of a typical able balance

One problem of assembled balance structures is an increase in creep, related to one piece balances. This creep is caused by imperfect contact between the different parts, through which micro-movements are possible. These movements cause friction that is seen as creep in the signal. Another is the above-mentioned difference in sensitivity for positive and negative loading. Both depend on the quality of the contact between the parts. In a finite element calculation the contact characteristics can be set to different values and so the influence of the contact on the balance signal can be analyzed. Also the requirements for a connection without influence can be formulated. One example of such an analysis and the results are given in [25].

The main reason for the asymmetric sensitivity behavior is found in the asymmetric wall thickness distribution of the inner and outer tubes (see Fig. 9.9). This asymmetry results in a difference in the deformation of tubes for positive and negative loading. The effect was detected by comparison of a balance with central and eccentric wall distributions [25].

The calculations show that a balance with eccentric wall thickness distribution does not only have different sensitivity for positive and negative loading, additionally a side force interaction is created that also has bi-directional sensitivity. The bi-directional behavior is a built-in abnormality that can be prevented by a symmetric design. In that case the extension of the balance matrix to a 6×96 matrix is not necessary.

9.1.6 Verification and Accuracy

The task of determining the uncertainties or the accuracy of a balance is an especially complex task, because of the large number of error sources. These sources can be broadly classified into the following four groups: the balance itself, the measurement equipment, the calibration procedure and finally, the applied loads.

All of these error sources can be subdivided into two groups. The first group consists of the uncertainties that are generated by the calibration procedure itself.

For example: A dead weight seems to be the best true calibration load, but even a precisely measured weight has some small uncertainties. When this load is applied to the calibration sleeve of the balance it has to be aligned to the geodetic axis system. This alignment must be controlled by another measurement device, which has its own uncertainties. For a moment, the distance to the reference center must be measured and this introduces a further uncertainty of the true load.

The second group consists of the uncertainties that originate from the balance and the measurement equipment.

To calculate an uncertainty from every single influence would be a rather complicated task, so usually the back calculated data are used to determine the uncertainty of a balance. The back calculated data are the loads that are calculated using the balance evaluation matrix and the signals measured during calibration.

The difference between the calibration load and the back calculated load is assumed to be representative of the uncertainty of the balance. These data include the uncertainties of the balance and the calibration process. Normally for a precise calibration, the uncertainties caused by the true loads and the calibration process itself are one order of magnitude lower than the uncertainties of the balance. If this is not the case, it is not possible to separate errors and to assign them either to the balance or to the calibration. It is sometimes possible to use a balance with a good repeatability to find a systematic error in the calibration. More precisely, if there are systematic errors in the process, they are not always generated by the balance.

The difference for every loading point is also called the residual load. These residuals contain both random errors and systematic errors (bias). So statistics applied to this data may not be the physical and mathematical correct way to describe the quality of the balance, but for the balance user they provide very useful information on how accurate at last the force measurement with his balance will be.

For a single calibration, some hundreds or thousands of calibration points may be performed and there is one residual load for every calibration loading. So the residual load data include a significant amount of information, but does not necessarily reveal the true quality of the balance. However, by using statistical procedures some significant quality factors can be extracted from the residual loads.

Widely accepted are the following:

1. Minimum and maximum residual load
2. Mean value of the residual loads
3. Standard deviation of the residual loads.

These quantities can be given either in engineering units or in percentage related to the full scale value of the force/moment component.

The mean value of the residual loads (ΔF_{rm}) is given as:

$$\Delta F_{rm} = \sum_{i=1}^{n} \frac{\Delta F_{ri}}{n} \tag{9.9}$$

The standard deviation can be calculated from the following equation:

$$S = \sqrt{\frac{\sum_{i=1}^{n} (\Delta F_{ri})^2}{n - 1}} \quad (9.10)$$

The analysis of the residual loads gives the balance user information about the accuracy class of the instrument under certain environmental conditions. All the conditions that are kept constant during calibration, like temperature and humidity, may have a further influence on the accuracy of the measurement in the wind tunnel. The effects of temperature must be corrected or calibrated separately.

Another source of error not taken into account is creep. Creep effects in the balance must also be tested separately.

All this data supplies information about the accuracy of the balance, but in the end the problem remains how to formulate the requirements for the accuracy in a specification and how to compare these requirements with the calibration data?

One suggestion to solve this problem was made by *Ewald* and *Graewe* in the early 1980s and briefly described in [6]. They formulate an equation that takes into account that the error is influenced by the number and the magnitude of the interactions.

$$\delta_i = A \cdot F_{\text{imax}} \cdot \left(a_i + b_i \cdot \sum_{n=1;n\neq i}^{6} \left| \frac{F_i}{F_{\text{imax}}} \right| \right) \quad (9.11)$$

In this equation δ_i is the allowed residual load and F_{imax} is the maximum combined load of component i. F_i are the loads applied during calibration. The factor A is the general accuracy factor and the accuracy factors a_i are for the individual components i. b_i is a weighting of the interaction. The customer can specify the factors A, a_i, b_i and after calibration the balance manufacturer has to verify whether all the residual loads are below the allowed value.

A more global verification is to calculate the factors A, a_i, b_i from the calibration data and compare them with the specified values.

All other influences on the accuracy of a balance can be accounted for using the Gauss relation for propagation of errors For the influence of the temperature Eq. (9.11) can be written as:

$$\delta_i = A \cdot F_{\text{imax}} \cdot \sqrt{\left(a_i + b_i \cdot \sum_{n=1;n\neq i}^{6} \left| \frac{F_i}{F_{\text{imax}}} \right| \right)^2 c_i{}^2 + (d_i + \Delta T_i)^2 + (e_i + \Delta T_2)^2} \quad (9.12)$$

9.1.7 Traceability

A fundamental requirement for a calibration is the traceability of the applied loads to some national standard. This implies that the applied loads are identical to the definition for force and moment of the national bureau of standards. The force is defined by the mass (m) and the local value of the acceleration due to gravity (g_{loc}). The dimension of the force is Newton [N]. To determine the force, the local value of acceleration due to gravity must be known precisely and of course the mass of the weight. Therefore, the determination of the mass is the key measure to obtain the value of a calibration load.

With good scales, the mass of a weight can be determined with an uncertainty of less than $\pm$ 0.005% of full scale. The scales must be calibrated themselves with standard weights of OIML (**O**rganisation **I**nternational de **M**étrologie **L**égale) class E1 or F1 to ensure that they work within their tolerance. By this chain of traceability, the magnitude of the applied weight corresponds to national standard.

If a load reference is used, for example a load cell, the calibration of the load cell itself with weights is necessary to insure the traceability. If the direction of the load is not exactly known, the orientation measurement system must be included into the process to determine the correct relation between applied load and the national standard.

The influence of hydrostatic lift (buoyancy) in air must be taken into account if the density of the standard weights is different to the density of the weights that are used for the calibration.

$$\Delta G = 100 \cdot \left(1 - \frac{1 - \frac{\rho_{Air}}{\rho_{Cal}}}{1 - \frac{\rho_{Air}}{\rho_{St}}}\right) \quad [\%] \tag{9.13}$$

In Eq. (9.13) ρ_{Air} is the density of air, ρ_{Cal} is the density of the material of the calibration weights and ρ_{St} is the density of the standard weights (OIML E1, F1). Normally steel is used for the standard weights. If steel weights for the calibration are used, the difference is zero. If weights with less density, like water tanks, are used, the error is 0.105%, which is significant for a normal calibration of a balance and cannot be neglected.

9.1.8 Realignment

For a direct calibration with weights, a realignment process as shown in Fig. 9.13 is required. To insure the correct leveling to the geodetic axis system, water levels, bubbles, theodolites or inclinometers must be used. The uncertainties of these devices add up to the overall uncertainty of the calibration load. Fortunately, the uncertainty is almost equal to the repeatability of the value that represents the exact horizontal position. To perform the realignment, a mechanical system that enables rotatory and lateral displacements is required.

9.1.9 Signal Conditioning

To introduce the requirements necessary for the signal conditioning system, first the sensor capabilities are analyzed. Here only the wire strain gauge is analyzed because this is the most frequently used sensor. For a wire strain gauge the natural limiting factor of the signal resolution for the resistance measurement is the thermal noise [14]. Therefore, the maximum signal-to-noise ratio is determined by this limit. This thermal noise can be calculated using the following equation:

$$U_n = \sqrt{4kTRB} \tag{9.14}$$

$$\begin{aligned} U_n &= \text{Thermal noise [V]} \\ k &= \text{Boltzmann constant (1.380662E} - 23\,\text{J/K)} \\ T &= \text{Absolute temperature(300 K)} \\ R &= \text{Resistance (350)} \\ B &= \text{Bandwidth observed (1 Hz relate to } \sqrt{1\,\text{Hz}}) \end{aligned}$$

The values for a typical strain gauge measurement are given above. Using these values for the calculation, the noise voltage is $U_n = 2.5\,\text{nV}/\sqrt{\text{Hz}}$ for a strain gauge. With a carrier frequency based system and a quasi-static measurement like in a wind tunnel, a bandwidth smaller than 1 Hz is realistic. For such a system the thermal noise lies in the same range as for the strain gauge resistance. For both, the noise is in the range of $U_{\text{RMS}} = 5\,\text{nV}/\sqrt{\text{Hz}}$. With direct current systems such small noise levels at a small bandwidth cannot be realized.

Typcically an uncertainty of a 2σ confidence level is realistic and so for a peak-to-peak noise $2 \cdot (2U_{\text{RMS}})$ the uncertainty is $U_{pp} = 20\,\text{nV}/\sqrt{\text{Hz}}$.

For a strain gauge transducer the typical full scale output is 2 mV/V and with an excitation of 5 or 10 V the full scale signal (U_{FS}) is then $U_{\text{FS}(5\,\text{V})} = 10\,\text{mV}$ or $U_{\text{FS}(10\,\text{V})} = 20\,\text{mV}$. The relation of the full scale signal (U_{FS}) versus the peak-to-peak noise (U_{pp}) is the maximum possible resolution in digits or the maximum signal-to-noise ratio.

Figure 9.10 shows the possible resolution for a strain gauge measurement versus the bandwidth. With high quality signal conditioning systems (e.g. HBM DMP 40), resolutions of 2,000,000 digits at low bandwidth of about 0.2 Hz is possible. Carrier frequency based systems also reduce the influence of other error sources like thermocouple voltage or high frequency electromagnetic interference (EMI) and the high resolution effectively can be used for the strain measurement.

After showing what is possible with modern signal conditioning equipment the question remains, what is needed for the measurement in the wind tunnel?

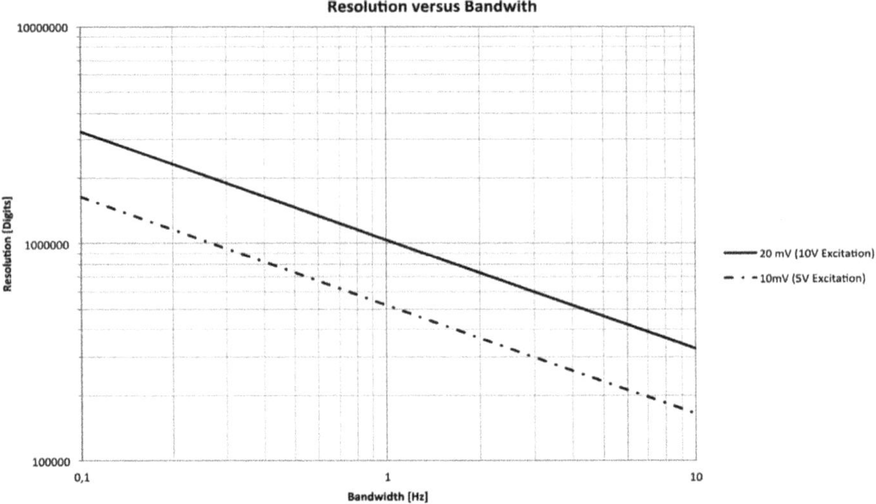

Fig. 9.10 Possible resolution versus bandwidth for 10 and 20 V excitation (HBM)

In the literature about the quality for wind tunnel measurements often one drag count is given as an accuracy requirement. This means that the drag coefficient should be measurable with an accuracy better than 1×10^{-4}; $\Delta c_D < \pm 1 \times 10^{-4}$. This requirement has to be transferred from the dimensionless aerodynamic coefficient to the force measurement [N] of the wind tunnel by using the tunnel and the balance data. Wind tunnels and their balances differ from each other, so only an approximation in the order of $\pm$ 0.15% for the accuracy requirement can be estimated here. The signal conditioning system should not influence the overall uncertainty significantly and therefore the uncertainty of the system should be one order lower than the required one. This means 0.015% of full scale of balance load range. Here it is assumed that the balance load range is perfectly adapted to the tunnel. Normally this is not the case and so the system quality should be below 0.01% of the system full scale output. Good systems offer this quality or are even better, but these systems also have their cost.

Most manufactures of measurement equipment provide an accuracy class of the equipment. This number gives the combined uncertainty for the strain measurement of the system and this value should be lower than the above estimated values. If only uncertainties are given for the individual characteristics, like sensitivity, zero, thermal drift, hysteresis etc., then the overall uncertainty must be calculated according to the test procedures. For example, if frequent wind off data are taken, the zero drift must not necessarily be taken into account, while the drift of the sensitivity plays a major role for the uncertainty.

9.2 Calibration Equipment

For a calibration many instruments are required and some aspects related to these components are now discussed.

9.2.1 Calibration Sleeve

It is very important to understand that the calibration axis system is fixed to this sleeve, not to the balance. As a consequence the calibration sleeve has to be very stiff so that almost no deformation under load will lead to different positions of the load related to the axis system without load. Deformations should be lower than a few hundreds of a millimeter and a hundredth of a degree for rotations. The correct values can be obtained by the following consideration. If a pure force (full scale) is applied to the calibration sleeve without generating an error larger than 0.01% in a component normal to the applied direction, the allowed misalignment in rotation is 0.006°, if the full scale value of the force normal to the loaded direction is the same as the loaded direction.

$$\Delta F_a = F_{n(100\%)} \cdot \sin \Delta\alpha \Rightarrow \Delta\alpha \leq \arcsin \frac{0.01}{100} \leq 0.006^\circ \qquad (9.15)$$

Unfortunately, in the combination of normal force and axial force the full scale of axial force normally is much lower than the full scale value of normal force. So a misalignment in rotation affects the axial force much more. When the ratio between full scale normal force and axial force is 2:1 than the allowed misalignment in rotation must be lower than 0.003° and the value decreases almost linearly with this ratio and makes the stiffness requirement more stringent.

The allowed lateral displacement depends on the ratio between the full scale value of two forces and the full scale value of the corresponding moment. For example, the sleeve is loaded with full scale normal force and full scale pitching moment. The result will be a lateral deformation in normal direction. In this situation an error in pitching moment will be generated by an applied axial force. It has to be lower than 0.01% of the full scale pitching moment.

Using such considerations the stiffness requirements for all deformations of the sleeve can be calculated. The assumption of allowed errors smaller than 0.01% are not very stringent for sleeve deformations. More demanding requirements will result in much lower deformations and a much stiffer sleeve. Such requirements will undoubtedly increase the sleeve volume or the sleeve weight. During calibration the sleeve volume is normally not restricted, but the sleeve weight should not exceed the weight of a normal model. On the contrary, the sleeve weight should be as small as possible, because sleeve weight and sleeve moment are unknown loads and load the balance to some extent (tare load). Calibration points smaller than these tare loadings cannot be performed in a manual calibration and so non-linearities of a balance output around zero cannot be detected by calibration.

9.2.2 Weights

Weights that are used for a calibration can either be trimmed to an exact mass or their exact mass can be measured. To trim weights to an exact value is an expensive task and such weights are usually used to close the chain of traceability to the national standard. For the calibration of balances usually weights are used that have almost the same mass and their mass is determined with a precision scale. It does not matter if this is done in the lab itself or by a calibration service. In the end there must be a number on the weight and a list where the correct mass is documented.

The advantage of trimmed weights is that there is no need to know the order of the weights, because every loading step corresponds to the same value. Using weights with an individual value for every step, the calibration staff has to carefully transcribe the correct value into the data, or make sure that the weight with the required number is applied.

For high loads the weight stack can reach a considerable height and some safety measures have to be considered so that the stack cannot topple over.

In the range up to 10 or 20 kN it is possible to conduct the loading manually with easy-to-handle weights. If one has to calibrate a balance with a load range of a few ten thousands of Newtons (10 kN) load stacks become a real problem. Manual handling of weights up to 250 N,are allowed depending on the local operation safety regulations. For 10 kN this represents 40 pieces, for 50 kN, 200 pieces are needed and it will require considerable time and space to store the pieces for the loading. In this case an automated loading with larger load pieces is necessary or another principle may be the better choice to deliver better results.

Usually steel is used as the material to manufacture calibration loads, sometimes water tanks are used to check the load for large capacity balances. The filling of the tank is straightforward, but this method requires a reference load cell in the loading hanger to ensure the correct value. A simple measurement of the water volume by an indicator gauge is not very precise and the influence of the hydrostatic lift is about 0.105%. A general problem of water tanks is that the water hardly settles down to a static value.

9.2.3 Position Measurement Equipment

If the loads are not aligned with the axis system of the calibration sleeve, the orientation of the loads relative to the axis system of the calibration sleeve must be measured with the same accuracy as the load itself. The uncertainty of the calibration will be dominated by the uncertainty of the worst measurement. If weights are used, normally the uncertainty of all other instrumentation is worse than that of the weights. So the requirements for the orientation measurement systems become very high.

9.2.4 Water Levels

The simple water level can be a very precise and inexpensive instrument for adjusting something horizontally and is often the best choice. The disadvantage is have no electric or digital output is available to connect to a data acquisition system. The accuracy can be checked easily by turning them by 180°. If the same inclination is indicated in both directions the level has been adjusted correctly. Water levels are available with a resolution of 0.01 mm/m. This corresponds to an angle of less than 0.001°.

9.2.5 Inclinometers; Accelerometers

For automatic control of horizontal level alignment electrical sensors must be used. Electrical inclinometers offer the best performance with high accuracy, but they are also expensive. A resolution of less than one thousandths of a degree is possible. The working principle is that a pendulum is kept exactly in place by an electromagnetic coil system. The current through these coils is a measure for the misalignment in relation to gravity. This principle only functions if there are no other accelerations beside gravity, but this is a prerequisite for all level measurement systems.

More recently electronic levels based on silicon sensors are becoming more accurate and less expensive. In such sensors the pendulum is made of a silicon wafer and the compensation of pendulum deflection is based on capacitive reactance. Resolutions of about a few thousandths of a degree are available and these sensors offer a repeatability of one hundredth of a degree. So in most cases electronic levels are sufficient for leveling in calibration applications. Additionally, they offer the possibility to measure small angular orientation changes in roll and pitch direction with sufficient accuracy.

9.2.6 Theodolite

Theodolites are widely used in civil engineering and thus excellent instruments are available for a relatively low price. They can be used to measure vertical and horizontal alignment of linking rods between calibration sleeves and applied loads. The precision of water levels is better, but they measure only at a single point and for the correct alignment of a rod the positions of the end points are relevant for the load vector.

Fig. 9.11 Linking rod under compression

The sketch in Fig. 9.11 shows a linking rod under compression force. The deformation is magnified for demonstration. The force and the reaction force are exactly at the same level, but water levels, if used for the alignment, will always indicate an inclination. The accuracy of a theodolite used in civil engineering is about 0.005°, which is sufficient for alignment. Using a plumb as a reference can always be used to check the correct adjustment of a theodolite.

9.2.7 Laser Based Position Measurement

Lateral deflections and angular changes in yaw direction cannot use gravity as a reference and so other principles must be used to obtain this information. Mechanical devices to measure lateral deflections must have contact with the object and the reference plane. This causes some unavoidable traction that affects the calibration results. To obtain this information non-contact optical devices must be used. A laser can be used as reference. To do so, the initial orientation must be defined by the relevant axis system and in this position the measurements of the optical system define the non-deformed situation. Long-term stability of laser and optical measurement systems is an ultimate requirement and is also the limiting factor for the accuracy of the system. Position sensitive devices (PSD) can be used as sensors. Beside two-axes lateral sensors, there are systems available that offer the measurement of two lateral and two angular deflections with a resolution of about $0.1\,\mu$m and $5\,\mu$rad, corresponding to an uncertainty for the absolute measurement better than $\pm\,10\ \mu$m and $\pm\,0.12$ mrad (0.007°). This is sufficient for a calibration.

While measuring some of the deflections with two sensors, one using gravity and one using a reference laser, the stability of the optical system can be checked assuming that it is the same for all directions.

9.2.8 Reference Load Cells

In calibration machines or loadings using a reference load cell, the load cells define the lowest uncertainty that is possible with the calibration system. The characteristics of most interest are long term stability of zero and sensitivity, low creep, low hysteresis and low thermal dependence in the required temperature range. The best load cells available on the market offer these characteristics:

$$Thermal\,stability\,of\,zero\colon\;0.001\%/K$$
$$Thermal\,stability\,of\,sensitivity\colon\;0.001\%/K$$
$$Hysteresis\colon\;0.002\%$$
$$Creep(20\text{ min})\colon\;0.01\%$$

For the long-term stability usually there are no guaranteed values because there are no specified test procedures and therefore this stability has to be checked by the user. Good strain gauge based load cells usually have a perfect long term stability and to control this a reference load system is required. It is sufficient to monitor one signal for one certain loading. If the signal for zero and the specific loading stays within the specified values, no re-calibration is required. Based on data given from manufacturers for reference load cells, an overall accuracy of better than 0.01% is possible and creep is usually the limiting factor. For calibration machines and the use of a reference force in wind tunnel applications, the highest available quality of load cells is required. These load cells are expensive but their accuracy is the limit for the repeatability of the system and so the load cell sets the ultimate limit for the achievable minimum uncertainty of the calibration. Related to the other costs, investments in load cells pay off in quality of measurement.

The second important load cell characteristic is stiffness. The deformation of load cells for full scale loading is in the range of 0.5–0.3 mm. If the load cell is used in a multi-component arrangement, deformations in this range lead to misalignment of the linking rods related to the given axis system. This is the reason for interactions even in a large external balance. The smaller the deformations of the load cells are, the smaller the interactions are, or the more compact an external balance can be designed.

9.2.9 Cantilevers

To transfer loads from one direction to the normal direction cantilevers are used. The levers can be manufactured with high geometrical precision and the dimension of the length of the arms can be measured with an uncertainty of about one hundredth of a millimeter. The uncertainty of the transfer load by geometry is therefore within a range lower than 0.005%. The critical component is the bearing of the lever. The ultimate requirements for the bearing are no deformation under load, no friction and no break-away force or break-away torque. The bearing that fulfills these requirements best is the knife-edge bearing. When push-pull cantilevers must be used for positive and negative loading, flexures are a practical solution. In some cases also carefully selected and dimensioned ball bearings can be used, but they are the poorest solution with respect to friction. Which bearing is the one with the lowest effect on the transformation of the load cannot be answered in general. It is always a compromise between the requirements and the specific disadvantages.

9.2.10 *Knife-Edge Bearings*

Knife-edges have been well developed over the years and are widely used in mechanical scales. They have the lowest friction and the lowest deformation under load. Usually the wedges and sockets are made of hardened tool steels. In special cases sockets can be made of agate stone. One knife-edge bearing can be used for each direction. In case of a push-pull application, two different knife-edge bearings are required. Additionally, the correct position of the wedge in the socket must be controlled after every change of a load, making the use of knife-edges rather complex. The problem mainly occurs when the bearing is low loaded and the self-centering of the wedge in the edge does not function very well. So knife-edge bearings work best when a certain pre-load can be applied.

9.2.11 *Ball-Socket Bearings*

Ball socket bearings are not recommended because of the high local stress in the contact area. The ball flattens there elastically and the consequence is a relatively high break-away torque.

9.2.12 *Ball Bearings*

The rolling elements inside a ball bearing also deform under load and this leads to break-away torque which results in hysteresis and friction. These unwanted effects can be reduced if a relatively high static load rating is chosen. Then the bearings are relatively large but the deformations of the rolling elements are small. Of course high quality bearings must be selected to minimize the inner resistance caused by manufacturing tolerances. No grease and no versions with dust sealing are allowed, because this increases friction. The bearings must be carefully de-greased with solvent and if oil is needed, only very thin oil should be used.

9.2.13 *Flexures*

Flexures are elastic bearings that allow a certain movement with almost no friction while allowing push-pull loading. The disadvantages of flexures are small reaction forces and moments in every deformation direction. The requirements of stiffness for small deformations under load and low elastic reaction loads of the bearing are contradictory requirements and a good compromise must be found in the design

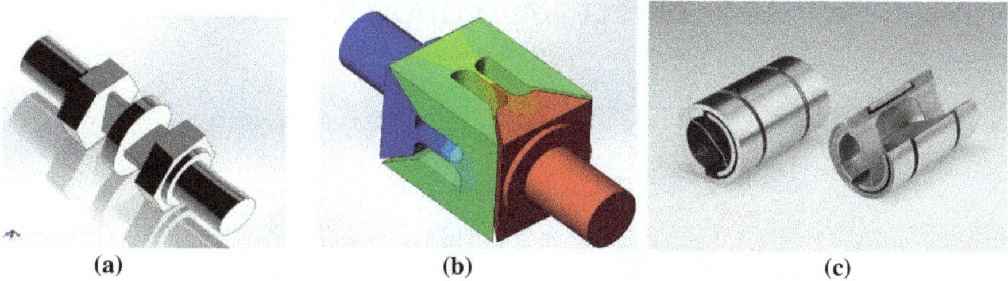

Fig. 9.12 Types of flexures **a** orthogonal separated; **b** Cardan joint (Ormond); **c** swivel joint (Riverhawk Co.)

of the flexure. In any case the flexure will deform under load and the deformation requires realignment.

In Fig. 9.12 some flexure types are shown. Figure 9.12a, b show flexures that can be used in linking rods as pivots to minimize the transfer of parasitic moments. Figure 9.12c. shows a swivel joint that can be used to support a lever or a bell crank.

The *Cardan* joint is a rather complicated device, but both orthogonal hinge axes intersect at almost one point. To build this kind of flexure requires a finite element calculation to find the best compromise between flexibility and stability and the manufacturing process is also rather complex and costly. If relatively long rods can be used, the orthogonal separated flexure type offers the best compromise between cost and flexibility. With a long rod between the flexures the relative movement of both axes does not result in significant parasitic transfer loads.

9.3 Calibration Principles

The introduction to this section describes in general the boundary conditions for a calibration. However, these boundary conditions allow a large variety of possibilities to calibrate a wind tunnel balance. The different principles are now characterized and described in more detail.

Manual or automatic calibration
The main differentiating factor is whether a calibration is performed manually or automatically. In this context automatic means the calibration process is completely carried out by a machine. Manual means loading and realignment is conducted completely manually or partly automatically. For example, for very high loads it is not practical to load the balance manually because the weights are too large and so some device must be used for this step, but the realignment and all the control processes are carried out manually.

Direct or indirect calibration
There are two possibilities to load a balance for calibration. The usual way is to fix the balance at the sting end and to load the metric end, because this is the configuration

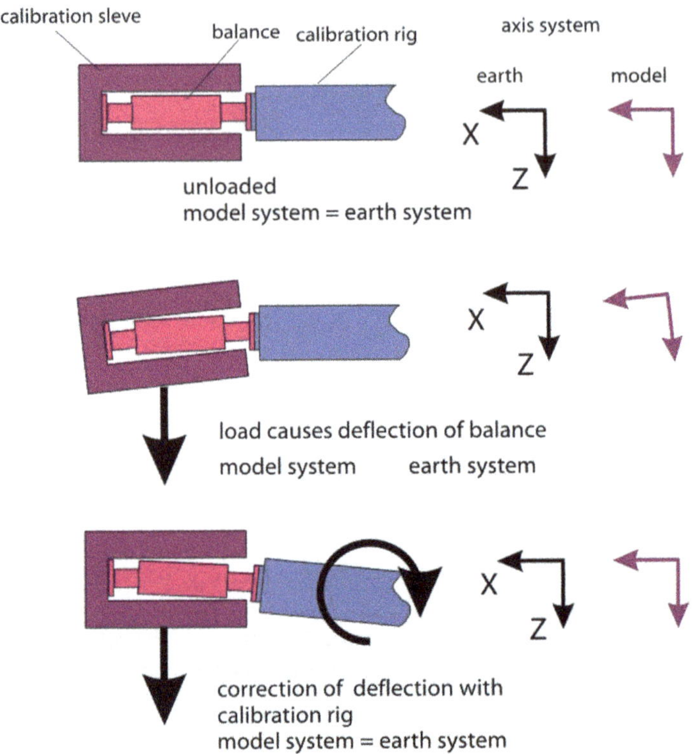

Fig. 9.13 Direct calibration principle

used during wind tunnel tests. This method is called direct calibration. However, it is also possible to fix the metric end and to load the sting end. This way is called indirect calibration.

Direct calibration

As already mentioned in Chap. 2, the reference system for the calibration is fixed to the model. During calibration there is no model on the balance so there must be something that represents the model to apply the calibration loads. Usually this device is called calibration sleeve and it is indeed something like a generic wind tunnel model so that the loads and load combinations can be applied through this model to the balance. During direct calibration the loads have to always be aligned to the axis system of the sleeve or their orientation relative to the sleeve axis system has to be known exactly. Beside the use of precise loads, either a realignment or an orientation measurement system is required for this direct calibration.

Indirect calibration

For an indirect calibration the balance is fixed on the metric end and the axis system of the calibration is identical with the axis system of the fixation. The orientation and the magnitude of the load must be now known exactly in relation to the axis system of the support. In other words, the real calibration loads are the reaction forces in the fixation, and that is why this is called an indirect calibration. A realignment system in this case is not necessary, because both axis systems—balance axis system and support axis system—are always identical.

When weights are used as calibration loads their correct position related to the calibration axis system must be measured. Especially the distance between the non-metric end of the balance and the calibration system changes by the deformation of the balance under load. If the fixation (support) is a load reference system, no further deformation or orientation equipment is needed (Figs. 9.14 and 9.15).

Using either the direct or the indirect method, several different principles can be realized.

Direct manual calibration with weights

As described above in a direct calibration the non-metric end of the balance is fixed to a calibration rig and the metric end is connected to a calibration sleeve that is loaded with the calibration loads.

Normal force and pitching moment can be easily applied by hanging weights on the sleeve at different positions. Side force and yawing moment can also be applied by weights, after the balance is rolled by 90° or 270° inside the sleeve. The only force that has to be applied horizontally is the axial force. To use weights a deflection lever is required for transforming the vertical force into the horizontal direction. A simple vertical arrangement of the balance for the axial force loading is not sufficient because in this position no combined loading to calibrate the interactions is possible. The interactions of normal force and moments on the axial component are the dominating ones. Therefore their calibration is the most important.

Direct automatic calibration

For the direct automatic calibration the balance is mounted with its non-metric end to a support and a calibration model covers the metric end. The loads are applied to the calibration model by load generator. Additional to the load generator a direction measurement device and a load transducer have to be implemented in every load line. Two methods are applicable.

1. Measure the orientation of the load vector relative to the axis system of the calibration model with the orientation measurement system and calculate the components in the calibration axis system from the vector. The sum of all components in the same direction is the calibration load (Fig. 9.16 left).
2. Use the orientation measurement system to realign the load axis exactly to the calibration axis system by moving the base support. Then the measured load in direction of the load generating is the calibration load.

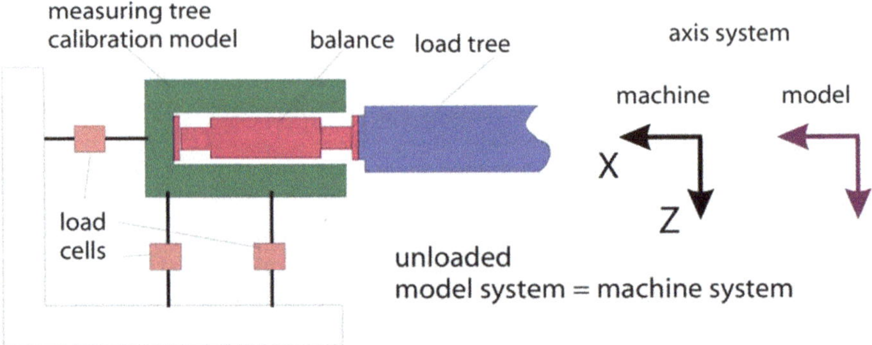

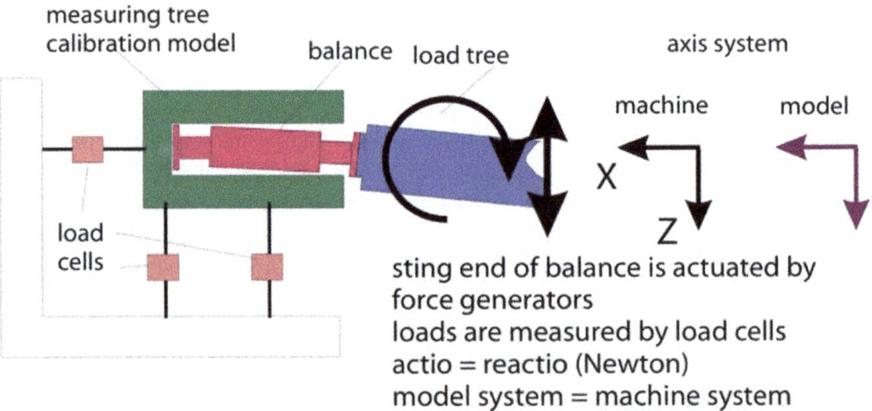

Fig. 9.14 Indirect calibration principle

Indirect manual calibration
Manually the indirect calibration method can be realized by using an external wind tunnel balance as a reference balance and mount the sample balance with the metric end to the metric end of the external balance. Loads are applied on the non-metric end of the sample balance manually and the external balance measures the calibration loads. This method has been successfully practiced to calibrate a half-model balance by using an external balance (see Fig. 9.17).

The major precondition for a successful calibration is that the overall accuracy of the external balance is much better than the required accuracy for the sample balance.

Indirect automatic calibration
For an indirect automatic calibration the metric end of the sample balance must be connected to a reference balance that measures exactly the loads. On the non-metric end a load generation system has to be added. To do this there are two possibilities.

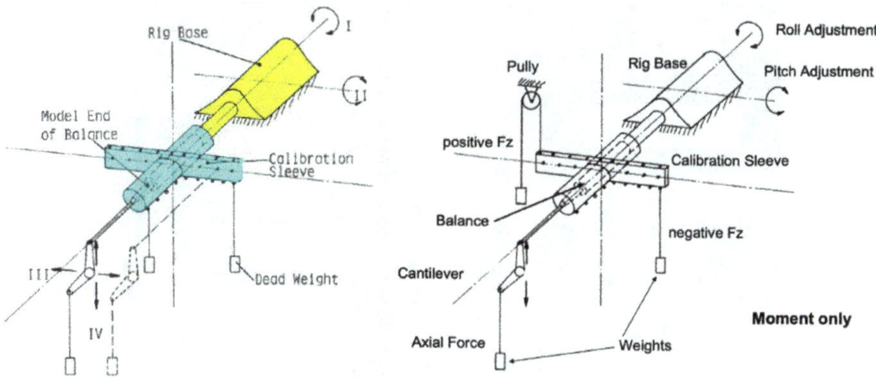

Fig. 9.15 Direct calibration manual calibration

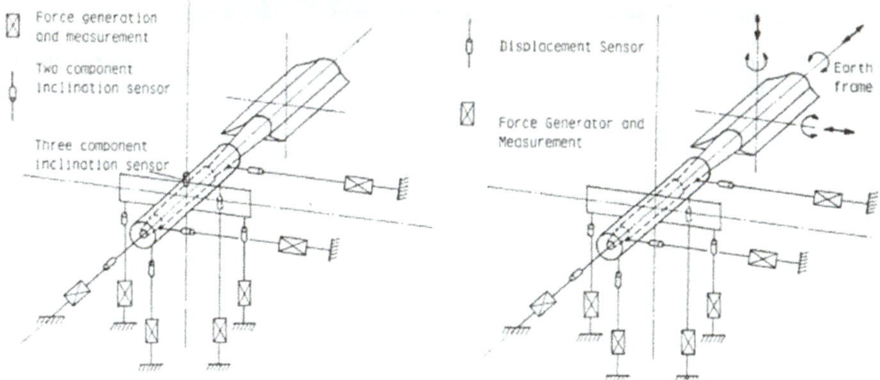

Fig. 9.16 Direct calibration with automatic load generation

1. Cover the sample balance with a sleeve and apply the loads to this sleeve. The advantage is that a force generator can apply pure loads if it is aligned with the direction of a calibration axis. The deformation of the balance will cause a change in the orientation of the sleeve and unavoidable small loads will affect the other components, but the reference balance measures these interactions.
2. Use two force generators in every plane to generate force and the corresponding moment (see Fig. 9.18 left diagram). If the load generators are arranged symmetrically to the calibration axis system, they can be dimensioned identical for every plane and the full scale capability of the generators can be smaller than the maximum required load. The mounting and dismounting of a sample balance always requires a complete disassembling of the load generation system and makes the use of the system inconvenient.

If the complete load generation system is attached to the non-metric end of the sample balance, as shown in right diagram of Fig. 9.18, the actuating system can be

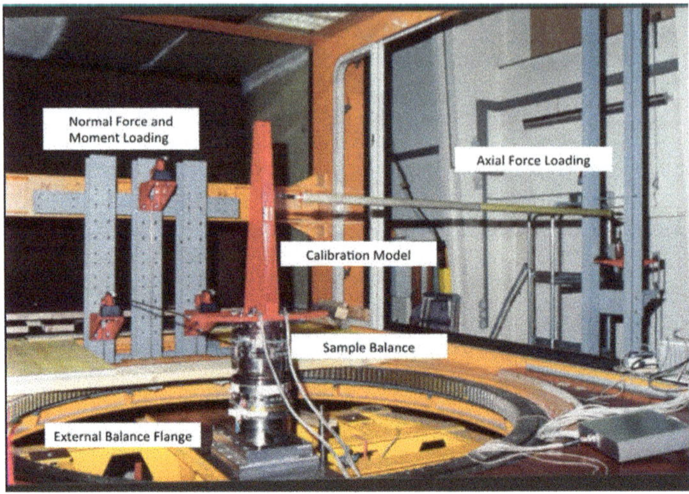

Fig. 9.17 Calibration of a half-model balance on the external wind tunnel balance in the TU Darmstadt low speed wind tunnel

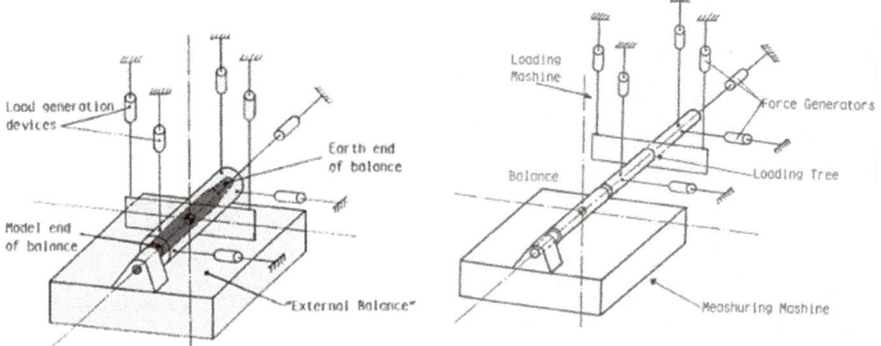

Fig. 9.18 Indirect automatic calibration

dismounted much easier. To produce a pure force in this arrangement requires a full scale capability for some generators that is higher than the required full scale force.

The calibration machines designed by TU Darmstadt follow more the principle of using a calibration sleeve because the loads in the force generators are much smaller. However, instead of using a sleeve a so-called loading tree was used (yellow in Fig. 9.19) around the space where the sample balance is located. In the calibration machine for the ETW a climatic chamber occupied most of the space inside the loading tree. This chamber was necessary to calibrate the balance throughout the operating temperature range of ETW.

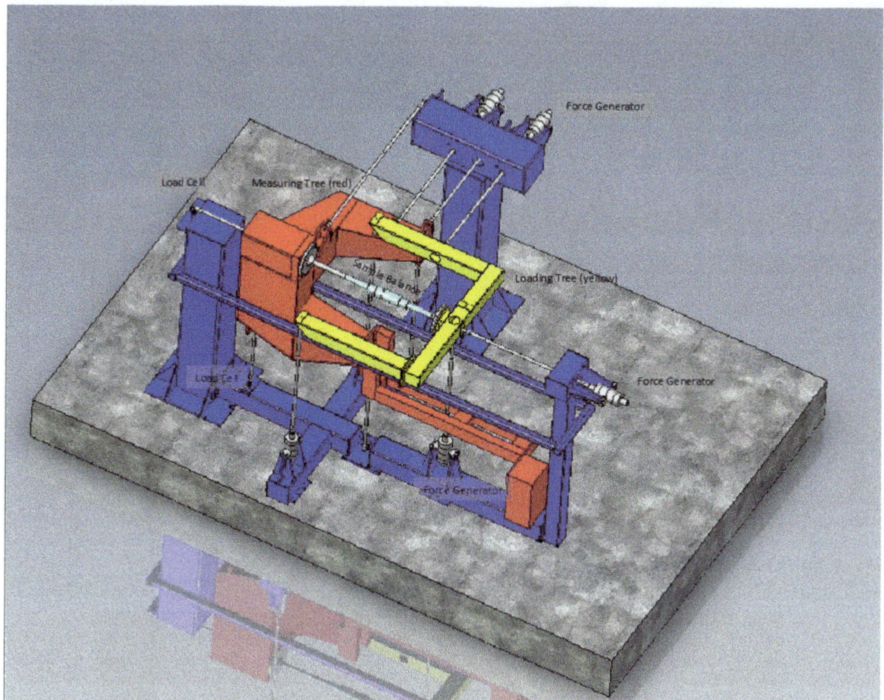

Fig. 9.19 TU Darmstadt balance calibration machine

9.4 Direct Manual Calibration

Manual direct or indirect calibration is possible as described in Sect. 9.3. Here the equipment and the process of a direct manual calibration, which is most widely used, will be described. A manual calibration rig is shown in Fig. 9.20. This rig was built for balances of small or medium size, but it still consists of all necessary hardware for arbitrary balance calibration. The rig consists of a main frame (Base Frame) that represents the earth side.

On top of the rig a support is mounted where the balances can be attached to a flange. The flange can be rotated by 90°, 180° and 270° to enable positive and negative side force loading, positive and negative normal force and the corresponding moments. To do this the fixation must be uncoupled and then the flange can be rotated. Fixation is only possible in the 0°, 90°, 180° and 270° position. After fixation, small roll variations with a spindle drive are possible (Fig. 9.21).

To adjust the calibration sleeve in pitch, the support can also be swiveled in pitch direction by a spindle drive.

On this rig normal force and side force are loaded separately from each other by weights. If normal force and side force interactions have to be calibrated, an additional bell crank support for side force loading is necessary. The corresponding

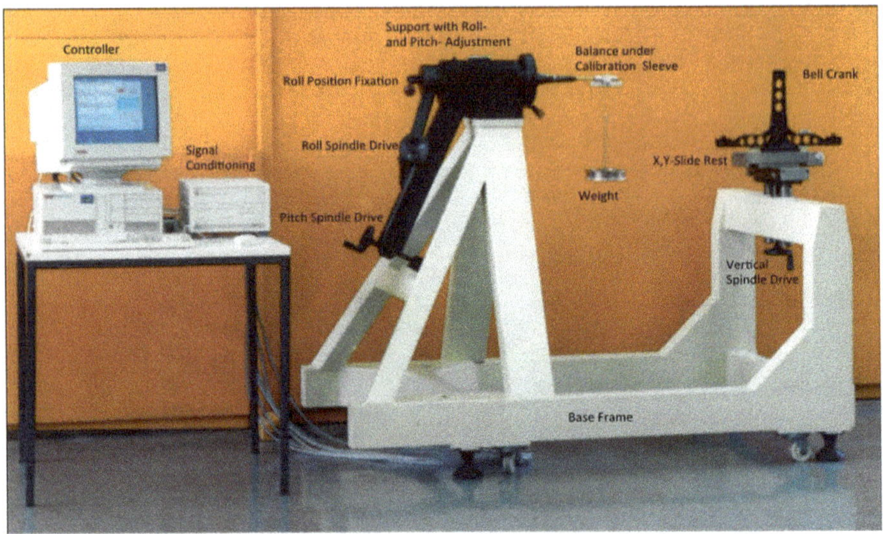

Fig. 9.20 Manual calibration rig

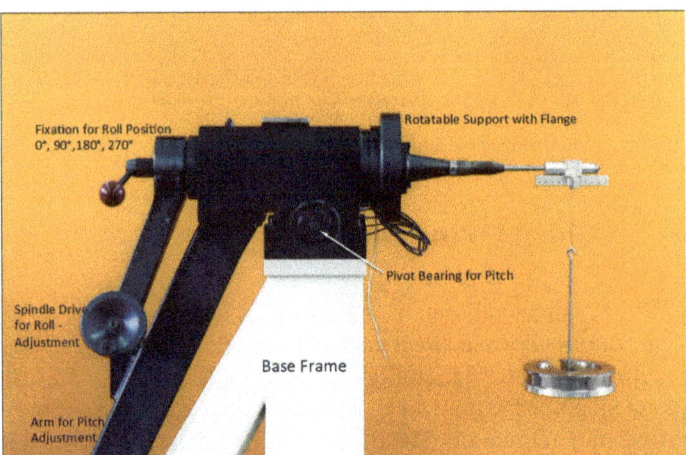

Fig. 9.21 Balance support

moments are applied by hanging the weights off balance center on the sleeve. The computer for data acquisition shown in this figure and the calibration program are rather old-fashioned, but nevertheless with new equipment the functionality will be the same. For signal conditioning an HBM device with 21 bit resolution was used which is sufficient for a precise calibration.

For axial force loading a support with bell crank is used to transfer the load from the vertical into the horizontal direction. This support can be adjusted in X and Y direction by a compound slide rest and vertical by a spindle drive. By moving the

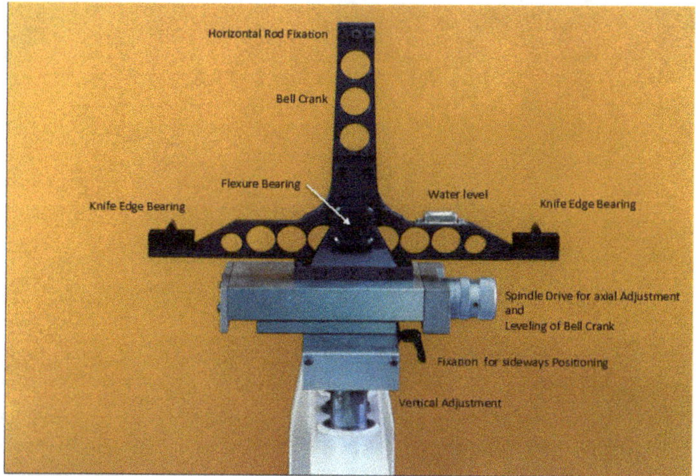

Fig. 9.22 Bell crank for horizontal loading

support out of the X axis in Y direction a yawing moment can be applied by an axial force. The bell crank is shown in detail in Fig. 9.22. The bell crank is pivot mounted by using elastic swivel joints. These bearings could be used because the loads on this rig were very low and they cause minimum hysteresis under these loads. The weights are applied via hangers on the knife-edge bearings. The knife-edge bearing can be initially adjusted in lengths such that the transformation ratio is exactly one to one. This ensures the best positioning repeatability and accuracy. When the load is applied, the bell crank will rotate and it has to be re-positioned by moving the support backward or forward with the spindle drive until the water level indicates the correct horizontal position (Fig. 9.23).

The calibration sleeve looks like a generic wind tunnel model that enables a variation in loading points to apply loads similar to that in a wind tunnel test. To enable loading in the different rotation positions the sleeve must be re-rotated to the horizontal position after every turn. The design of the calibration sleeve has to fulfill this requirement with a high repeatability. Additional to this requirement, the sleeve has to be so stiff that the sleeve deformation does not falsify the calibration load. This is in contradiction to the requirement of a light calibration sleeve to keep the tare loading low. Calculations should be performed to find a good compromise.

After description of all the hardware for the calibration now the basic procedure of a manual calibration is described and recommendations are made for the calibration program used to control the process.

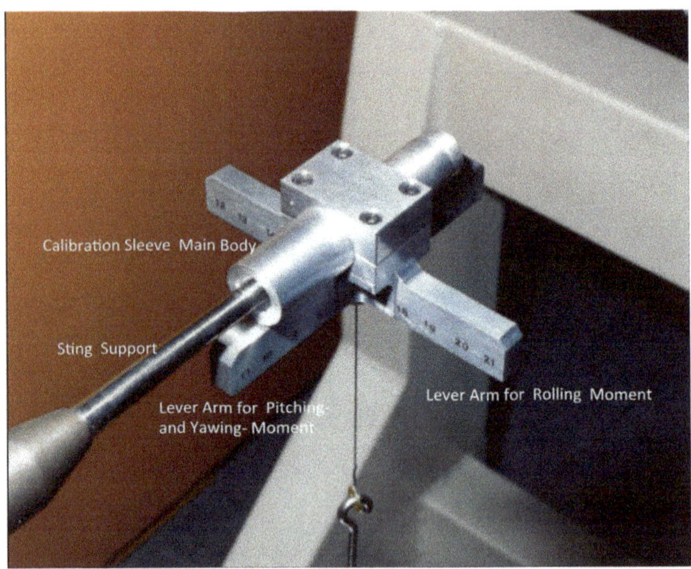

Fig. 9.23 Calibration sleeve

9.4.1 Calibration Program

In the program that controls the calibration process, a large number of parameters must be set and a large amount of data is to be acquired. It is practical to separate the data into a parameter file that contains all the data that is needed for a certain calibration, and measurement data files that are subsequently used for the data evaluation and matrix generation. The advantage of this approach is that the measurement data files are not overloaded with repetitious parameter data and inspection of the data when trouble-shooting is much easier. It should be ensured that corresponding parameter and measurement files are always stored at the same place.

The quantities to be documented in the parameter file are:

Balance Type:

Internal; external; semi span; etc.

Signal conditioning:

Measurement equipment

Number of measured bridges

Bridge configuration (full; half; ...)

Excitation voltage

Range

Filters

Gain

Sampling rate

Integration rate and time

Settling time
Number of temperature measurements
Type of temperature sensor
Characteristic of temperature sensors
Characteristic of inclinometers
Characteristic of alignment sensors

Loads:
Local acceleration due to gravity
Mass of weights
Primary load, combined load

Calibration sleeve:
Geometric data of the calibration sleeve
Orientation

Ambient conditions:
Air temperature
Air density
Humidity

Limits:
Allowed signal drift with time
Allowed signal difference between measurements
Allowed standard deviation for data recording
Allowed deviation of temperature
Maximum signal for security warnings
Allowed deviation of inclination
Allowed deviation of alignment

All the parameters must be recorded in the parameter file for the calibration of a certain balance so that it is possible to reuse or check them.
In the data file for the calibration itself the following data must be recorded:
Name of parameter file
Time
Primary load, combined load
Balance signals
Loads
Temperatures
Humidity
Sleeve orientation
Loading point

During the loading process the quality and stability of the data can be automatically controlled. For example, if the standard deviation of averaged measuring points exceeds a given value (allowed standard deviation) the program waits with data storage until the required standard deviation occurs. The reason for a high standard deviation can be a small, almost invisible movement of the load or vibrations in the rig. Such an automatic control prevents taking poor data and in the end saves time. The control of air temperature prevents taking data with a large temperature influence, and so on.

Another useful check is to calculate a sensitivity after the first two loadings. With this sensitivity the signal of the next loading can be predicted. If the measured signal of the next loading is outside a certain allowable window related to the estimated signal, than probably something is wrong with the load. For example, load numbers are mixed up or a weight was not applied or water levels for checking the alignment were left on the balance. Such mistakes in the process can be immediately detected and the measurement can be locked until the mistake is eliminated.

The process itself consists of two main parts, first the loading process and second the data evaluation and matrix generation.

Before a loading process is started the whole measurement chain must be calibrated. Nowadays this can be done by automatic processes but these processes must be checked by a measurement standard to guarantee the traceability to the national standard.

Over the entire duration of the process the ambient conditions (temperature and humidity) have to be recorded. Even if the temperature in the calibration lab is not stabilized to a certain value, this is a basic requirement for calibration laboratories (see EN/ISO 17025).

The loading process itself can be performed according to the following flowchart (Fig. 9.24).

The first two steps are obvious. After the balance is functional, some tests should be made. The calibration program should provide the possibility to take the signals in this adjusted position without the sleeve, or the signal has to be recorded manually. If these signals are logged over the balance lifetime, differences in these signals are the first to indicate if something is wrong with the balance.

The next step in this position is a check of the correct signs by loading the balance manual at the metric end. The signs and the functionality of lift and pitch, yaw and side force and axial force can be checked easily. To check roll with a simple twist by hand is not that easy on larger balances and possibly a simple tool is necessary for that.

If these first checks are satisfactory, the sleeve can be mounted to the balance. When this is finished the sleeve has to be aligned to the geodetic axis system. If there are large discrepancies between the no sleeve and the sleeve alignment, this is the point where the geometric accuracy of the model adapter and the calibration sleeve must be rechecked and eventually corrected. If inaccuracies are accepted, the misalignment will increase the interactions. After the alignment of the sleeve to the axis system the tare load signal has to be taken. In the program these signals can be used to check automatically the correct condition (alignment, hysteresis, left water levels and left loading devices) on the balance after de-loading.

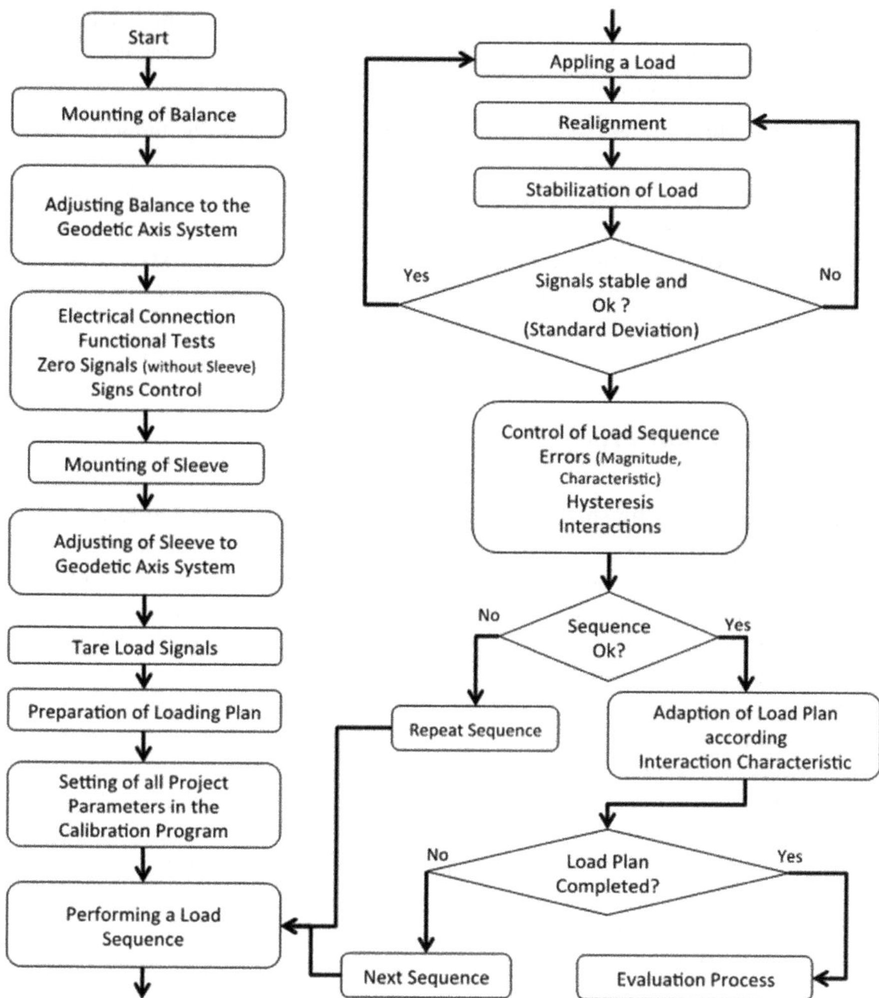

Fig. 9.24 Flowchart of loading sequences

Now the balance is ready for loading. At the beginning of the load sequences some consideration for the loading plan has to be made. As already stated, there is a philosophy for the type of loads and load combinations that have to be performed.

On the one hand the varieties of loads and load combinations that can be calibrated depend first on the rig that is being used. For example, on the rig described above no single moment calibration is possible. Moments can only be calibrated in combination with a force. This is not a major disadvantage, because in a tunnel test there are no single moment load cases.

On the other hand, the load combinations that have to be calibrated depend on the magnitude of the interactions inside the balance. These differ from balance type to balance type. Manufacturers mostly know the characteristics of the balances they build very well, so the philosophy of the load plan differs from manufacturer to manufacturer. Some general rules for the load plan are:

1. Calibrate only loads and load combinations that can occur in the tunnel or during a test.
2. Calibrate single loads even though they do not occur during test, to determine the major sensitivities.
3. Perform load combinations that have large interactions and are expected to occur during the test. Significant interactions deliver signals in the order of the allowed error.

9.4.2 Evaluation Process

After all loadings are finished the evaluation process can begin. Up to now every single loading has been checked for quality, but there is no information whether there are discrepancies between the different loadings. For example, if a combined loading is performed there are some loadings of the primary load variation and changes in the combined load from sequence to sequence. To check problems and error areas all data are stored in one file and plotted as a "visual matrix" as shown in Fig. 9.25. The first diagram of this visual matrix is a plot of all axial force signals versus axial force. The next diagram (first column second row) is a plot of the side force signals versus axial load and so on.
In this matrix irregularities can be readily recognized. For example, the offset in the diagram of the signals of side force versus side force (second column, second row) which is caused by the tare load, or a sign error in normal force (third column, third row). This information can be used either to correct the data files or to repeat the calibration sequence to find the reason for the inconsistency.

After all major problems are eliminated, a first matrix is calculated and a verification calculation is performed. The data calculated with this process are used to plot what is known as the 'visual error matrix', as depicted in Fig. 9.26. The first diagram of this visual matrix is a plot of all axial force signal errors versus axial force. The next diagram (first column second row) is a plot of the side force signal errors versus axial load, and so on.

In this matrix, areas with large errors can be easily detected. To do this, either the scale of the ordinate must be checked, or if the same scale is used for all, the width of the dot distribution indicates areas or large errors. After the analysis of some of these patterns for different balances, there will be some experience about how a pattern for a good balance should look like and if the errors are systematic errors or random errors.

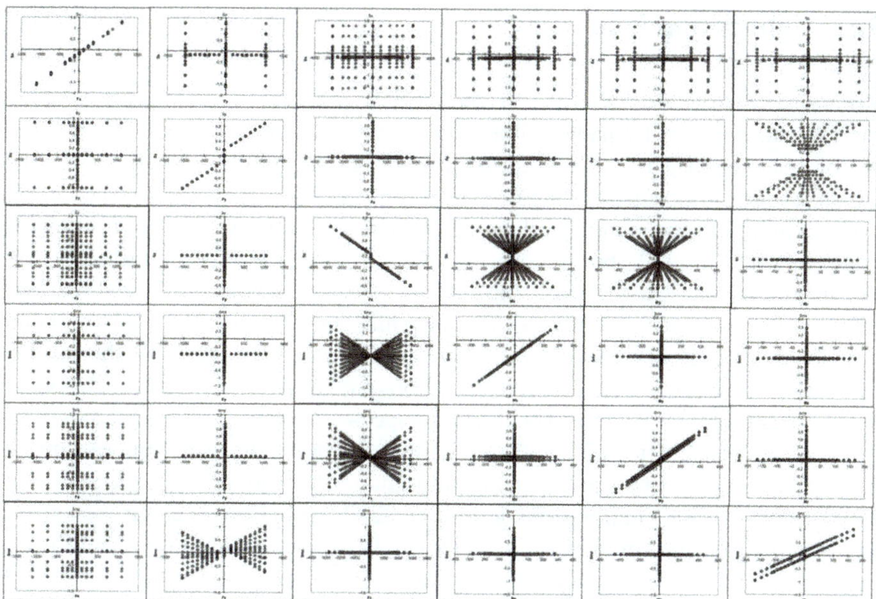

Fig. 9.25 Visual matrix

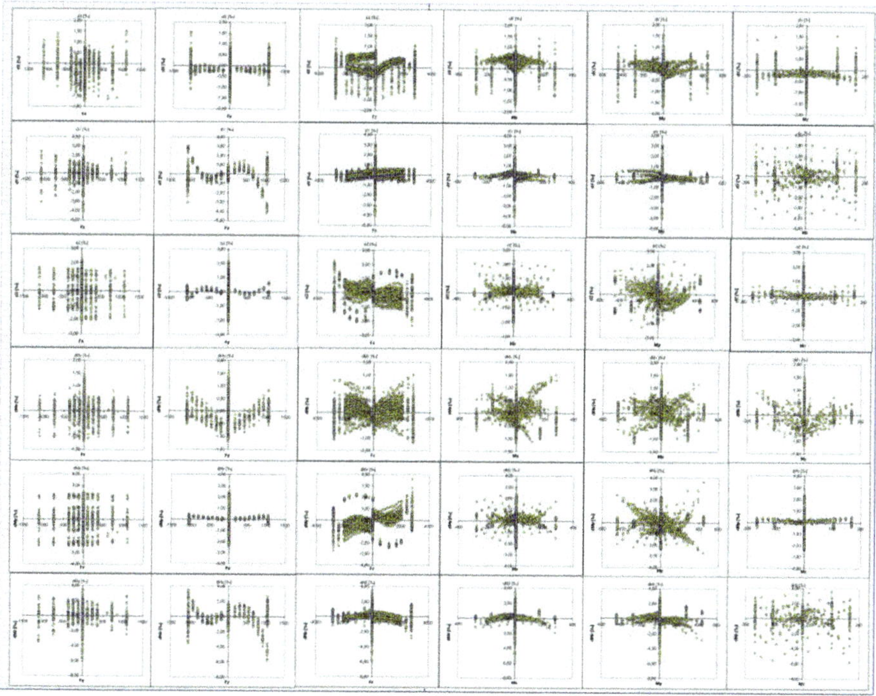

Fig. 9.26 Visual error matrix

9.5 Calibration Machines

Normally every wind tunnel center should have an automatic calibration facility, available to perform frequent calibrations of their balances. Unfortunately this is seldom the case because the machines are very expensive and therefore, only few machines exist in a country or region. The major task in the specification of a calibration machine is to specify the load ranges of the machine that should cover numerous wind tunnel balances. To build such a machine for a wide range is contradictory to the high precision that is required for the machine. The uncertainty for the calibration is given by the full scale uncertainty of the machine and therefore, as a rule of thumb, a machine should only be used for balances with more than 25% of the machine full scale load ranges. To be able to re-calibrate a balance frequently is the advantage of a calibration machine. The disadvantage is a possible higher uncertainty of a machine because the machine has to be calibrated itself. And so the uncertainty can be greater than that of a good manual calibration. The European Union once sponsored a comparison of calibrations at different facilities in Europe using one balance. The outcome of this comparison was that the best manual calibration and the best machine calibration have uncertainties of the same order of magnitude and that the advantage of the machine is that it can be used for frequent calibrations without a greater uncertainty.

9.5.1 Indirect Automatic Calibration

The indirect method is used mostly in the calibration machines designed by TU Darmstadt. Up to now four inverse machines in use at different institutes have been built in cooperation with commercial vendors. The basic principle is described in Sect. 9.3. Here the machines are described in more detail.

The main part of the machine is the reference balance that measures the applied calibration loads. In principle this is an external balance. To the metric end of this balance (measuring tree, red) the metric end of the sample balance (light blue, grey) is connected. For good separation of the components and small interactions, this measuring tree is connected via long rods to the load cells (red). At every connection between the rods there are elastic flexible joints so that no friction can cause hysteresis in the measurement. The great advantage of this principle is that no realignment or deflection measurement process is necessary during operation of the machine.

On the non-metric end of the sample balance a loading tree (yellow) is mounted and this loading tree is connected to the force generators via long rods. The connections of these rods can be ball joints because friction on this side does not affect the result of the measurement.

For the vertical loading the force generators are arranged in an equilateral triangle. By this arrangement it is possible to place the force generators beside the load cells and so it is possible to have access to the generators for maintenance. If all vertical force generators act in the same direction they generate a normal force Z. If the force generators beside the X axis act in the opposite direction to the force generator in the X axis they generate moment around the Y axis, M_y. If the two force generators beside the X axis act in opposite direction, they generate a moment around the X axis, M_x.

In the horizontal plane there are also three force generators. One is directly in the X axis and generates axial force, F_x. The other two are arranged equidistant to the side force load cell Y. If they act in the same direction at the same time they generate side force F_y and if they act in opposite direction they generate the moment around the Z axis M_z.

The load cells measure the amount of the load and determine the repeatability and the signal stability of the machine, so they must be of the best quality that is available on the market. To determine the load components exactly, the exact orientation of the loads has to also be known. The orientation of the components is defined by the machine axis system. This axis system is an orthogonal axis system that is fixed in the reference point of the machine. The reference point of the machine is the intersection of the normal force Z, axial force X and side force Y direction. At first this is a virtual point and axis system. Practically, the reference point is defined by adapting an orthogonal axis system so that the each axis intersects the connecting point of the rod at the measuring tree. To define an exact axis system, the positions of these attachment points must be manufactured very precisely. In addition it is necessary to bring the connecting rods and the load cells precisely in line with the virtual exact orthogonal axis system. For the orientation of the measuring tree the contact flange for the balances is the practical reference surface.

The force generators must be able to generate the maximum required loads and to maintain the set loads absolutely constant during the time that is needed to take the measurement of the load cells and the balance signals. Another requirement for the force generators is that sufficiently small increments are possible to calibrate also the balance with the smallest allowed load ranges. These requirements can be realized by pneumatic force generators, which were specifically developed by TU Darmstadt.

The working principle of the generator is shown in Fig. 9.28. To generate a force one of the cylinders must be filled with air and the air pressure in the cylinder acts against the flexibility of the balance. Balances with different load ranges will deform different under the same pressure inside the cylinder. Air instead of hydraulic oil enables the best adaption to this requirement.

A rolling diaphragm closes the gap between the cylinder and the piston. The whole cylinder must be absolutely air-tight to enable the stability of the set pressure and the commanded force. Not shown in Fig. 9.28 is a central guide bar that holds the piston in the middle of the cylinder. This guide bar moves in linear ball bearings, which are mounted in the cylinder. A precision controller that is usually used for pressure sensor calibration controls the pressure in the cylinder. After a certain stability is established, all valves are closed and the load is constant during the measurement interval. To speed up the process, at first the set values are controlled within a tolerance of $\pm 1\%$ of the set value. After closing of the valves, the load is kept constant within $\pm 0.01\%$ of the last pressure controlled value. With that the absolute load values only match

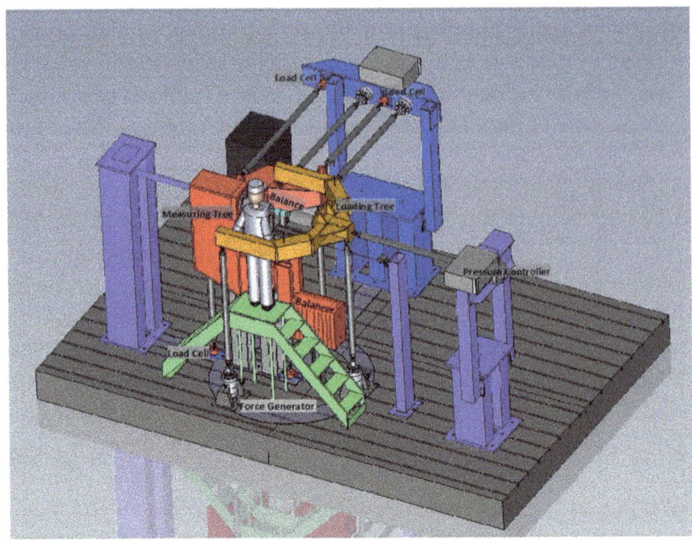

Fig. 9.27 TU Darmstadt machine in operation mode

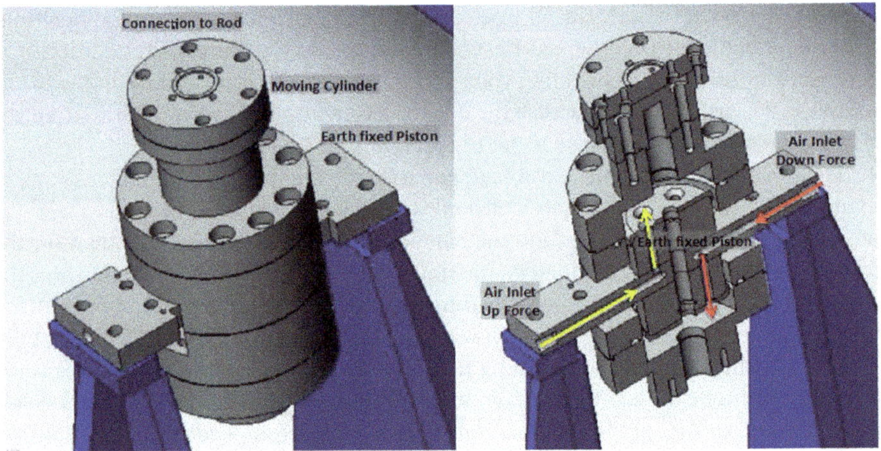

Fig. 9.28 Pneumatic force generator

the set value with an uncertainty of ± 1%, but within this tolerance the stability of the load is rather precise. For the whole calibration process it is not necessary to apply loads of a fixed value. It is only necessary to apply loads with a high stability and a reasonable number of different loads within the full load range.

The design of the measuring tree was improved related to stiffness. The load cells of the vertical loads (Z, M_x, M_y) must carry the load of the weighted construction (measuring tree, rod, flexures) and this causes a certain pre-load on these load cells. The load ranges of the load cells were selected so that they are pre-loaded with half

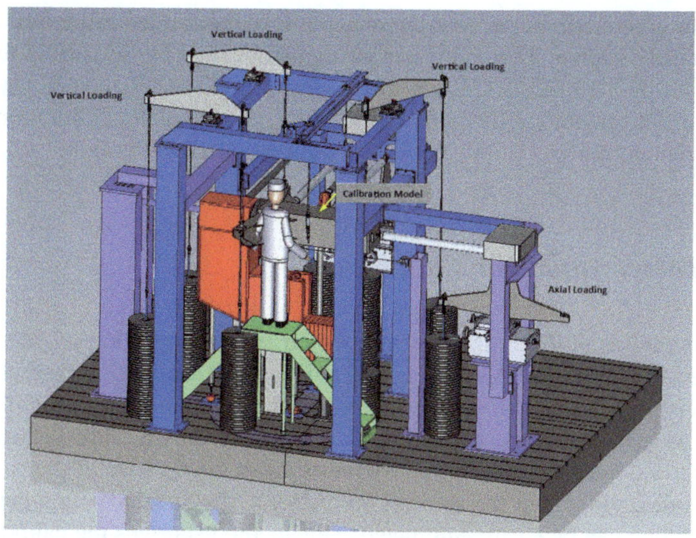

Fig. 9.29 Master calibration setup of the TU Darmstadt machine

the full scale necessary output. As a consequence, a mass corresponding to these values can be used for the stiffness optimization of the measuring tree. The normal force load cell has to carry most of the weight. It is not necessarily placed in the center of gravity of the measuring tree. A balance (see Fig. 9.27) is necessary to distribute the pre-load, so that every load cell is pre-loaded to half its range. The maximum deformation of the measuring tree, especially at the connecting flange, should be below 1/500 mm.

The deformations of the loading tree are not of interest because they do not affect the calibration. The dynamic behavior of the control system instead is heavily influenced by the mass of the loading tree so that in this case the mass should be kept as low as possible.

After the machine is built up it must be calibrated to determine the relation between the load cell signals and the loads acting in the machine reference point. This calibration is called master calibration. For a master calibration weights must be used to enable the traceability of the calibration loads to the national standard. The setup for the machine in the master calibration mode is shown in Fig. 9.29.

For the master calibration the whole force generating equipment is removed from the machine. Instead of these devices rods, levers and hangers are installed to enable a loading of every component with weights. In place of the sample balance a calibration model is mounted to the metric flange of the machine and the corresponding points on the model are connected to the loading devices. Now the axis system of the machine is defined by the geometry and position of the calibration model. As a consequence, the model has to be manufactured precisely to match the axis system of the machine and the rods have to be aligned exactly in the machine axis system to apply the calibration loads in the correct direction. Also the magnitude of the applied calibration loads has to be precisely known. This value can easily be obtained by measuring the masses exactly. During the loading realignment to the machine axis system is necessary to keep the exact orientation of the load. In contrast to the direct calibration method, this realignment process is only necessary during master calibration.

9.5.2 Direct Automatic Calibration

One semi-automatic direct calibration method is the Single Vector Method [13]. It was developed by NASA Langley and it is a direct calibration method where a weight is applied to the calibration sleeve and the orientation of the earth mount is varied. It is semi-automatic because with one weight only a range of loads can be generated. In the current version servomotors move the rig base of the balance into different orientations, but the settings are performed manually with a control unit. This procedure can be easily automated. This method shortens the calibration time significantly in comparison to total manual calibration.

During calibration the weight is always aligned vertical so that the value of load is known precisely. The calibration sleeve orientation has to be measured exactly to calculate the calibration loads from the orientation information and the weight.

Applying different weights varies the magnitude of the load and hanging the weight on the sleeve away from the center axis generates moments. To cover the balance load ranges, several weights at different sleeve positions must be applied manually: this is the reason why this method cannot be fully automated.

For this method a high accuracy orientation measurement system is required to calculate the components of the loads. To determine the orientation, two separate accelerometer systems are used, one on the balance sleeve and one on the hanger. From these systems the relative orientation of the sleeve to the weight is measured.

Other sensitive parts affecting accuracy of the method are the roll and ball bearings (see Fig. 9.30) that enable the vertical alignment of the yoke on which the weights are placed. Friction and breakaway effects can cause misalignment of the yoke. In case of the NASA rig, the position of the yoke related to the acceleration due to gravity is measured by accelerometers and if there are misalignments, the data of these accelerometers can be used for correction.

Another calibration machine working according the direct principle is built by *Israel Aircraft Industries* (IAI) and was sold in different versions to various institutes throughout the world. The basic principle of this machine is shown in the left diagram of Fig. 9.16.

The balance to be calibrated is mounted to a rigid frame with its non-metric side and the metric side is connected to a load adapter. Six load cells and six force generators are mounted between the load adapter and the rigid frame via rods to close the force loop. These systems generate the calibration loads. With this arrangement the force that is acting in the direction of the rod can be measured precisely.

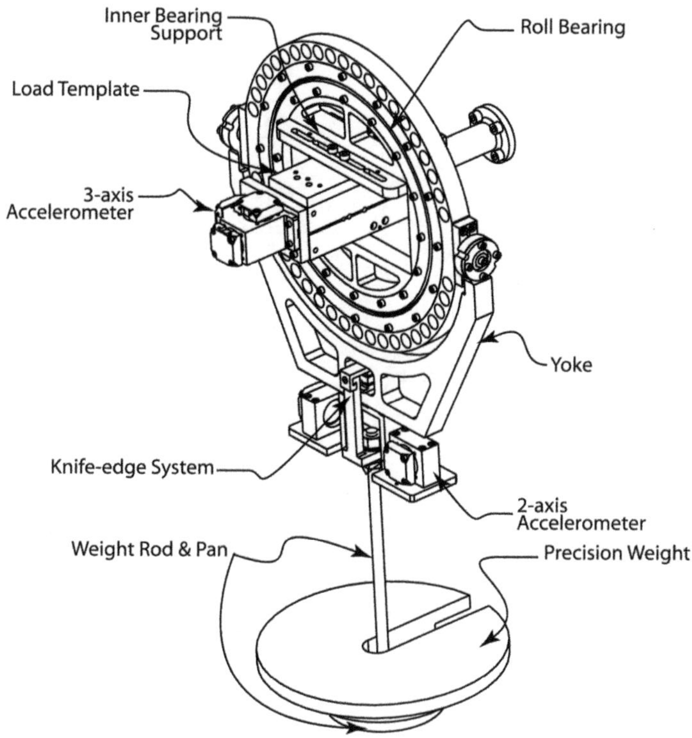

Fig. 9.30 Direct calibration using single vector principle of NASA Langley

The reference axis system for the calibration is the axis system of the load adapter, which is identical to the reference system of the metric end of the balance. Due to balance deformations the generated forces are not acting in the direction of the reference axis system and so the exact direction of the force vector related to the reference system has to be known. This is achieved by measuring the orientation and the movement of the load adapter, is measured by a six-axis optical linear displacement measurement system. Using the measured load and the relative movements the calibration forces and moments acting in the reference axis system have to be calculated (see [16]) (Fig. 9.31).

The advantage of this machine design is that the calibration of the entire machine is reduced to the single calibration of each load cell and the calibration of each sensor of the orientation measurement system. This offers also the possibility to change the load ranges of the machine by exchanging load generators and load cells. The only adjustment that has to be made is the balancing of the new load system using the existing balancing system. The disadvantage is that always two measurement systems, force and orientation measurement systems, are required to determine the calibration loads. The overall uncertainty of the machine is in the order of 0.1% of full scale of the load range.

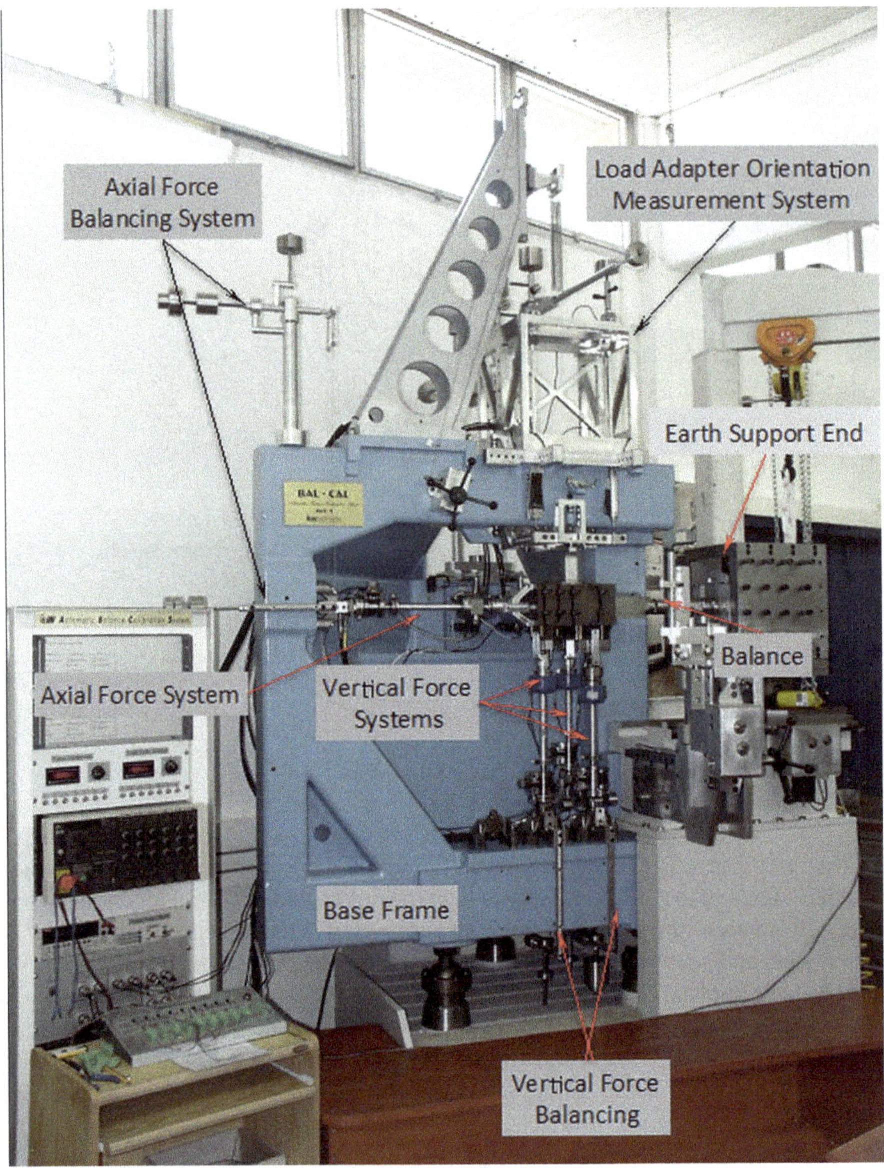

Fig. 9.31 IAI automatic balance calibration system (ABCS)

References

1. AIAA: Calibration and use of internal strain gage balances with application to wind tunnel testing. AIAA Recommended Practice, AIAA R-091-2017
2. AIAA: Assessment of experimental uncertainty with application to wind tunnel testing. AIAA S-017A Washington DC **6**, 1–15 (1999)
3. Bergmann, R., Philipsen, I.: An experimental comparison of different load tables for balance calibration. In: 27th AIAA Aerodynamic Measurement Technology and Ground Testing Conference, p. 4544 (2010)
4. Bergmann, R., Philipsen, I.: Some contemplations on a proposed definition of uncertainty for balances. In: 27th AIAA Aerodynamic Measurement Technology and Ground Testing Conference (2010)
5. Cook, T.: A note on the calibration of strain gauge balances for wind tunnel models. RAE Techn. Note AERO **2631** (1959)
6. Ewald, B.: The accuracy of internal wind tunnel balances for wind tunnel force measurements, the problem of definition and verification. Aeronaut. J. **105**(1050), 443–449 (2001)
7. Ewald, B.: Standard-Verfahren zur Eichung und Messauswertung von DMS-Waagen. VFW-Fokker, Bericht Ex 1-450 (Aug 1974)
8. Galway, R.: A comparison of methods for calibration and use of multi-component strain gauge wind tunnel balances. NRCA Aeronautical Report (LR-6001980) (1980)
9. Galway, R.D.: A consideration of tare weight effects in calibration and use of wind tunnel strain-gauge balances. In: The Second International Symposium on Strain Gauge Balances (Mai 1999, Bedford,UK)
10. Holmes, J.E.: Static Multiple-Load Measurement Technique as Utilized in the Naval Surface Weapons Center's Wind Tunnels. Technical Report, Naval Surface Weapons Center White Oak Lab Silver Spring MD (1976)
11. Hufnagel, K.: The truth about true loads. In: 5th International Symposium on Strain Gauge Balances (9-12 May 2006 Modane; France 2006)
12. ISO: Guide to the Expression of Uncertainty in Measurement (GUM). International Organization for Standardization, Genève, Switzerland (1995)
13. Jones, S.M., Rhew, R.D.: Recent developments and status of the Langley single vector balance calibration system (SVS). In: 4th Symposium on Strain Gauge Balances (May 2004)
14. Kreuzer, M.: High-Precision Measuring Technique for Strain Gauge Transducers. Internal Publication of Hottinger Baldwin Messtechnik, GmbH, Darmstadt, Germany (1999)
15. Leung, S.Y., Link, Y.Y.: Comparison and Analysis of Strain Gauge Balance Calibration Matrix Mathematical Models. Technical Report DSTO Tech-Report TR0857 (1999)
16. Levkovitch, M.: Accuracy Analysis of the IAI Automatic Balance Calibration System. Technical Report, IAI Company Report (Tel Aviv, Apr 2016)
17. Philipsen, I., Bergmann, R.: The evaluation of balance calibration data and the regression procedure at DNW. In: 27th AIAA Aerodynamic Measurement Technology and Ground Testing Conference, p. 4205. Chicago (July 2010)
18. Smith, D.L.: An Efficient Algorithm Using Matrix Methods to Solve Wind Tunnel Force-Balance Equations. Technical Report NASA TN D-6860, NASA Langley (1972)
19. Ulbrich, N.: Combined load diagram for a wind tunnel strain-gage balance. In: 27th AIAA Aerodynamic Measurement Technology and Ground Testing Conference, p. 4203. Chicago (2010)
20. Ulbrich, N.: Iterative strain-gage balance calibration data analysis for extended independent variable sets. In: 49th AIAA Aerospace Sciences Meeting Including the New Horizons Forum and Aerospace Exposition, p. 949. Orlando (2011)
21. Ulbrich, N.: A universal tare load prediction algorithm for strain-gage balance calibration data analysis. In: 47th AIAA/ASME/SAE/ASEE Joint Propulsion Conference & Exhibit, p. 6090. San Diego (2011)

22. Ulbrich, N.: Comparison of iterative and non-iterative strain-gage balance load calculation methods. In: 27th AIAA Aerodynamic Measurement Technology and Ground Testing Conference, p. 4202. Chicago (July 2010)
23. Ulbrich, N., Volden, T.: Regression model term selection for the analysis of strain-gage balance calibration data. In: 27th AIAA Aerodynamic Measurement Technology and Ground Testing Conference, p. 4545. Chicago (July 2010)
24. van der Veer, J.: User Guide of the Program EXBSCH. Memorandum WN-79-044; NLR, NLR (1979)
25. Xiang, G.: Mechanical Simulation of Bi-Directional Behavior of Multi Piece Strain Gauge Balances. Technical Report, TU Darmstadt (2017)
26. Zhai, J.: Analyse und Optimierung der internen Windkanalwaagen mit FEM. Ph.D. thesis, TU Darmstadt (1996)

Chapter 10
Utilization of Balances in the Wind Tunnel

10.1 Rigging and Test Preparation

Despite the detail outlined in the previous chapters regarding design, construction, calibration and mounting of internal wind tunnel balances, the integration of balances into wind tunnel text campaigns remains a formidable task, usually with very specific challenges in each wind tunnel. Thus, there is no general procedure to be followed; nevertheless, the flow chart discussed in this section (Fig. 10.1) can be used as a guideline.

Once the balance is mounted to the tunnel support, connected to the data system and adjusted to the geodetic axis system, first a check of the balance zero signals must be made. If all balance zeros are very near to the zeros of calibration, the installation can be continued. There are two avenues for the next step:

1. Mount the sleeve that is used for calibration and perform a check loading to confirm that the sensitivity of the balance is within the calibration limits.
2. Mount the model and perform a load check with the model. This is only possible if the model provides points to apply loads at certain well-known positions.

If this loading check is successful, the next step is the recording of the weight polar or wind off polar. To perform this the model must be moved in all test relevant positions so that it is possible to calculate functions that can be used to generate wind off data for all positions to be encountered in the test. Now the balance is in principle ready for the test and this is the opportunity to set the parameters for the signal processing, like integration time, filtering etc. If all is complete, the test can be performed and data can be taken in any position. If a model configuration change is foreseen, the test must be ended with wind off data in the zero position. After the modification of the model the entire process must be repeated, starting with the wind off polar.

The signals taken during a test are transformed first into forces using the balance matrix. It is important, especially for nonlinear behavior, to transfer the signals always into loads before a tare correction is made, because the balance matrix is only valid for the position leveled to the geodetic axis system. These calculated loads are the

K. Hufnagel, *Wind Tunnel Balances*, Experimental Fluid Mechanics,
https://doi.org/10.1007/978-3-030-97766-5_10

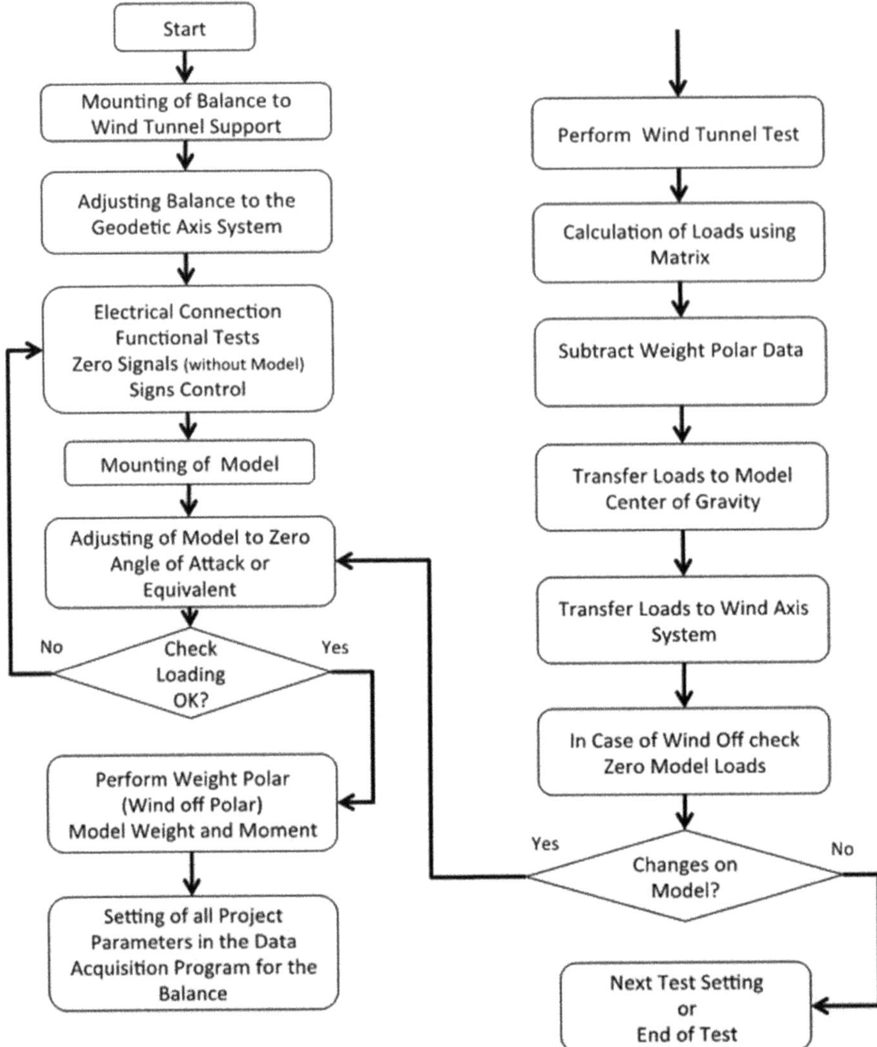

Fig. 10.1 Flow chart for balance use in wind tunnel test campaigns

loads related to the position of the balance sleeve reference point. From these loads the wind off loads must be subtracted and then the loads can be transferred to the model center of gravity. Now the loads can be transferred to the wind axis system using the information of the model orientation.

10.2 Damping Systems

Model, balance and sting form together a mass-spring system that can potentially oscillate. The excitation for the oscillation can be the unsteady flow initiated by flow separation on the model at high angle of attack or other instationary flow effects. Depending on the total elasticity distribution, the model starts to oscillate in the lowest eigenfrequency in the direction of the lowest stiffness. Oscillation is possible around every axis. Important is to realize that it is not always the balance being the weakest part of this system; however, it is probably the most expensive part that can be damaged or destroyed by such oscillations. It is likely that the sting and the support system have the lowest eigenfrequency, simply because of their length. To protect the entire system from resonant oscillations it is necessary to know the eigenfrequencies prior to the test. They can be measured with accelerometers that are already on board the model if attitude of the model is measured by an onboard system, or if there is no onboard system, additional accelerometers must be installed temporarily. To determine the eigenfrequencies in the different directions, the model must be carefully deflected in the different directions and suddenly released. The system than oscillates at the eigenfrequency. By performing a Fourier transformation of the measured data the eigenfrequency can be determined.

After the character of the most sensitive vibration is known, damping systems can be installed on the sting or the support system. In the low speed wind tunnel of TU Darmstadt a simple system that uses two external oil dampers has been used (see Fig. 10.2).

The sting is connected via steel cables to the dampers so that changes in angle of attack and yaw angle can be made while the dampers are active. The total height of the oil reservoir and the lateral distance of the two dampers are determined by the total necessary deflection for yaw and pitch displacement. With such an arrangement pitch and yaw oscillations can be effectively damped.

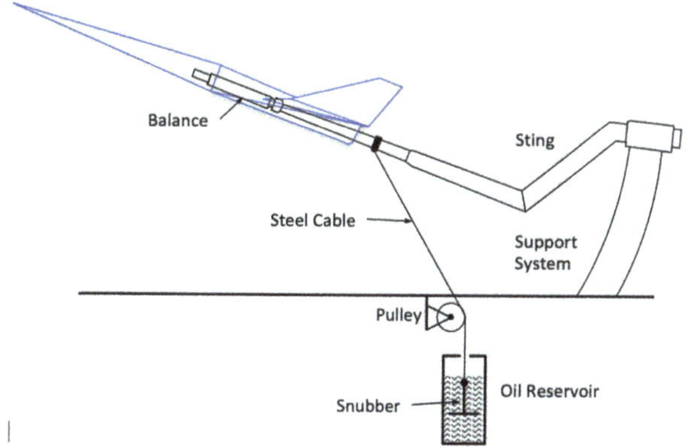

Fig. 10.2 Sting damping system of TU Darmstadt low speed wind tunnel1

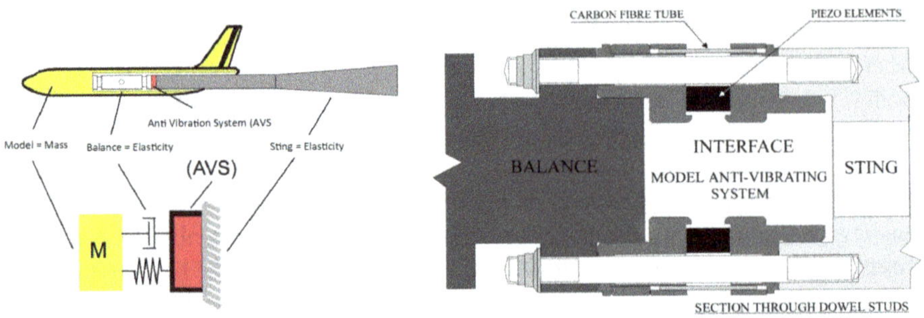

Fig. 10.3 ETW sting integrated anti-vibration system (ERAS company, Göttingen)

In wind tunnels where the access to external dampers is not possible, sting integrated damping systems must be used. An example for such a system is the active anti-vibration system (AVS) of ETW (Fig. 10.3).

Between balance and sting a system out of radially arranged piezo-actuators work with a phase shift of 180° relative to the vibrations coming from the balance and thus, cancel the excitation. The balance signals are used as control signals for the active damping control circuit. With this system the damping of vibrations in all lateral directions are effected [1].

References

1. Fehren, H., Gnauert, U., Wimmel, R., Hefer, G., Schimanski, D.: Aktive Schwingungskompensation für den Europäischen Transsonischen Windkanal. Adaptronic Congress; April, Berlin (2001)
2. Yoder, D.L.: Dynamic Monitoring and Life Prediction of Internal Strain-Gage Balances. Master Thesis University of Tennesee (2016)

GPSR Compliance
The European Union's (EU) General Product Safety Regulation (GPSR) is a set of rules that requires consumer products to be safe and our obligations to ensure this.

If you have any concerns about our products, you can contact us on

ProductSafety@springernature.com

In case Publisher is established outside the EU, the EU authorized representative is:

Springer Nature Customer Service Center GmbH
Europaplatz 3
69115 Heidelberg, Germany

www.ingramcontent.com/pod-product-compliance
Ingram Content Group UK Ltd.
Pitfield, Milton Keynes, MK11 3LW, UK
UKHW021009290726
14059UKWH00001BA/49